REED'S
NAVAL ARCHITECTURE
FOR
MARINE ENGINEERS

REED'S NAVAL ARCHITECTURE FOR MARINE ENGINEERS

By

E. A. STOKOE

C.Eng., F.R.I.N.A., F.I.Mar.E., M.N.E.C.Inst.,
Principal Lecturer in Naval Architecture at South Shields Marine and Technical College

PUBLISHED BY THOMAS REED PUBLICATIONS LIMITED
SUNDERLAND AND LONDON

First Edition – 1963
Second Edition – 1967
New Edition – 1973 (*in SI units*)
Reprint – 1975
Reprint – 1977

ISBN 0 901281 39 5

PRINTED BY THOMAS REED AND COMPANY LIMITED
SUNDERLAND, GLASGOW AND LONDON

PREFACE

This book is intended to cover the theoretical work in the Syllabus for Naval Architecture in Part B of the Department of Trade Examinations for Second and First Class Engineers.

In each section the work progresses from an elementary stage to the standard required for First Class Examinations. Parts of the subject matter and the attendant Test Examples are marked with the prefix "f" to indicate that they are normally beyond the syllabus for the Second Class Examination and so can be temporarily disregarded by such candidates. Throughout the book emphasis is placed on basic principles, and the profusely illustrated text, together with the worked examples, assists the student to assimilate these principles more easily.

All students attempting Part B of their certificate will have covered the work required for Part A, and several of the principles of Mathematics and Mechanics are used in this volume. Where a particularly important principle is required, however, it is revised in this book. Fully worked solutions are given for all Test Examples and Examination Questions. In several cases shorter methods are available and acceptable in the examination, but the author has attempted to use a similar method for similar problems, and to avoid methods which may only be used in isolated cases. It should be noted that a large proportion of the worked solutions include diagrams and it is suggested that the students follow this practice. The typical Examination Questions are intended as a revision of the whole of the work, and should be treated as such by attempting them in the order in which they are given. The student should avoid attempting a number of similar types of questions at the same time. A number of Examination Questions have been selected from Department of Trade papers and are reproduced by kind permission of the Controller of Her Majesty's Stationery Office.

An engineer who works systematically through this volume will find that his time is amply repaid when attending a course of study at a college and his chance of success in the Examination will be greatly increased.

CONTENTS

INTRODUCTION TO SI UNITS

SI is the abbreviation for Système International d'Unités, the metric system of measurement now coming into international use and being adopted by British Industry.

BASIC UNITS

There are six basic units in the system:

QUANTITY	UNIT	SYMBOL
length	metre	m
mass	kilogramme	kg
time	second	s
temperature	kelvin	K
electric current	ampere	A
luminous intensity	candela	cd

DERIVED UNITS

It is possible to obtain derived units from these basic units. The system has been designed in such a way that the basic units are used without numerical multipliers to obtain the fundamental derived units. The system is therefore said to be *coherent*.

unit area $= m^2$
unit volume $= m^3$
unit velocity $= m/s$
unit acceleration $= m/s^2$

The unit of *force* is the *newton* N

Now force = mass × acceleration
Hence 1 newton $= 1\ kg \times 1 m/s^2$
N $= kg\ m/s^2$

The unit of *work* is the *joule* J

Now work done = force × distance.
1 joule = 1 N × 1 m
J = N m

The unit of *power* is the *watt* W

Now power = work done per unit time
1 watt = 1 J ÷ 1 s
W = J/s

The unit of *pressure* is the *pascal* Pa

Now pressure = force per unit area
1 Pa $= 1N \div 1\ m^2$
Pa $= N/m^2$

MULTIPLES AND SUB-MULTIPLES

In order to keep the number of names of units to a minimum, multiples and sub-multiples of the fundamental units are used. In each case powers of ten are found to be most convenient and are represented by prefixes which are combined with the symbol of the unit.

MULTIPLICATION FACTOR	STANDARD FORM	PREFIX	SYMBOL
1 000 000 000 000	10^{12}	tera	T
1 000 000 000	10^{9}	giga	G
1 000 000	10^{6}	mega	M
1 000	10^{3}	kilo	k
100	10^{2}	hecto	h
10	10^{1}	deca	da
0·1	10^{-1}	deci	d
0·01	10^{-2}	centi	c
0·001	10^{-3}	milli	m
0·000 001	10^{-6}	micro	μ
0·000 000 001	10^{-9}	nano	n
0·000 000 000 001	10^{-12}	pico	p

Only one prefix may be used with each symbol. Thus a thousand kilogrammes would be expressed as a Mg and not kkg. When a prefix is attached to a unit, it becomes a new unit symbol on its own account and this can be raised to positive or negative powers of ten.

Multiples of 10^3 are recommended but others are recognised because of convenient sizes and established usage and custom. A good example of this convenient usage lies in the calculation of volumes. If only metres or millimetres are used for the basic dimensions, the volume is expressed in m^3 or mm^3.

now $1 m^3 = 10^9 mm^3$

i.e. the gap is too large to be convenient. If, on the other hand, the basic dimensions may be expressed in decimetres or centimetres in addition to metres and millimetres, the units of volume change in 10^3 intervals.

i.e.
$$1\ m^3 = 1000\ dm^3$$
$$1\ dm^3 = 1000\ cm^3$$
$$1\ cm^3 = 1000\ mm^3$$

Several special units are introduced, again because of their convenience. A megagramme, for instance, is termed a *tonne* which is approximately equal to an imperial ton mass. Pressure may be expressed in *bars* (b) of value $10^5 N/m^2$. A bar is approximately equal to one atmosphere. Stresses may be expressed in *hectobars* ($10^7 N/m^2$) of about $\frac{2}{3}$tonf/in^2.

It is unwise, however, to consider comparisons between imperial and SI units and it is probable that the pressure and stress units will revert to the basic unit and its multiples.

CHAPTER 1

HYDROSTATICS

DENSITY ρ of a substance is the mass of a unit volume of the substance and may be expressed in grammes per millilitre (g/ml), kilogrammes per cubic metre (kg/m^3) or tonnes per cubic metre (t/m^3). The numerical values of g/ml are the same as t/m^3. The density of fresh water may be taken as 1·000 t/m^3 or 1000 kg/m^3 and the density of sea water 1·025 t/m^3 or 1025 kg/m^3.

RELATIVE DENSITY or specific gravity of a substance is the density of the substance divided by the density of fresh water, i.e. the ratio of the mass of any volume of the substance to the mass of the same volume of fresh water. Thus the relative density (rd) of fresh water is 1·000 while the relative density of sea water is 1025 ÷ 1000 or 1·025. It is useful to know that the density of a substance expressed in t/m^3 is numerically the same as the relative density. If a substance has a relative density of x, then one cubic metre of the substance will have a mass of x tonnes. V cubic metres will have a mass of Vx tonnes or $1000Vx$ kilogrammes.

Thus:

$$\text{mass of substance} = \text{volume} \times \text{density of substance}$$

Example. If the relative density of lead is 11·2, find
(a) its density
(b) the mass of 0·25 m^3 of lead.

Density of lead = relative density of lead × density of fresh water
= 11·2 t/m^3
Mass of lead = 0·25 × 11·2
= 2·8 t

Example. A plank 6 m long, 0·3 m wide and 50 mm thick has a mass of 60 kg. Calculate the density of the wood.

Volume of wood $= 6{\cdot}0 \times 0{\cdot}3 \times 0{\cdot}050$
$= 90 \times 10^{-3}\ \text{m}^3$

Density of wood $= \dfrac{\text{mass}}{\text{volume}}$

$= \dfrac{60}{90 \times 10^{-3}} \quad \dfrac{\text{kg}}{\text{m}^3}$

$= 667 \quad \text{kg/m}^3$

PRESSURE EXERTED BY A LIQUID

Liquid pressure is the load per unit area exerted by the liquid and may be expressed in multiples of N/m^2.

e.g. $10^3\ \text{N/m}^2 = 1\ \text{kN/m}^2$
$10^5\ \text{N/m}^2 = 1\ \text{bar}$

This pressure acts equally in all directions and perpendicular to the surface of any immersed plane.

Consider a trough containing liquid of density ρ kg/m³
Let A = cross-sectional area of a cylinder of this liquid in m²
and h = height of cylinder in m (Fig. 1).

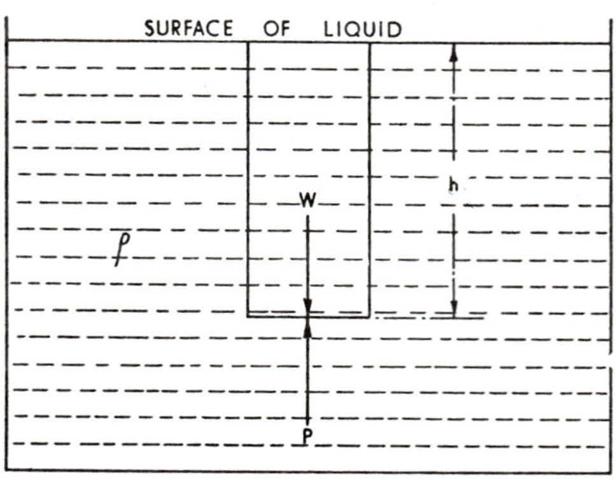

Fig. 1

The cylinder is in equilibrium under the action of two vertical forces:—

(a) the gravitational force W acting vertically down
(b) the upthrust P exerted by the liquid on the cylinder.

Thus $P = W$
but $P = pA$
where p = liquid pressure at a depth h m
and $W = \rho gAh$
$\therefore pA = \rho gAh$
$p = \rho gh$

Thus it may be seen that the liquid pressure depends upon the density ρ and the vertical distance h from the point considered to the surface of the liquid. Distance h is known as the *head.*

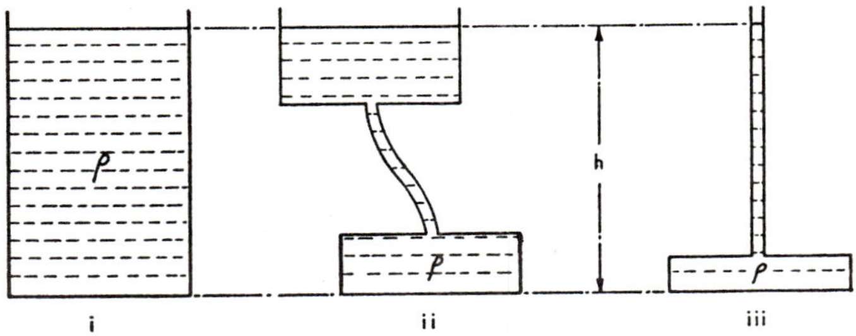

Fig. 2

The pressure at the base of each of the containers shown in Fig. 2 is ρgh although it may be seen that the total mass of the liquid is different in each case. Container (ii) could represent a supply tank and header tank used in most domestic hot water systems. The pressure at the supply tank depends upon the height of the header tank.

Container (iii) could represent a double bottom tank having a vertical overflow pipe. The pressure inside the tank depends upon the height to which the liquid rises in the pipe.

The total load exerted by a liquid on a *horizontal* plane is the product of the pressure and the area of the plane.

$$P = p\,A$$

Example. A rectangular double bottom tank is 20 m long, 12 m wide and 1·5 m deep, and is full of sea water having a density of 1·025 tonne/m³.

Calculate the pressure in kN/m³ and the load in MN on the top and bottom of the tank if the water is:

(a) at the top of the tank
(b) 10 m up the sounding pipe above the tank top.

(a) Pressure on top $= \rho g h$
$= 1{\cdot}025 \times 9{\cdot}81 \times 0$
$= 0$

Load on top $= 0$

Pressure on bottom $= 1{\cdot}025 \times 9{\cdot}81 \times 1{\cdot}5 \quad \frac{\text{Mg}}{\text{m}^3} \times \frac{\text{m}}{\text{s}^2} \times \text{m}$
$= 15{\cdot}09 \text{ kN/m}^2$

Load on bottom $= 15{\cdot}09 \times 20 \times 12$
$= 3622 \text{ kN}$
$= 3{\cdot}622 \text{ MN}$

(b) Pressure on top $= 1{\cdot}025 \times 9{\cdot}81 \times 10$
$= 100{\cdot}6 \text{ kN/m}^2$

Load on top $= 100{\cdot}6 \times 20 \times 12$
$= 24\ 144 \text{ kN}$
$= 24{\cdot}144 \text{ MN}$

Pressure on bottom $= 1{\cdot}025 \times 9{\cdot}81 \times 11{\cdot}5$
$= 115{\cdot}6 \text{ kN/m}^2$

Load on bottom $= 115{\cdot}6 \times 20 \times 12$
$= 27\ 744 \text{ kN}$
$= 27{\cdot}744 \text{ MN}$

This example shows clearly the effect of a head of liquid. It should be noted that a very small volume of liquid in a vertical pipe may cause a considerable increase in load.

LOAD ON AN IMMERSED PLANE

The pressure on any horizontal plane is constant, but if the plane is inclined to the horizontal there is a variation in pressure over the plane due to the difference in head. The total load on such a plane may be determined as follows.

Consider an irregular plane of area A, totally immersed in a liquid of density ρ and lying at an angle θ to the surface of the liquid as shown in Fig. 3.

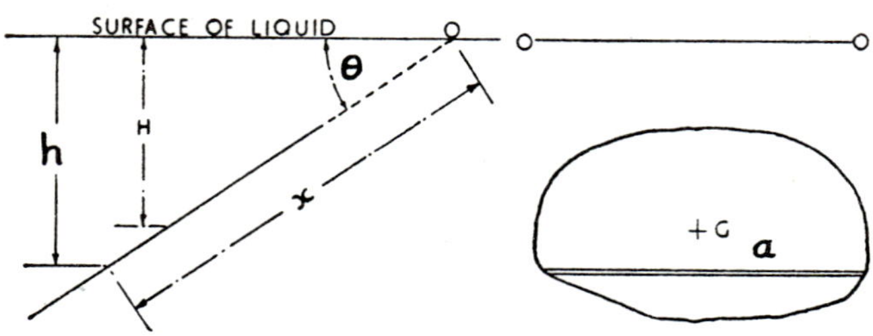

Fig. 3

Divide the plane into thin strips parallel to the surface of the liquid. Let one such strip, distance h below the surface of the liquid, have an area a. Since the strip is thin, any variation in pressure may be ignored.

$$\text{Load on strip} = \rho gah$$
$$\text{Load on plane} = \rho g(a_1 h_1 + a_2 h_2 + a_3 h_3 + \ldots) = \rho g \Sigma ah$$

But Σah is the first moment of area of the plane about the surface of the liquid.

If H is the distance of the centroid of the plane from the liquid surface, then:

$$\Sigma ah = AH$$
$$\therefore \text{Load on plane} = \rho g AH$$

Example. A rectangular bulkhead is 10 m wide and 8 m deep. It is loaded on one side only with oil of relative density 0·8.

Calculate the load on the bulkhead if the oil is:

(a) just at the top of the bulkhead.
(b) 3 m up the surrounding pipe.

(a) Load on bulkhead

$$= \rho g AH = 0{\cdot}8 \times 1{\cdot}0 \times 9{\cdot}81 \times 10 \times 8 \times \frac{8}{2} = 2511 \text{ kN}$$

(b) Load on bulkhead

$$= 0{\cdot}8 \times 1{\cdot}0 \times 9{\cdot}81 \times 10 \times 8 \times \left(\frac{8}{2}+3\right) = 4395 \text{ kN}$$

CENTRE OF PRESSURE

The centre of pressure on an immersed plane is the point at which the whole liquid load may be regarded as acting.

Consider again Fig. 3.

Let the strip be distance x from the axis 0-0.

Then $h = x \sin \theta$

$$\text{Load on strip} = \rho gah = \rho gax \sin \theta$$

$$\text{Load on plane} = \rho g \sin \theta\, (a_1x_1 + a_2x_2 + a_3x_3 + \ldots\ldots) = \rho g \sin \theta\, \Sigma ax$$

Taking moments about axis 0-0:

$$\text{Moment of load on strip} = x \times \rho g a x \sin\theta$$
$$= \rho g a x^2 \sin\theta$$
$$\text{Moment of load on plane} = \rho g \sin\theta\,(a_1 x_1{}^2 + a_2 x_2{}^2 + \ldots\ldots)$$
$$= \rho g \sin\theta\,\Sigma a x^2$$

$$\text{Centre of pressure from 0-0} = \frac{\text{moment}}{\text{load}}$$
$$= \frac{\rho g \sin\theta\,\Sigma a x^2}{\rho g \sin\theta\,\Sigma\, a x}$$
$$= \frac{\Sigma a x^2}{\Sigma a x}$$

But Σax is the first moment of area of the plane about 0-0 and Σax^2 is the second moment of area of the plane about 0-0. If the plane is vertical, then 0-0 represents the surface of the liquid, and thus:

Centre of pressure from surface

$$= \frac{\text{second moment of area about surface}}{\text{first moment of area about surface}}$$

The second moment of area may be calculated using the *theroem of parallel axes*.

If I_{NA} is the second moment about an axis through the centroid (the neutral axis), then the second moment about an axis 0-0 parallel to the neutral axis and distance H from it is given by

$$I_{OO} = I_{NA} + AH^2$$

where A is the area of the plane
Thus Centre of pressure from 0-0

$$= \frac{I_{OO}}{AH}$$
$$= \frac{I_{NA} + AH^2}{AH}$$
$$= \frac{I_{NA}}{AH} + H$$

I_{NA} for a rectangle is $\frac{1}{12}\ BD^3$
I_{NA} for a triangle is $\frac{1}{36}\ BD^3$
I_{NA} for a circle is $\frac{\pi}{64}\ D^4$

The following examples show how these principles may be applied.

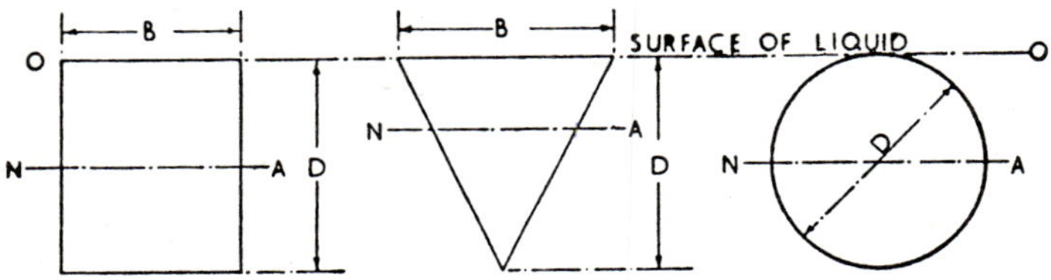

Fig. 4

(a) RECTANGULAR PLANE WITH EDGE IN SURFACE

$$\text{Centre of pressure from 0-0} = \frac{I_{NA}}{AH} + H$$

$$= \frac{\frac{1}{12}\ BD^3}{BD \times \frac{1}{2}D} + \frac{D}{2}$$

$$= \frac{D}{6} + \frac{D}{2}$$

$$= \tfrac{2}{3}\ D$$

(b) TRIANGULAR PLANE WITH EDGE IN SURFACE

$$\text{Centre of pressure from 0-0} = \frac{I_{NA}}{AH} + H$$

$$= \frac{\frac{1}{36}\ BD^3}{\frac{1}{2}BD \times \frac{1}{3}D} + \frac{D}{3}$$

$$= \frac{D}{6} + \frac{D}{3}$$

$$= \tfrac{1}{2}D$$

(c) CIRCULAR PLANE WITH EDGE IN SURFACE

$$\text{Centre of pressure from 0-0} = \frac{I_{NA}}{AH} + H$$

$$= \frac{\frac{\pi}{64} D^4}{\frac{\pi}{4} D^2 \times \frac{1}{2} D} + \frac{D}{2}$$

$$= \frac{D}{8} + \frac{D}{2}$$

$$= \tfrac{5}{8} D$$

If the top edge of the plane is below the surface of the liquid, these figures change considerably.

Example. A peak bulkhead is in the form of a triangle, apex down, 6 m wide at the top and 9 m deep. The tank is filled with sea water. Calculate the load on the bulkhead and the position of the centre of pressure relative to the top of the bulkhead if the water is;

(a) at the top of the bulkhead
f(b) 4 m up the sounding pipe.

(a) Load on bulkhead

$$= \rho g A H$$

$$= 1{\cdot}025 \times 9{\cdot}81 \times \frac{6 \times 9}{2} \times \frac{9}{3}$$

$$= 814{\cdot}5 \text{ kN}$$

Centre of pressure from top

$$= \frac{D}{2}$$

$$= \frac{9}{2}$$

$$= 4{\cdot}5 \text{ m}$$

(b) Load on bulkhead

$$= 1{\cdot}025 \times 9{\cdot}81 \times \frac{6 \times 9}{2} \times \left(\frac{9}{3} + 4\right)$$

$$= 1901 \text{ kN}$$

Centre of pressure from surface

$$= \frac{I_{NA}}{AH} + H$$

$$= \frac{\frac{1}{36} \times 6 \times 9^3}{\frac{1}{2} \times 6 \times 9 \times 7} + 7$$

$$= 0{\cdot}624 + 7$$

$$= 7{\cdot}642 \text{ m}$$

Centre of pressure from top

$$= 7{\cdot}642 - 4$$

$$= 3{\cdot}642 \text{ m}$$

LOAD DIAGRAM

If the pressure at any point in an immersed plane is multiplied by the width of the plane at this point, the load per unit depth of plane is obtained. If this is repeated at a number of points, the resultant values may be plotted to form the load diagram for the plane.

The *area* of this load diagram represents the *load* on the plane, while its *centroid* represents the position of the *centre of pressure.*

For a rectangular plane with its edge in the surface, the load diagram is in the form of a triangle.

For a rectangular plane with its edge below the surface, the load diagram is in the form of a trapezoid.

The load diagrams for triangular planes are parabolic.

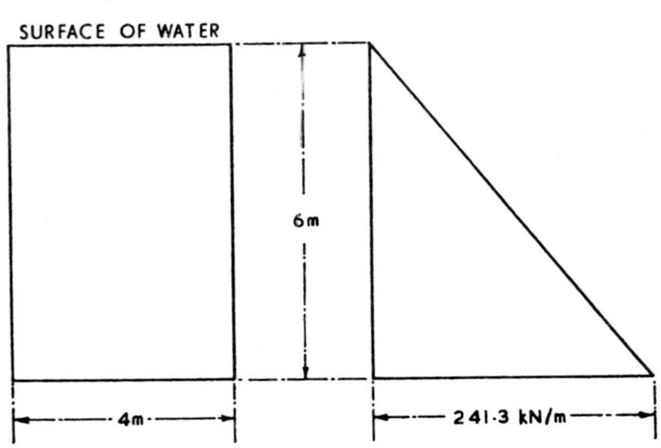

Fig. 5

Consider a rectangular bulkhead 4 m wide and 6 m deep loaded to its top edge with sea water.

Load/m at top of bulkhead $= \rho g h \times \text{width}$
$= 1{\cdot}025 \times 9{\cdot}81 \times 0 \times 4$
$= 0$

Load/m at bottom of bulkhead
$= 1{\cdot}025 \times 9{\cdot}81 \times 6 \times 4$
$= 241{\cdot}3$ kN

Load on bulkhead = area of load diagram
$= \frac{1}{2} \times 6 \times 241{\cdot}3$
$= 723{\cdot}9$ kN

Centre of pressure = centroid of load diagram
$= \frac{2}{3} \times 6$
$= 4$ m from top

Check: Load on bulkhead $= \rho g A H$
$= 1{\cdot}025 \times 9{\cdot}81 \times 4 \times 6 \times \frac{1}{2} \times 6$
$= 724{\cdot}0$ kN

and Centre of pressure $= \frac{2}{3} D$
$= 4$ m from top

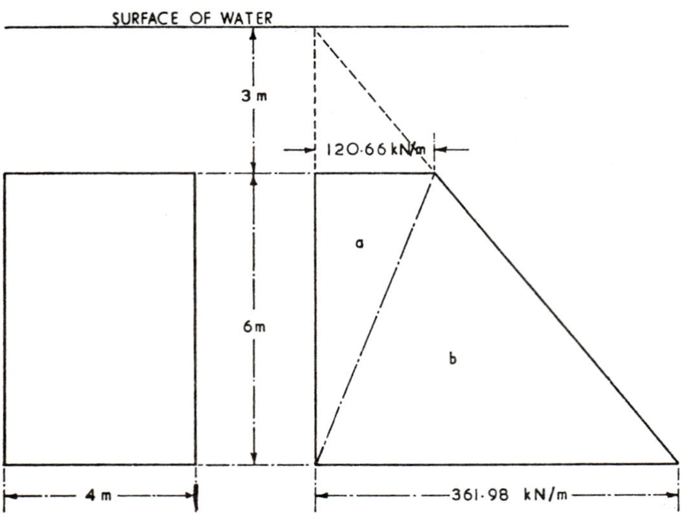

Fig. 6

If, in the above example, there is a 3 m head of water above the bulkhead, then:

Load/m at top of bulkhead $= 1{\cdot}025 \times 9{\cdot}81 \times 3 \times 4$
$= 120{\cdot}66$ kN

Load/m at bottom of bulkhead
$= 1{\cdot}025 \times 9{\cdot}81 \times 9 \times 4$
$= 361{\cdot}98$ kN

The load diagram may be divided into two triangles *a* and *b*

Area *a* $= \frac{1}{2} \times 6 \times 120{\cdot}66$
$= 361{\cdot}98$ kN

Area *b* $= \frac{1}{2} \times 6 \times 361{\cdot}98$
$= 1085{\cdot}94$ kN

Total load $= 361{\cdot}98 + 1085{\cdot}94$
$= 1447{\cdot}92$ kN

Taking moments about the top of a bulkhead

$$\text{Centre of pressure} = \frac{361{\cdot}98 \times \frac{1}{3} \times 6 + 1085{\cdot}94 \times \frac{2}{3} \times 6}{361{\cdot}98 + 1085{\cdot}94}$$

$$= \frac{723{\cdot}96 + 4343{\cdot}76}{1447{\cdot}92}$$

$= 3{\cdot}5$ m from top of bulkhead

These results may again be checked by calculation.

Load on bulkhead $= 1{\cdot}025 \times 9{\cdot}81 \times 4 \times 6 \times (\frac{1}{2} \times 6 + 3)$
$= 1448$ kN

Centre of pressure from surface

$$= \frac{I_{NA}}{AH} + H$$

$$= \frac{\frac{1}{12} \times 4 \times 6^3}{4 \times 6 \times (\frac{1}{2} \times 6 + 3)} + (\frac{1}{2} \times 6 + 3)$$

$= 0{\cdot}5 + 6$
$= 6{\cdot}5$ m

Centre of pressure $= 6{\cdot}5 - 3$
$= 3{\cdot}5$ m from top of bulkhead.

SHEARING FORCE ON BULKHEAD STIFFENERS

A bulkhead stiffener supports a rectangle of plating equal to the length of the stiffener times the spacing of the stiffeners. If the bulkhead has liquid on one side to the top edge, the stiffener supports a load which increases uniformly from zero at the top to a maximum at the bottom (Fig. 7).

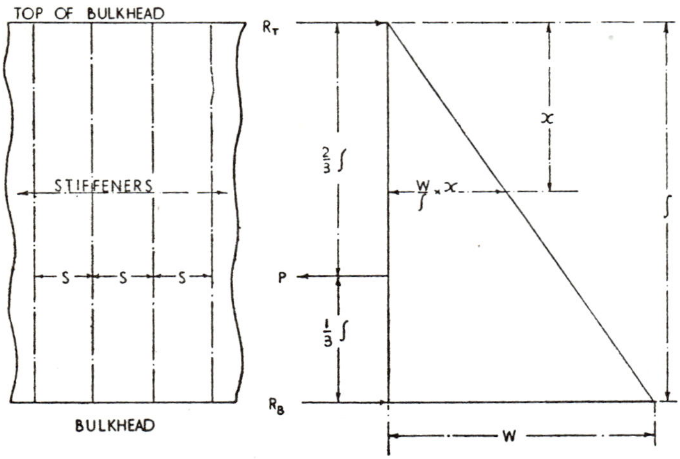

Fig. 7

Let l = length of stiffener in m
s = spacing of stiffeners in m
ρ = density of liquid in kg/m³
P = load on stiffener
W = load/m at bottom of stiffener

Then
$$P = \rho g l s \times \frac{l}{2}$$
$$= \tfrac{1}{2}\rho g l^2 s$$
$$W = \rho g l s$$
$$\therefore \quad P = W \times \frac{l}{2}$$

The load P acts at the centre of pressure which is $\frac{2}{3}l$ from the top. Reactions are set up by the end connections at the top (R_T) and at the bottom (R_B).

Taking moments about the top,

$$R_B \times l = P \times \tfrac{2}{3}l$$
$$R_B = \tfrac{2}{3}P$$
$$\text{and } R_T = \tfrac{1}{3}P$$

The shearing force at a distance x from the top will be the reaction at the top, less the area of the load diagram from this point to the top.

$$\text{i.e. } SF_x = R_T - \frac{W\,x}{l} \times \frac{x}{2}$$

$$= \tfrac{1}{3}P - \frac{Wx^2}{2l}$$

$$= \frac{Wl}{6} - \frac{Wx^2}{2l}$$

Let $x = 0$

$$\text{SF at top} = \frac{Wl}{6}$$

$$= \tfrac{1}{3}P$$

Let $x = l$

$$\text{SF at bottom} = \frac{Wl}{6} - \frac{Wl^2}{2l}$$

$$= \frac{Wl}{6} - \frac{Wl}{2}$$

$$= -\frac{Wl}{3}$$

$$= -\tfrac{2}{3}P$$

Since the shearing force is positive at the top and negative at the bottom, there must be some intermediate point at which the shearing force is zero. This is also the position of the maximum bending moment.

Let SF $= 0$

$$O = \frac{Wl}{6} - \frac{Wx^2}{2l}$$

$$\frac{Wx^2}{2l} = \frac{Wl}{6}$$

$$x^2 = \frac{l^2}{3}$$

Position of zero shear $x = \frac{l}{\sqrt{3}}$ from the top

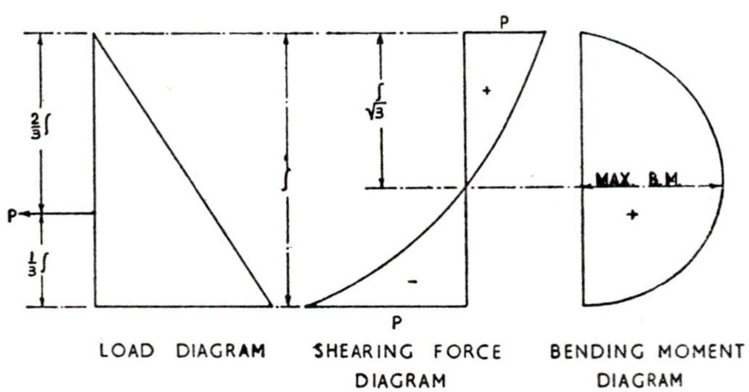

Fig. 8.

Example. A bulkhead 9 m deep is supported by vertical stiffeners 750 mm apart. The bulkhead is flooded to the top edge with sea water on one side only. Calculate:

(a) shearing force at top
(b) shearing force at bottom
(c) position of zero shear.

$$\text{Load on stiffener } P = \rho g A H$$

$$= 1{\cdot}025 \times 9{\cdot}81 \times 9 \times 0{\cdot}75 \times \frac{9}{2}$$

$$= 305{\cdot}4 \text{ kN}$$

(a) Shearing force at top $= \frac{1}{3}P$

$$= \frac{305{\cdot}4}{3}$$

$$= 101{\cdot}8 \text{ kN}$$

(b) Shearing force at bottom $= \frac{2}{3}P$

$$= \tfrac{2}{3} \times 305{\cdot}4$$

$$= 203{\cdot}6 \text{ kN}$$

(c) Position of zero shear $= \dfrac{l}{\sqrt{3}}$

$$= \frac{9}{\sqrt{3}}$$

$$= 5{\cdot}197 \text{ m from the top}$$

TEST EXAMPLES 1

1. A piece of aluminium has a mass of 300 g and its volume is 42 cm^3. Calculate:
(a) its density in kg/m^3
(b) its relative density
(c) the mass of 100 cm^3 of aluminium.

2. A rectangular double bottom tank 12 m long and 10 m wide is full of sea water. Calculate the head of water above the tank top if the load due to water pressure on the tank top is 9·6 MN.

3. A double bottom tank is 1·2 m deep and has a sounding pipe extending 11 m above the tank top. The tank is filled with oil (rd 0·89) to the top of the sounding pipe. The double bottom floors are spaced 750 mm apart and are connected to the tank top by riveted angles, the rivets having a pitch of 7 diameters. If the maximum allowable stress in the rivets is 30 MN/m^2, calculate the pressure in kN/m^2 on the outer bottom and the diameter of the rivets.

4. A ballast tank is 15 m long, 12 m wide and 1·4 m deep and is filled with fresh water. Calculate the load on the top and short side, if:
(a) the tank is just completely full
(b) there is a head of 7m of water above the tank top.

5. A vertical bulkhead 9 m wide and 8 m deep has sea water on one side only to a depth of 6 m. Calculate the pressure in kN/m^3 at the bottom of the bulkhead and the load on the bulkhead.

6. A bulkhead is in the form of a trapezoid 9 m wide at the deck, 5 m wide at the bottom and 8 m deep. Find the load on the bulkhead if it has oil (rd 0·85) on one side only:
(a) to a depth of 6 m
(b) with 4 m head to the top edge.

*f*7. The end bulkhead of an oil fuel bunker is in the form of a rectangle 10 m wide and 12 m high. Find the total load and the position of the centre of pressure relative to the top of the bulkhead if the tank is filled with oil (rd 0·9):
(a) to the top edge
(b) with 3 m head to the top edge.

*f*8. A dock gate 6 m wide and 5 m deep has fresh water on one side to a depth of 3 m and sea water on the other side to a depth of 4 m. Calculate the resultant load and position of the centre of pressure.

9. A triangular bulkhead is 5 m wide at the top and 7 m deep. It is loaded to a depth D with sea water, when it is found that the load on the bulkhead is 190 kN. Find the depth D and the distance from the top of the bulkhead to the centre of pressure.

*f*10. A triangular bulkhead is 7 m wide at the top and has a vertical depth of 8 m. Calculate the load on the bulkhead and the position of the centre of pressure if the bulkhead is flooded with sea water on only one side:
(a) to the top edge
(b) with 4 m head to the top edge.

11. A watertight bulkhead is 8 m high and is supported by vertical stiffeners 700 mm apart, connected at the tank top by brackets having 10 rivets 20 mm diameter. The bulkhead is flooded to its top edge with sea water. Determine:
(a) shearing force at top of stiffeners,
(b) shear stress in the rivets,
(c) position of zero shear.

12. A bulkhead is supported by vertical stiffeners. The distance between the stiffeners is one ninth of the depth of the bulkhead. When the bulkhead is flooded to the top with sea water on one side only, the maximum shearing force in the stiffeners is 200 kN. Calculate:
(a) the height of the bulkhead,
(b) the shearing force at the top of the stiffeners,
(c) the position of zero shear.

CHAPTER 2

DISPLACEMENT, TPC, COEFFICIENTS OF FORM

ARCHIMEDES' PRINCIPLE

If a solid body is immersed in a liquid there is an apparent loss in weight. This loss in weight is the upthrust exerted by the liquid on the body and is equal to the weight of the volume of liquid which the body displaces.

If a solid body is suspended in fresh water, completely immersed, the upthrust is the weight of fresh water having the same volume as the body. Using this principle it is possible to determine the relative density of an irregular body

$$\text{Relative density} = \frac{\text{mass of body}}{\text{mass of equal volume of fresh water}}$$

$$= \frac{\text{weight of body in air}}{\text{weight in air—weight in fresh water}}$$

$$= \frac{\text{weight of body in air}}{\text{upthrust in fresh water}}$$

$$= \frac{\text{mass of body}}{\text{apparent loss in mass in fresh water}}$$

It will be noticed here that either mass or weight may be used as long as the units of the numerator are the same as those of the denominator. This is acceptable since the value of g used to obtain the weight in air is the same as the value of g used to obtain the upthrust in water.

Example. A solid block of cast iron has a mass of 500 kg. When it is completely immersed in fresh water the mass appears to be reduced to 430 kg. Calculate the relative density of cast iron.

$$\text{Mass of cast iron} = 500 \text{ kg}$$
$$\text{Apparent loss of mass in fresh water} = 500 - 430$$
$$= 70 \text{ kg}$$
$$\text{Relative density} = \frac{500}{70}$$
$$= 7{\cdot}143$$

Example. A piece of brass (rd 8·4) 0·06 m³ in volume is suspended in oil of rd 0·8. Calculate the apparent mass of the brass.

$$\text{Mass of brass} = 1000 \times 8{\cdot}4 \times 0{\cdot}06$$
$$= 504 \text{ kg}$$
$$\text{Mass of equal volume of oil} = 1000 \times 0{\cdot}8 \times 0{\cdot}06$$
$$= 48 \text{ kg}$$
$$\therefore \quad \text{Apparent mass in oil} = 504 - 48$$
$$= 456 \text{ kg}$$

FLOATING BODIES

If a solid body, having a relative density less than 1, is completely immersed in fresh water, the upthrust exerted by the water on the body will exceed the weight of the body. The body will then rise until part of its volume emerges, i.e. it will *float*. The upthrust will then be reduced to the weight of the body. Thus:

(a) A body of rd less than 1 will float in fresh water.
A body of rd of less than 13·6 will float in mercury whose rd is 13·6.

(b) The *weight* of a floating body is equal to the *weight* of the volume of liquid it displaces, and since g is constant, the *mass* of a floating body is equal to the *mass* of the volume of liquid it displaces.

(c) The percentage of volume of a floating body which remains immersed depends upon the relative density of the body and the relative density of the liquid; e.g. a body of rd 0·8 will float in fresh water with 80 per cent of its volume immersed, and in sea water with $\frac{1{\cdot}000}{1{\cdot}025} \times 80$ per cent of its volume immersed.

Example, A block of wood 4 m long, 0·3 m wide and 0·25 m thick floats at a draught of 0·15 m in sea water. Calculate the mass of the wood and its relative density.

$$\begin{aligned}\text{mass of wood} &= \text{mass of water displaced}\\ &= 1025 \times 4 \times 0{\cdot}3 \times 0{\cdot}15\\ &= 184{\cdot}5 \text{ kg}\end{aligned}$$

$$\begin{aligned}\text{mass of equal volume of fresh water} &= 100 \times 4 \times 0{\cdot}3 \times 0{\cdot}25\\ &= 300 \text{ kg}\end{aligned}$$

$$\begin{aligned}\text{Relative density of wood} &= \frac{184{\cdot}5}{300}\\ &= 0{\cdot}615\end{aligned}$$

Example. A box barge 40 m long and 9 m wide floats in sea water at a draught of 3·5 m. Calculate the mass of the barge.

$$\begin{aligned}\text{mass of barge} &= \text{mass of water displaced}\\ &= 1025 \times 40 \times 9 \times 3{\cdot}5\\ &= 1292 \times 10^3 \text{ kg}\\ &= 1292 \text{ tonne}\end{aligned}$$

DISPLACEMENT

When a ship is floating freely at rest the mass of the ship is equal to the mass of the volume of water displaced by the ship and is therefore known as the *displacement* of the ship. Thus if the volume of the underwater portion of the ship is known, together with the density of the water, it is possible to obtain the displacement of the ship.

It is usual to assume that a ship floats in sea water of density 1025 kg/m³ or 1·025 t/m³. Corrections may then be made if the

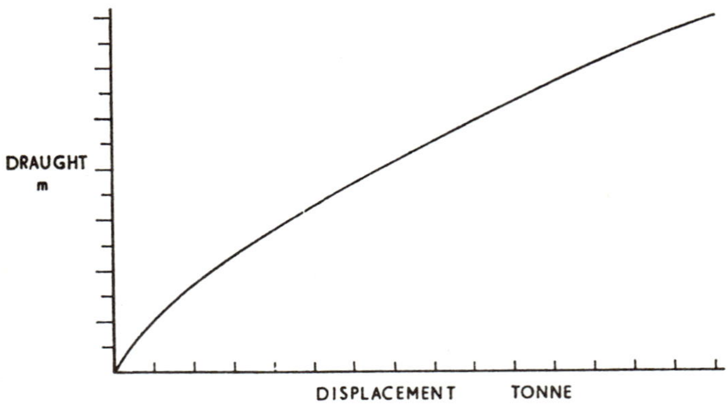

Fig. 9

vessel floats in water of any other density. Since the volume of water displaced depends upon the draught, it is useful to calculate values of displacement for a range of draughts. These values may then be plotted to form a *displacement curve*, from which the displacement may be obtained at any intermediate draught.

The following symbols will be used throughout the text:

$\triangle$ = displacement in tonne

∇ = volume of displacement in m^3

Thus for sea water $\triangle = \nabla \times 1{\cdot}025$

Some confusion exists between the *mass* of the ship and the *weight* of the ship. This confusion may be reduced if the displacement is always regarded as a mass. The gravitational force acting on this mass—the weight of the ship—will then be the product of the displacement $\triangle$ and the acceleration due to gravity g.

Hence mass of ship (displacement) = $\triangle$ tonne

weight of ship = $\triangle\ g$kN

Example. A ship displaces 12 240 m^3 of sea water at a particular draught.

(a) Calculate the displacement of the ship.

(b) How many tonnes of cargo would have to be discharged for the vessel to float at the same draught in fresh water?

(a) Displacement in sea water = 12 240 × 1·025

= 12 546 tonne

(b) Displacement at same draught in fresh water

= 12 240 × 1·000

= 12 240 tonne

∴ Cargo to be discharged = 12 546 — 12 240

= 306 tonne

BUOYANCY

Buoyancy is the term given to the upthrust exerted by the water on the ship. If a ship floats freely, the buoyancy is equal to the weight of the ship.

The force of buoyancy acts at the *centre of buoyancy*, which is the centre of gravity of the underwater volume of the ship.

The *longitudinal* position of the centre of buoyancy (LCB) is usually given as a distance forward or aft of midships and is represented by the longitudinal centroid of the curve of immersed cross-sectional areas (see Chapter 3).

The *vertical* position of the centre of buoyancy (VCB) is usually given as a distance above the keel. This distance is denoted by *KB* and is represented by the vertical centroid of the waterplane area curve (see Chapter 3). The distance from the waterline to the VCB may be found by two other methods:

(a) from the displacement curve (Fig. 10)

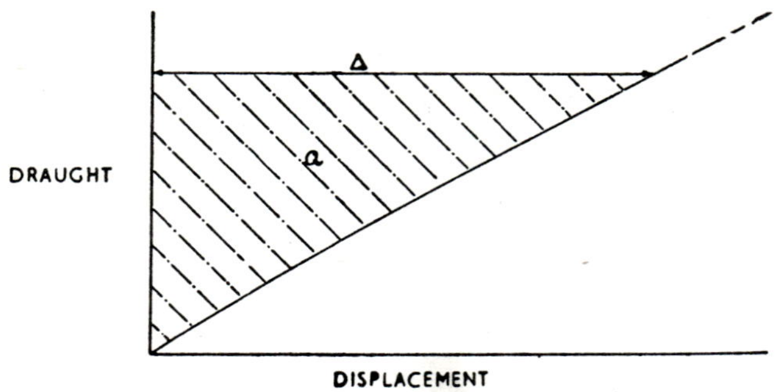

Fig. 10

VCB below waterline

$$= \frac{\text{area between displacement curve and draught axis}}{\text{displacement}}$$

$$= \frac{a}{\Delta}$$

(b) by Morrishes approximate formula

$$\text{VCB below the waterline} = \frac{1}{3}\left(\frac{d}{2} + \frac{\nabla}{A_w}\right)$$

where d = draught in m
∇ = volume of displacement in m^3
A_w = waterplane area in m^2

TONNE PER CENTIMETRE IMMERSION

The tonne per centimetre immersion (TPC) of a ship at any given draught is the mass required to increase the mean draught by 1 cm.

Consider a ship floating in water of density ρ t/m³.
If the mean draught is increased by 1 cm, then:

$$\text{Increase in volume of displacement} = \frac{1}{100} \times \text{waterplane area}$$

$$= \frac{A_w}{100} \quad \text{m}^3$$

$$\text{Increase in displacement} = \frac{A_w}{100} \times \rho \quad \text{t}$$

$$\text{Thus} \qquad \text{TPC} = \frac{A_w \times \rho}{100}$$

$$\text{For sea water } \rho = 1{\cdot}025 \text{ t/m}^3$$

$$\therefore \quad \text{TPC sw} = 0{\cdot}01025\, Aw$$

At different draughts, variations in waterplane area cause variations in TPC. Values of TPC may be calculated for a range of draughts and plotted to form a TPC curve, from which values of TPC may be obtained at intermediate draughts.

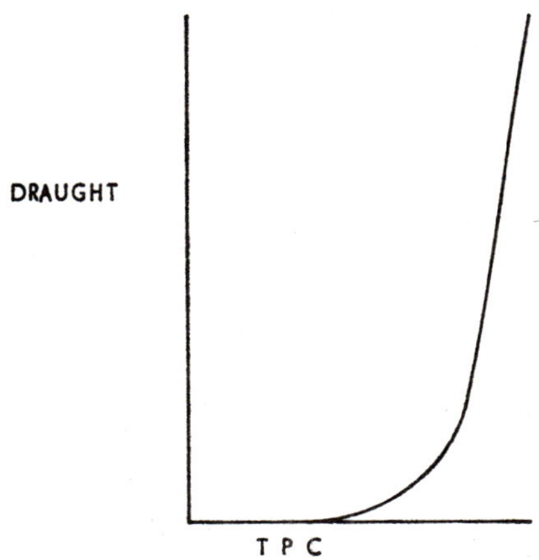

Fig. 11

The area between the TPC curve and the draught axis to any given draught represents the displacement of the ship at that draught, while its centroid represents the vertical position of the centre of buoyancy.

It may be assumed for small alterations in draught, that the ship is wall-sided and therefore TPC remains constant. If the change in draught exceeds about 0·5 m, then a mean TPC value should be used. If the change in draught is excessive, however, it is more accurate to use the area of the relevant part of the TPC curve.

Example. The waterplane area of a ship is 1730 m². Calculate the TPC and the increase in draught if a mass of 270 tonne is added to the ship.

$$\begin{aligned} \text{TPC} &= 0{\cdot}01025 \times 1730 \\ &= 17{\cdot}73 \end{aligned}$$

$$\begin{aligned} \text{Increase in draught} &= \frac{\text{mass added}}{\text{TPC}} \\ &= \frac{270}{17{\cdot}73} \\ &= 15{\cdot}23 \text{ cm} \end{aligned}$$

COEFFICIENTS OF FORM

Coefficients of form have been devised to show the relation between the form of the ship and the dimensions of the ship.

WATERPLANE AREA COEFFICIENT C_w is the ratio of the area of the waterplane to the product of the length and breadth of the ship. (Fig. 12).

$$\begin{aligned} C_w &= \frac{\text{waterplane area}}{\text{length} \times \text{breadth}} \\ &= \frac{A_w}{L \times B} \end{aligned}$$

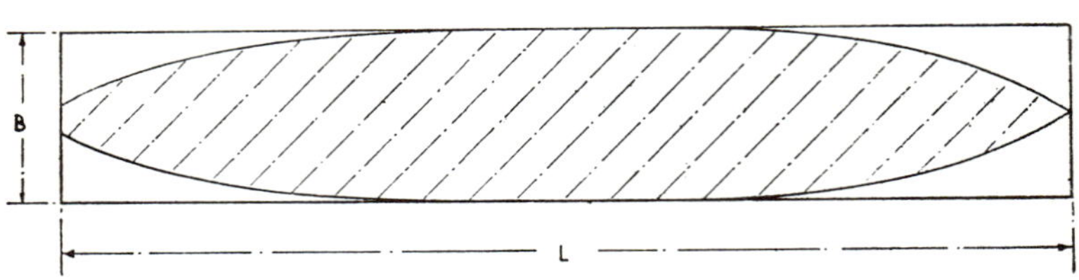

Fig. 12

MIDSHIP SECTION AREA COEFFICIENT C_m is the ratio of the area of the immersed portion of the midship section to the product of the breadth and the draught (Fig. 13).

$$C_m = \frac{\text{area of immersed midship section}}{\text{breadth} \times \text{draught}}$$

$$= \frac{A_m}{B \times d}$$

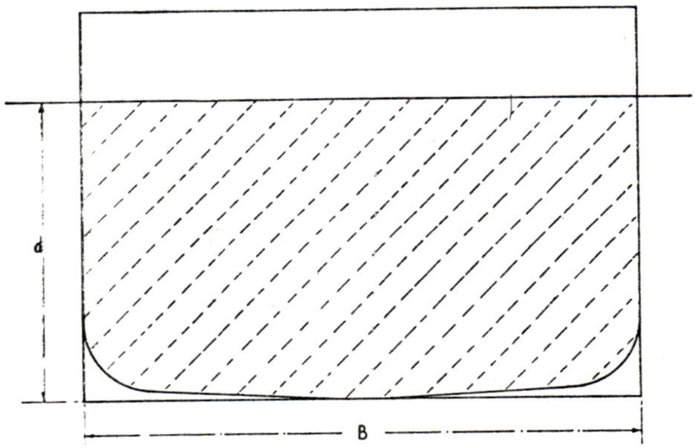

Fig. 13

BLOCK COEFFICIENT OR COEFFICIENT OF FINENESS C_b is the ratio of the volume of displacement to the product of the length, breadth and draught (Fig. 14).

$$C_b = \frac{\text{volume of displacement}}{\text{length} \times \text{breadth} \times \text{draught}}$$

$$= \frac{\nabla}{L \times B \times d}$$

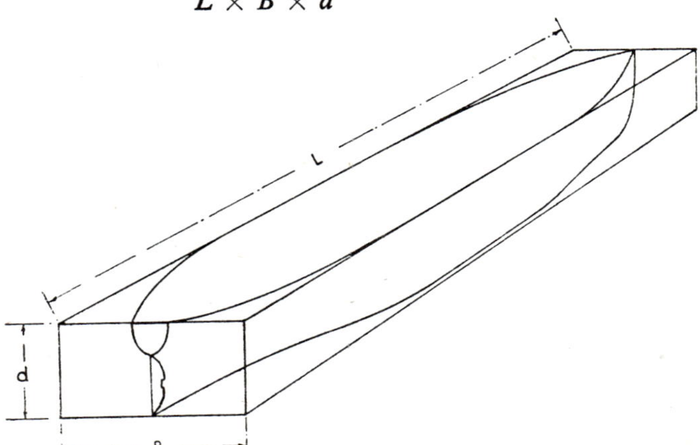

Fig. 14

PRISMATIC COEFFICIENT C_p is the ratio of the volume of displacement to the product of the length and the area of the immersed portion of the midship section (Fig. 15).

$$C_p = \frac{\text{volume of displacement}}{\text{length} \times \text{area of immersed midship section}}$$

$$= \frac{\nabla}{L \times A_m}$$

But $\nabla = C_b \times L \times B \times d$

and $A_m = C_m \times B \times d$

Substituting these in the expression for C_p;

$$C_p = \frac{C_b \times L \times B \times d}{L \times C_m \times B \times d}$$

$$C_p = \frac{C_b}{C_m}$$

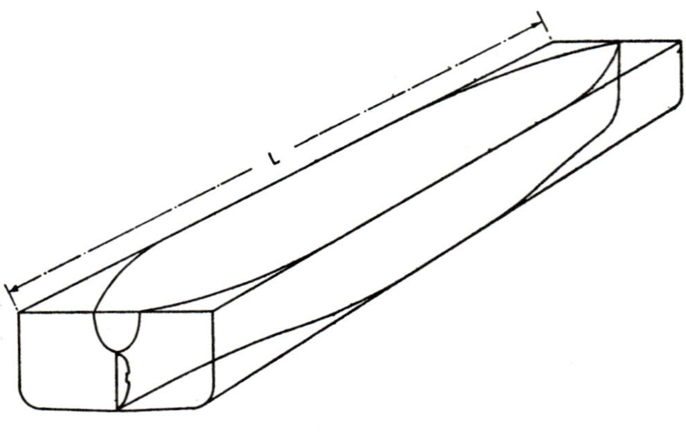

Fig. 15

Example. A ship 135 m long, 18 m beam and 7·6 m draught has a displacement of 14 000 tonne. The area of the load waterplane is 1925 m² and the area of the immersed midship section 130 m² Calculate (a) C_w; (b) C_m; (c) C_b; (d) C_p.

(a)
$$C_w = \frac{1925}{135 \times 18}$$
$$= 0{\cdot}792$$

(b)
$$C_m = \frac{130}{18 \times 7{\cdot}6}$$
$$= 0{\cdot}950$$

(c)
$$\nabla = \frac{14\,000}{1{\cdot}025}$$
$$= 13\,658 \text{ m}^3$$
$$C_b = \frac{13\,658}{135 \times 18 \times 7{\cdot}6}$$
$$= 0{\cdot}740$$

(d)
$$C_p = \frac{13\,658}{135 \times 130}$$
$$= 0{\cdot}778$$

Alternatively
$$C_p = \frac{0{\cdot}740}{0{\cdot}950}$$
$$= 0{\cdot}778$$

WETTED SURFACE AREA

The wetted surface area of a ship is the area of the ship's hull which is in contact with the water. This area may be found by putting the transverse girths of the ship, from waterline to waterline, through Simpson's Rule and adding about ½ per cent to allow for the longitudinal curvature of the shell. To this area should be added the wetted surface area of appendages such as cruiser stern, rudder and bilge keels.

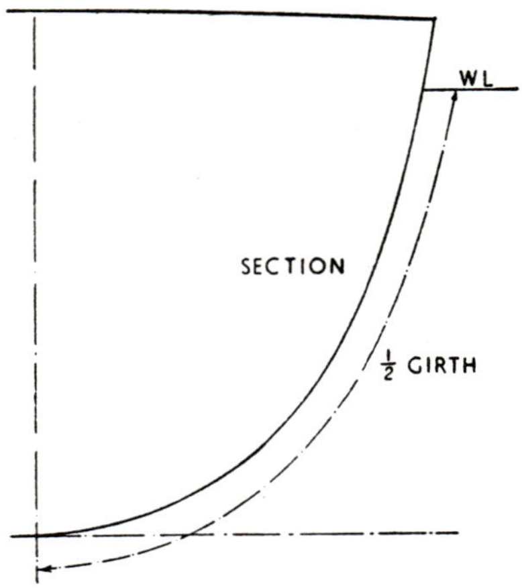

Fig. 16

Several approximate formulae for wetted surface area are available, two of which are:

Denny $$S = 1{\cdot}7Ld + \frac{\nabla}{d}$$

Taylor: $$S = c\sqrt{\Delta L}$$

where S = wetted surface area in m^2
L = length of ship in m
d = draught in m
∇ = volume of displacement in m^3
Δ = displacement in tonne
c = a coefficient of about 2·6 which depends upon the shape of the ship.

Example. A ship of 5000 tonne displacement, 95 m long, floats at a draught of 5·5 m. Calculate the wetted surface area of the ship:

(a) Using Denny's formula

(b) Using Taylor's formula with c = 2·6

(a) $$S = 1{\cdot}7Ld + \frac{\nabla}{d}$$

$$= 1{\cdot}7 \times 95 \times 5{\cdot}5 + \frac{5000}{1{\cdot}025 \times 5{\cdot}5}$$

$$= 888{\cdot}2 + 886{\cdot}9$$
$$= 1775{\cdot}1 \text{ m}^2$$

(b) $$S = c\sqrt{\triangle L}$$
$$= 2{\cdot}6\sqrt{5000 \times 95}$$
$$= 1793 \text{ m}^2$$

SIMILAR FIGURES

Two planes or bodies are said to be similar when their linear dimensions are in the same ratio. This principle may be seen in a projector where a small image is projected from a slide onto a screen. The height of the image depends upon the distance of the screen from the light source, but the *proportions and shape* of the image remain the same as the image on the slide. Thus the image on the screen is a scaled-up version of the image on the slide.

The *areas* of similar figures vary as the *square* of their corresponding dimensions. This may be shown by comparing two circles having diameters D and d respectively.

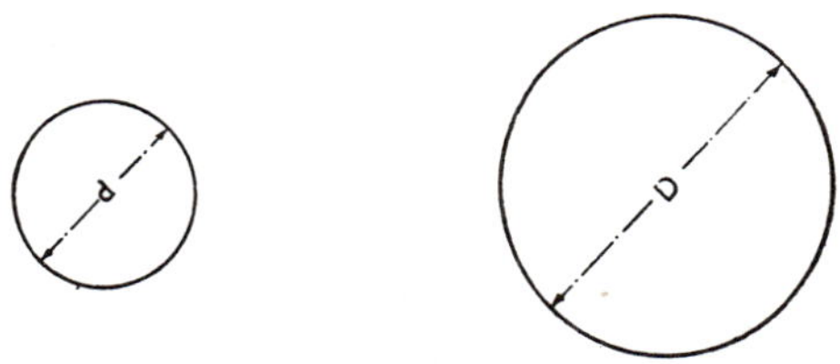

Fig. 17

$$\text{Area of large circle} = \frac{\pi}{4} D^2$$

$$\text{Area of small circle} = \frac{\pi}{4} d^2$$

Since $\frac{\pi}{4}$ is constant:

$$\text{ratio of areas} = \frac{D^2}{d^2}$$

Thus if the diameter D is *twice* diameter d, the area of the former is *four times* the area of the latter.

$$\frac{A_1}{A_2} = \left(\frac{L_1}{L_2}\right)^2 = \left(\frac{B_1}{B_2}\right)^2$$

The *volumes* of similar bodies vary as the *cube* of their corresponding dimensions. This may be shown by comparing two spheres of diameters D and d respectively.

$$\text{Volume of large sphere} = \frac{\pi}{6} D^3$$

$$\text{Volume of small sphere} = \frac{\pi}{6} d^3$$

Since $\frac{\pi}{6}$ is constant:

$$\text{ratio of volumes} = \frac{D^3}{d^3}$$

Thus if diameter D is *twice* diameter d, the volume of the former is *eight times* the volume of the latter.

$$\frac{V_1}{V_2} = \left(\frac{L_1}{L_2}\right)^3 = \left(\frac{B_1}{B_2}\right)^3 = \left(\frac{D_1}{D_2}\right)^3$$

These rules may be applied to any similar bodies no matter what their shape, and in practice are applied to ships.

Thus if L = length of ship
S = wetted surface area
$\triangle$ = displacement,

then $S \propto L^2$
or $S^{\frac{1}{2}} \propto L$
and $\triangle \propto L^3$
or $\triangle^{\frac{1}{3}} \propto L$
$\therefore$ $S^{\frac{1}{2}} \propto \triangle^{\frac{1}{3}}$
and $S \propto \triangle^{\frac{2}{3}}$
or $\triangle \propto S^{\frac{3}{2}}$

Example. A ship 110 m long displaces 9000 tonne and has a wetted surface area of 2205 m². Calculate the displacement and wetted surface area of a 6 m model of the ship.

$$\frac{\triangle_1}{\triangle_2} = \left(\frac{L_1}{L_2}\right)^3$$

$$\triangle_2 = 9000\left(\frac{6}{110}\right)^3$$

Displacement of model = 1·46 tonne

$$\frac{S_1}{S_2} = \left(\frac{L_1}{L_2}\right)^2$$

$$S_2 = 2205\left(\frac{6}{110}\right)^2$$

Wetted surface area of model
= 6·56 m²

f SHEARING FORCE AND BENDING MOMENT

Consider a loaded ship lying in still water. The upthrust over any unit length of the ship depends upon the immersed cross-sectional area of the ship at that point. If the values of upthrust at different positions along the length of the ship are plotted on a base representing the ship's length, a *bouyancy curve* is formed (Fig. 18). This curve increases from zero at each end to a maximum value in way of the parallel midship portion. The area of this curve represents the total upthrust exerted by the water on the ship.

The total weight of a ship consists of a number of independent weights concentrated over short lengths of the ship, such as cargo, machinery, accommodation, cargo handling gear, poop and forecastle, and a number of items which form continuous material over the length of the ship, such as decks, shell and tank top. A *curve of weights* is shown in the diagram.

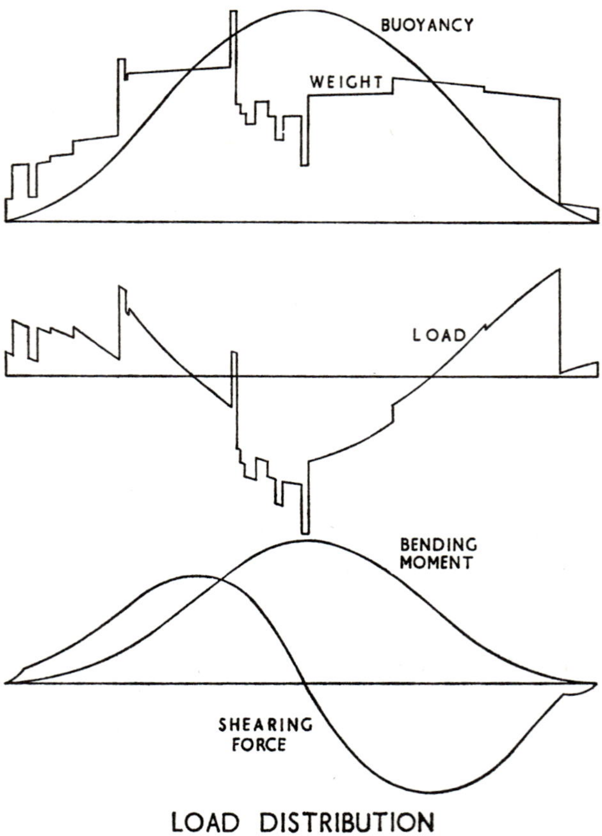

Fig. 18

The difference between the weight and buoyancy at any point is the *load* at that point. In some cases the load is an excess of weight over buoyancy and in other cases an excess of buoyancy over weight. A load diagram is formed by plotting these differences. Because of this unequal loading, however, shearing forces and bending moments are set up in the ship.

The *shearing force* at any point is represented by the *area* of the load diagram on one side of that point. A *shearing force diagram* may be formed by plotting these areas on a base of the ship's length.

The *bending moment* at any point is represented by the *area* of the shearing force diagram on one side of that point. A *bending moment diagram* may be formed by plotting such areas on a base of the ship's length.

The maximum bending moment occurs where the shearing force is zero and this is usually near amidships.

Example. A box barge 200 m long is divided into five equal compartments. The weight is uniformly distributed along the vessel's length.

500 tonne of cargo are added to each of the end compartments. Sketch the shearing force and bending moment diagrams and state their maximum values.

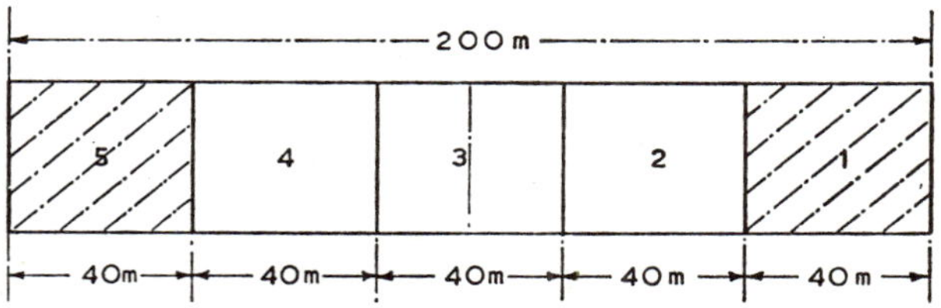

Fig. 19

Before adding the cargo, the buoyancy and weight were equally distributed and produced no shearing force or bending moment. It is therefore only necessary to consider the added cargo and the additional buoyancy required.

$$\text{Additional buoyancy/m} = \frac{1000\,g}{200}$$

$$= 5\,g\text{ kN}$$

Compartments 1 and 5

$$\text{Additional weight/m} = \frac{500g}{40}$$

$$= 12{\cdot}5\,g \text{ kN}$$

$$\text{Load/m} = 12{\cdot}5\,g - 5\,g$$

$$= 7{\cdot}5\,g \text{ kN excess weight}$$

Compartments 2, 3 and 4

$$\text{Load/m} = 5\,g \text{ kN excess buoyancy}$$

These values may be plotted to form a load diagram.

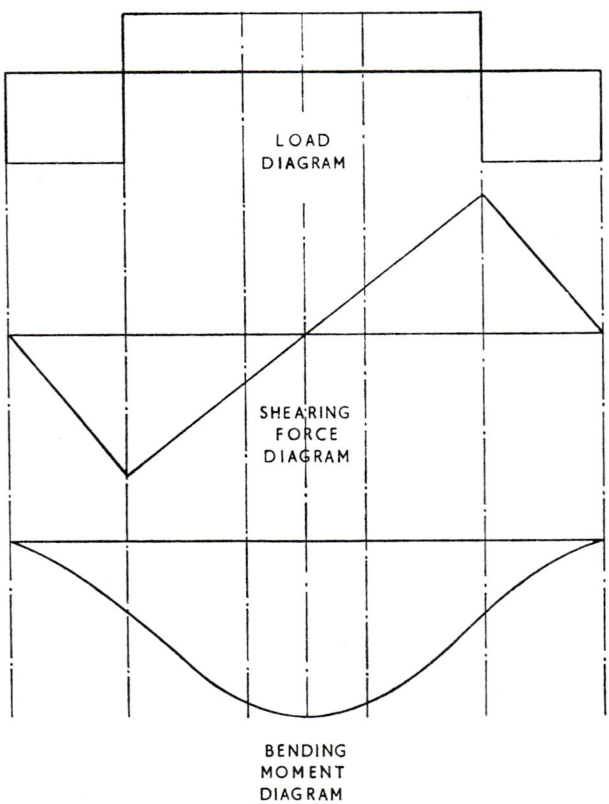

Fig. 20

$$
\begin{aligned}
\text{Shearing force at A} &= 0 \\
\text{Shearing force at B} &= -\ 7{\cdot}5\, g \times 40 \\
&= -\ 300\, g = 2943 \text{ kN } \textit{max.} \\
\text{Shearing force at C} &= -\ 300\, g + 5\, g \times 40 \\
&= -\ 100\, g = 981 \text{ kN} \\
\text{Shearing force at D} &= -300\, g + 5\, g \times 60 \\
&= 0 \\
\text{Shearing force at E} &= +\ 100\, g \\
\text{Shearing force at F} &= +\ 300\, g \\
\text{Shearing force at G} &= 0
\end{aligned}
$$

$$\text{Bending moment at A and G} = 0$$

$$
\begin{aligned}
\text{Bending moment at B and F} &= -300\, g \times \frac{40}{2} \\
&= -\ 6000\, g = 58{\cdot}86 \text{ MN m}
\end{aligned}
$$

$$
\begin{aligned}
\text{Bending moment at C and E} &= -15\,000\, g + 100\, g \times \frac{20}{2} \\
&= -\ 14\,000\, g = 137{\cdot}34 \text{ MN m}
\end{aligned}
$$

$$
\begin{aligned}
\text{Bending moment at D} &= -\ 300\, g \times \frac{100}{2} \\
&= -15\,000\, g \\
&= 147{\cdot}15 \text{ MN m } \textit{max.}
\end{aligned}
$$

TEST EXAMPLES 2

1. A piece of metal 250 cm^3 in volume is attached to the bottom of a block of wood 3·5 dm^3 in volume and having a relative density of 0·6. The system floats in fresh water with 100 cm^3 projecting above the water. Calculate the relative density of the metal.

2. A raft 3 m long and 2 m wide is constructed of timber 0·25 m thick having a relative density of 0·7. It floats in water of density 1018 kg/m^3. Calculate the minimum mass which must be placed on top of the raft to sink it.

3. A box barge 65 m long and 12 m wide floats at a draught of 5·5 m in sea water. Calculate:
   (a) the displacement of the barge,
   (b) its draught in fresh water.

4. A ship has a constant cross-section in the form of a triangle which floats apex down in sea water. The ship is 85 m long, 12 m wide at the deck and has a depth from keel to deck of 9 m. Draw the displacement curve using 1·25 m intervals of draught from the keel to the 7·5 m waterline. From this curve obtain the displacement in fresh water at a draught of 6·50 m.

5. A cylinder 15 m long and 4 m outside diameter floats in sea water with its axis in the waterline. Calculate the mass of the cylinder.

6. Bilge keels of mass 36 tonne and having a volume of 22 m^3 are added to a ship. If the TPC is 20, find the change in mean draught.

7. A vessel 40 m long has a constant cross-section in the form of a trapezoid 10 m wide at the top, 6 m wide at the bottom and 5 m deep. It floats in sea water at a draught of 4 m. Calculate its displacement.

8. The waterplane areas of a ship at 1·25 m intervals of draught, commencing at the 7·5 m waterline, are 1845, 1690, 1535, 1355 and 1120 m^2. Draw the curve of tonne per cm immersion and determine the mass which must be added to increase the mean draught from 6·10 m to 6·30 m.

9. A ship 150 m long and 20·5 m beam floats at a draught of 8 m and displaces 19 500 tonne. The TPC is 26·5 and midship section area coefficient 0·94. Calculate the block, prismatic and waterplane area coefficients.

10. A ship displaces 9450 tonne and has a block coefficient of 0·7. The area of immersed midship section is 106 m^2.

If a beam $= 0{\cdot}13 \times$ length $= 2{\cdot}1 \times$ draught, calculate the length of the ship and the prismatic coefficient.

11. The length of a ship is 18 times the draught, while the breadth is 2·1 times the draught. At the load waterplane, the waterplane area coefficient is 0·83 and the difference between the TPC in sea water and the TPC in fresh water is 0·7. Determine the length of the ship and the TPC in fresh water.

12. The $\frac{1}{2}$ girths of a ship 90 m long are as follows: 2·1, 6·6, 9·3, 10·5, 11·0, 11·0, 11·0, 9·9, 7·5, 3·9 and 0 m respectively. The wetted surface area of the appendages is 30 m^2 and $\frac{1}{2}\%$ is to be added for longitudinal curvature. Calculate the wetted surface area of the ship.

13. A ship of 14 000 tonne displacement, 130 m long, floats at a draught of 8 m. Calculate the wetted surface area of the ship using:
(a) Denny's formula
(b) Taylor's formula with $c = 2{\cdot}58$

14. A box barge is 75 m long, 9 m beam and 6 m deep. A similar barge having a volume of 3200 m^2 is to be constructed. Calculate the length, breadth and depth of the new barge.

15. The wetted surface area of a ship is twice that of a similar ship. The displacement of the latter is 2000 tonne less than the former. Determine the displacement of the latter.

16. A ship 120 m long displaces 11 000 tonne and has a wetted surface area of 2500 m^2. Calculate the displacement and wetted surface area of a 6 m model of the ship.

CHAPTER 3

CALCULATION OF AREA, VOLUME, FIRST AND SECOND MOMENTS

SIMPSON'S FIRST RULE

Simpson's First Rule is based on the assumption that the curved portion of a figure forms part of a parabola ($y = ax^2 + bx + c$), and gives the area contained between *three* consecutive, equally-spaced ordinates.

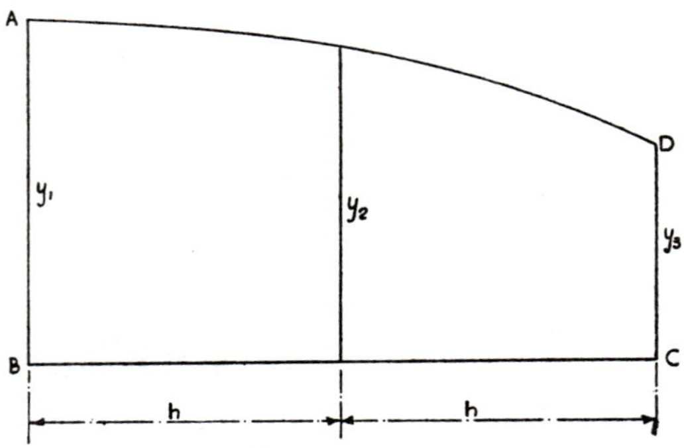

Fig 21.

$$\text{Area ABCD} = \frac{h}{3}(1y_1 + 4y_2 + 1y_3)$$

This rule may be applied repeatedly to determine the area of a larger plane such as EFGH (Fig. 22).

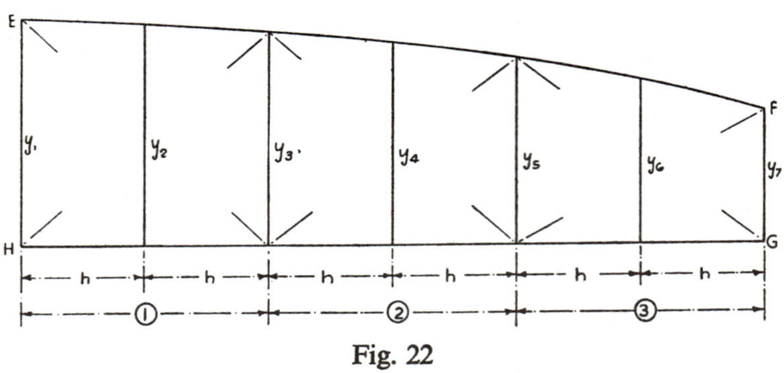

Fig. 22

$$\text{Area } 1 = \frac{h}{3}(1y_1 + 4y_2 + 1y_3)$$

$$\text{Area } 2 = \frac{h}{3}(1y_3 + 4y_4 + 1y_5)$$

$$\text{Area } 3 = \frac{h}{3}(1y_5 + 4y_6 + 1y_7)$$

Area EFGH
= Area 1 + Area 2 + Area 3

$$= \frac{h}{3}[(1y_1+4y_2+1y_3)+(1y_3+4y_4+1y_5)+(1y_5+4y_6+1y_7)]$$

$$= \frac{h}{3}[1y_1+4y_2+2y_3+4y_4+2y_5+4y_6+1y_7]$$

It should be noted at this stage that it is necessary to apply the whole rule and thus an *odd* number of equally-spaced ordinates is necessary. Greater speed and accuracy is obtained if this rule is applied in the form of a table. The distance h is termed the common interval and the numbers 1, 4, 2, 4, etc. are termed Simpson's multipliers.

When calculating the area of a waterplane it is usual to divide the length of the ship into about 10 equal parts, giving 11 sections. These sections are numbered from 0 at the after end to 10 at the fore end. Thus amidships will be section number 5. It is convenient to measure distances from the centreline to the ship side, giving half ordinates. These half ordinates are used in conjunction with Simpson's Rule and the answer multiplied by 2.

Example. The equally-spaced half ordinates of a watertight flat 27 m long are 1·1, 2·7, 4·0, 5·1, 6·1, 6·9 and 7·7 m respectively.

Calculate the area of the flat.

½ Ordinate	Simpson's Multipliers	Product for Area
1·1	1	1·1
2·7	4	10·8
4·0	2	8·0
5·1	4	20·4
6·1	2	12·2
6·9	4	27·6
7·7	1	7·7
		87·8 = Σ_A

Since there are 7 ordinates there will be 6 spaces

$$\therefore \text{ Common interval} = \frac{27}{6} = 4 \cdot 5 \text{ m}$$

$$\text{Area} = \frac{h}{3} \Sigma_A \times 2 = \frac{4 \cdot 5}{3} \times 87 \cdot 8 \times 2$$

$$= 263 \cdot 4 \text{ m}^2$$

APPLICATION TO VOLUMES

Simpson's Rule is a mathematical rule which will give the area under any continuous curve, no matter what the ordinates represent. If the immersed cross-sectional areas of a ship at a number of positions along the length of the ship are plotted on a base representing the ship's length (Fig. 23), the area under the resulting curve will represent the volume of water displaced by the ship and may be found by putting the cross-sectional areas through Simpson's Rule. Hence the displacement of the ship at any given draught may be calculated. The longitudinal centroid of this figure represents the longitudinal centre of buoyancy of the ship.

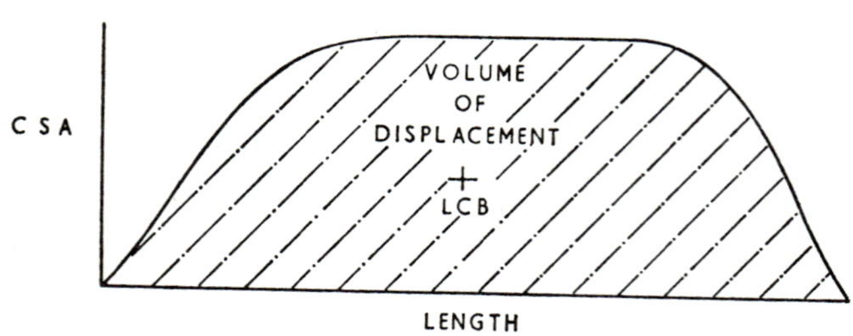

Fig. 23

It is also possible to calculate the displacement by using ordinates of waterplane area or tonne per cm immersion, with a common interval of draught (Fig. 24). The vertical centroids of these two curves represent the vertical centre of buoyancy of the ship.

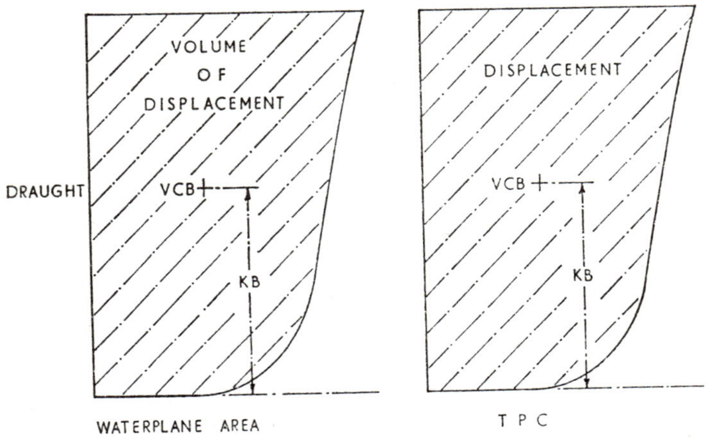

Fig. 24

Similar methods are used to determine hold and tank capacities.

Example. The immersed cross-sectional areas through a ship 180 m long, at equal intervals, are 5, 118, 233, 291, 303, 304, 304, 302, 283, 171, and 0 m^2 respectively. Calculate the displacement of the ship in sea water of 1·025 tonne/m^3.

CSA	SM	Product for Volume
5	1	5
118	4	472
233	2	466
291	4	1164
303	2	606
304	4	1216
304	2	608
302	4	1208
283	2	566
171	4	684
0	1	0
		6995 $= \Sigma_\nabla$

$$\text{Common interval} = \frac{180}{10} = 18 \text{ m}$$

$$\text{Volume of displacement} = \frac{h}{3}\Sigma_{\nabla}$$

$$= \frac{18}{3} \times 6995$$

$$= 41\ 970 \text{ m}^3$$

$$\text{Displacement} = \text{vol of displacement} \times \text{density}$$
$$= 41\ 970 \times 1{\cdot}025$$
$$= 43\ 019 \text{ tonne}$$

Example. The TPC values for a ship at 1·2 m intervals of draught commencing at the keel, are 8·2, 16·5, 18·7, 19·4, 20·0, 20·5 and 21·1 respectively. Calculate the displacement at 7·2 m draught.

Waterplane	TPC	SM	Product for Displacement
0	8·2	1	8·2
1·2	16·5	4	66·0
2·4	18·7	2	37·4
3·6	19·4	4	77·6
4·8	20·0	2	40·0
6·0	20·5	4	82·0
7·2	21·1	1	21·1
			332·3 = Σ_{Δ}

$$\text{Common interval} = 1{\cdot}2 \text{ m or } 120 \text{ cm}$$

$$\text{Displacement} = \frac{h}{3}\Sigma_{\Delta}$$

$$= \frac{120}{3} \times 332{\cdot}3$$

$$= 13\ 292 \text{ tonne}$$

Note: The common interval must be expressed in *centimetres* since the ordinates are tonne per *centimetre* immersion.

USE OF INTERMEDIATE ORDINATES

At the ends of the ship, where the curvature of a waterplane is considerable, it is necessary to reduce the spacing of the ordinates to ensure an accurate result. Intermediate ordinates are introduced to reduce the spacing to half or quarter of the normal spacing. While it is possible to calculate the area of such a waterplane by dividing it into separate sections, this method is not considered advisable. The following method may be used.

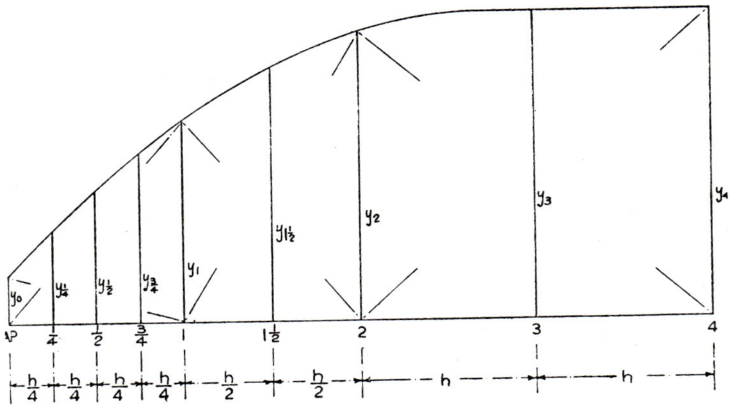

Fig. 25

If the length of the ship is divided initially into 10 equal parts, then:

$$\text{Common interval} = h = \frac{L}{10}$$

It is proposed to introduce intermediate ordinates at a spacing of $\frac{h}{4}$ from section 0 to section 1 and at a spacing of $\frac{h}{2}$ from section 1 to section 2. The ½ ordinates at sections AP, ¼, ½, ¾, etc. will be denoted by yo, $y\frac{1}{4}$, $y\frac{1}{2}$, $y\frac{3}{4}$, etc. respectively.

$$\text{Area from 0 to 1} = \frac{2}{3}\,\frac{h}{4}\,(1y_0 + 4y\tfrac{1}{4} + 2y\tfrac{1}{2} + 4y\tfrac{3}{4} + 1y_1)$$

$$= \tfrac{2}{3}\,h\,(\tfrac{1}{4}y_0 + 1y\tfrac{1}{4} + \tfrac{1}{2}y\tfrac{1}{2} + 1y\tfrac{3}{4} + \tfrac{1}{4}y_1)$$

$$\text{Area from 1 to 2} = \frac{2}{3}\,\frac{h}{2}(1y_1 + 4y1\tfrac{1}{2} + 1y_2)$$

$$= \tfrac{2}{3}\,h\,(\tfrac{1}{2}y_1 + 2y1\tfrac{1}{2} + \tfrac{1}{2}y_2)$$

$$\text{Area from 2 to 4} = \tfrac{2}{3}\,h\,(1y_2 + 4y_3 + 1y_4)$$

Thus Area from 0 to 4

$$= \tfrac{2}{3}\,h\,[(\tfrac{1}{4}y_0 + 1y\tfrac{1}{4} + \tfrac{1}{2}y\tfrac{1}{2} + 1y\tfrac{3}{4} + \tfrac{1}{4}y_1) + (\tfrac{1}{2}y_1 + 2y1\tfrac{1}{2} + \tfrac{1}{2}y_2) + (1y_2 + 4y_3 + 1y_4)]$$

$$= \tfrac{2}{3}\,h\,[\tfrac{1}{4}y_0 + 1y\tfrac{1}{4} + \tfrac{1}{2}y\tfrac{1}{2} + 1y\tfrac{3}{4} + \tfrac{3}{4}y_1 + 2y1\tfrac{1}{2} + 1\tfrac{1}{2}y_2 + 4y_3 + 1y_4]$$

When building up a system of multipliers it is wise to ignore the ordinates and concentrate only on the spacing and the multipliers. The following example shows how these multipliers may be determined

Example. The half ordinates of a cross-section through a ship are as follows:

WL	keel	0·25	0·50	0·75	1·0	1·5	2·0	2·5	3·0	4·0	5·0	6·0	7·0	m
½ ord	2·9	5·0	5·7	6·2	6·6	6·9	7·2	7·4	7·6	7·8	8·1	8·4	8·7	m

Calculate the area of the cross-section to the 7 m waterline.

Let the common interval $= h = 1$ m

Then interval from keel to 1m waterline $= \dfrac{h}{4}$

interval from 1 m to 3 m waterline $= \dfrac{h}{2}$

interval from 3 m to 7 m waterline $= h$

Multipliers from keel to 1 m with common interval $\frac{h}{4}$

$$= 1 : 4 : 2 : 4 : 1$$

or with common interval of h

$$= \tfrac{1}{4} : 1 : \tfrac{1}{2} : 1 : \tfrac{1}{4}$$

Multipliers from 1 m to 3 m with common interval $\frac{h}{2}$

$$= 1 : 4 : 2 : 4 : 1$$

or with common interval of h

$$= \tfrac{1}{2} : 2 : 1 : 2 : \tfrac{1}{2}$$

Multipliers from 3 m to 7 m with common interval h

$$= 1 : 4 : 2 : 4 : 1$$

Adding the respective multipliers we have:

$$\begin{array}{l} \tfrac{1}{4} : 1 : \tfrac{1}{2} : 1 : \tfrac{1}{4} \\ \qquad\qquad\quad\; \tfrac{1}{2} : 2 : 1 : 2 : \tfrac{1}{2} \\ \qquad\qquad\qquad\qquad\qquad\quad\; 1 : 4 : 2 : 4 : 1 \\ \hline \tfrac{1}{4} : 1 : \tfrac{1}{2} : 1 : \tfrac{3}{4} : 2 : 1 : 2 : 1\tfrac{1}{2} : 4 : 2 : 4 : 1 \end{array}$$

Waterline	½ ordinate	SM	Product for Area
Keel	2·9	¼	0·73
0·25	5·0	1	5·00
0·50	5·7	½	2·85
0·75	6·2	1	6·20
1·0	6·6	¾	4·95
1·5	6·9	2	13·80
2·0	7·2	1	7·20
2·5	7·4	2	14·80
3·0	7·6	1½	11·40
4·0	7·8	4	31·20
5·0	8·1	2	16·20
6·0	8·4	4	33·60
7·0	8·7	1	8·70
			156·63

Area of cross-section $= \frac{2}{3} \times 1{\cdot}0 \times 156{\cdot}63$

$= 104{\cdot}42 \text{ m}^2$

f APPLICATION OF SIMPSON'S RULE TO FIRST AND SECOND MOMENTS OF AREA

It is often found necessary to determine the centroid of a curved plane such as a waterplane and the second moment of area of a waterplane.

Consider a plane ABCD (Fig. 26).

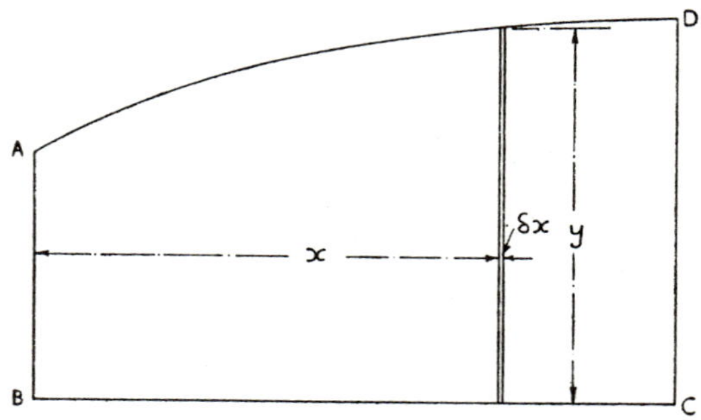

Fig. 26

Divide the plane into thin strips of length δx
Let one such strip, distance x from AB, have an ordinate y.

$$\text{Area of strip} = y \times \delta x$$
$$\therefore \text{Total area of plane} = (y_1 + y_2 + y_3 + \ldots)\delta x$$
$$= \Sigma\, y\, \delta x$$

But the area of the plane may be found by putting the ordinates y through Simpson's Rule.

$$\text{First moment of area of strip about AB} = x \times y\, \delta x$$
$$= x\, y\, \delta x$$
$$\therefore \text{First moment of area of plane about AB} =$$
$$(x_1y_1 + x_2y_2 + x_3y_3 + \ldots)\delta x$$
$$= \Sigma\, xy\, \delta x$$

Now it was mentioned earlier that Simpson's Rule may be used to find the area under any continuous curve, no matter what the ordinates represent. Such a curve may be drawn on a base equal to BC, with ordinates of xy, and the area under this curve may be found by putting the values of xy through Simpson's Rule

Second moment of area of strip about AB

$$= I_{NA} + A\bar{x}^2$$
$$= \tfrac{1}{12}y\,(\delta x)^3 + x^2y\,\delta x$$

This may be reduced to ($x^2y\,\delta x$) since δx is very small

∴ Second moment of area of plane about AB

$$= (x_1{}^2y_1 + x_2{}^2y_2 + x_3{}^2y_3 + \ldots)\delta x$$
$$= \Sigma\, x^2y\,\delta x$$

This may be found by putting the values of x^2y through Simpson's Rule.

First moment of area of strip about BC

$$= \tfrac{1}{2}y \times y\,\delta x$$
$$= \tfrac{1}{2}y^2\,\delta x$$

First moment of area of plane about BC

$$= (y_1{}^2 + y_2{}^2 + y_3{}^2 + \ldots)\tfrac{1}{2}\,\delta x$$
$$= \Sigma\,\tfrac{1}{2}\,y^2\,\delta x$$

This may be found by putting $\frac{1}{2}\,y^2$ through Simpson's Rule.

Second moment of area of strip about BC

$$= \tfrac{1}{12}y^3\,\delta x + (\tfrac{1}{2}y^2)y\,\delta x$$
$$= \tfrac{1}{3}\,y^3\,\delta x$$

Second moment of area of plane about BC

$$= (y_1{}^3 + y_2{}^3 + y_3{}^3 + \ldots)\tfrac{1}{3}\,\delta x$$
$$= \Sigma\,\tfrac{1}{3}\,y^3\,\delta x$$

This may be found by putting $\frac{1}{3}\,y^3$ through Simpson's Rule.

It is usually necessary to calculate area and centroid when determining the second moment of area of a waterplane about a transverse axis. Since the centroid is near amidships it is preferable to take moments about amidships. The following calculation shows the method used to determine area, centroid and second moment of area about the centroid for a waterplane having half ordinates of $y_0, y_1, y_2, \ldots y_{10}$ spaced h m apart commencing from aft.

The positive sign indicates an ordinate aft of midships.

The negative sign indicates an ordinate forward of midships.

Section	½ ord	SM	Product for area	Lever	Product for 1st moment	Lever	Product for 2nd moment
AP	y_0	1	$1y_0$	$+5h$	$+5y_0h$	$+5h$	$+25y_0h^2$
1	y_1	4	$4y_1$	$+4h$	$+16y_1h$	$+4h$	$+64y_1h^2$
2	y_2	2	$2y_2$	$+3h$	$+6y_2h$	$+3h$	$+18y_2h^2$
3	y_3	4	$4y_3$	$+2h$	$+8y_3h$	$+2h$	$+16y_3h^2$
4	y_4	2	$2y_4$	$+1h$	$+2y_4h$	$+1h$	$+2y_4h^2$
5	y_5	4	$4y_5$	$0h$	$\Sigma_{MA} \times h$	$0h$	—
6	y_6	2	$2y_6$	$-1h$	$-2y_6h$	$-1h$	$+2y_6h^2$
7	y_7	4	$4y_7$	$-2h$	$-8y_7h$	$-2h$	$+16y_7h^2$
8	y_8	2	$2y_8$	$-3h$	$-6y_8h$	$-3h$	$+18y_8h^2$
9	y_9	4	$4y_9$	$-4h$	$-16y_9h$	$-4h$	$+64y_9h^2$
FP	y_{10}	1	$1y_{10}$	$-5h$	$-5y_{10}h$	$-5h$	$+25y_{10}h^2$
			Σ_A		$\Sigma_{MF} \times h$		$\Sigma_I \times h^2$

$$\text{Area of waterplane } A = \tfrac{2}{3}\, h\, \Sigma_A$$

First moment of area of waterplane about amidships

$$= \tfrac{2}{3}\, h\, (\Sigma_{MA} + \Sigma_{MF})\, h$$

(added algebraically)

$$\text{Centroid from midships } x = \frac{\tfrac{2}{3}\, h^2\, (\Sigma_{MA} + \Sigma_{MF})}{\tfrac{2}{3}\, h\, \Sigma_A}$$

$$= h\, \frac{(\Sigma_{MA} + \Sigma_{MF})}{\Sigma_A} \text{ aft of midships}$$

(NOTE: If Σ_{MF} is greater than Σ_{MA}, the centroid will be forward of midships).

Second moment of area of waterplane about amidships

$$I_m = 2\, h\, \Sigma_I \times h^2$$
$$= \tfrac{2}{3}\, h^3\, \Sigma_I$$

Second moment of area of waterplane about the centroid

$$I_F = I_m - Ax^2$$

It should be noted that h remains in the table as a constant and left to the end of the calculation. It may be omitted from the table as will be seen from the worked examples.

Example. The half ordinates of a waterplane 180 m long are as follows:

section	AP	$\frac{1}{2}$	1	2	3	4	5	6	7	8	9	$9\frac{1}{2}$	FP
$\frac{1}{2}$ ord	0	5·0	8·0	10·5	12·5	13·5	13·5	12·5	11·0	7·5	3·0	1·0	0 m

Calculate:

(a) area of waterplane

(b) distance of centroid from midships

(c) second moment of area of waterplane about a transverse axis through the centroid.

Section	$\frac{1}{2}$ ordinate	SM	Product for Area	Lever	Product for 1st moment	Lever	Product for 2nd moment
AP	0	$\frac{1}{2}$	—	+5	—	+5	—
$\frac{1}{2}$	5·0	2	10·0	$+4\frac{1}{2}$	+45·0	$+4\frac{1}{2}$	+202·5
1	8·0	$1\frac{1}{2}$	12·0	+4	+48·0	+4	+192·0
2	10·5	4	42·0	+3	+126·0	+3	+378·0
3	12·5	2	25·0	+2	+50·0	+2	+100·0
4	13·5	4	54·0	+1	+54·0	+1	+ 54·0
					+323·0		
5	13·5	2	27·0	0		0	—
6	12·5	4	50·0	—1	—50·0	—1	+ 50·0
7	11·0	2	22·0	—2	—44·0	—2	+ 88·0
8	7·5	4	30·0	—3	—90·0	—3	+270·0
9	3·0	$1\frac{1}{2}$	4·5	—4	—18·0	—4	+ 72·0
$9\frac{1}{2}$	1·0	2	2·0	$-4\frac{1}{2}$	—9·0	$-4\frac{1}{2}$	+ 40·5
FP	0	$\frac{1}{2}$	—	—5	—	—5	—
			278·5		—211·0		+1447·0

$$\text{Common interval} = \frac{180}{10}$$

$$= 18 \text{ m}$$

(a) $$\text{Area of waterplane} = \tfrac{2}{3} \times 18 \times 278{\cdot}5$$
$$= 3342{\cdot}0 \text{ m}^2$$

(b) $$\text{Centroid from midships} = \frac{18\,(323 - 211)}{278{\cdot}5}$$

$$= 7{\cdot}238 \text{ m aft}$$

(c) Second moment of area of waterplane about midships
$$= \tfrac{2}{3} \times 18^3 \times 1447$$
$$= 5{\cdot}626 \times 10^6 \text{ m}^4$$

Second moment of area of waterplane about centroid

$$= 5{\cdot}626 \times 10^6 - 3342 \times 7{\cdot}238^2$$
$$= 5{\cdot}626 \times 10^6 - 0{\cdot}175 \times 10^6$$
$$= 5{\cdot}451 \times 10^6 \text{ m}^4$$

To determine the second moment of area of the waterplane about the centreline of the ship, the half ordinates must be *cubed* and then put through Simpson's Rule.

½ ordinate	(½ ordinate)³	SM	Product for 2nd moment
y_0	y_0^3	1	$1y_0^3$
y_1	y_1^3	4	$4y_1^3$
y_2	y_2^3	2	$2y_2^3$
y_3	y_3^3	4	$4y_3^3$
y_4	y_4^3	2	$2y_4^3$
y_5	y_5^3	4	$4y_5^3$
y_6	y_6^3	2	$2y_6^3$
y_7	y_7^3	4	$4y_7^3$
y_8	y_8^3	2	$2y_8^3$
y_9	y_9^3	4	$4y_9^3$
y_{10}	y_{10}^3	1	$1y_{10}^3$
			ΣI_{CL}

Second moment of area of waterplane about the centreline.

$$= \tfrac{2}{3} \times h \times \Sigma I_{CL} \times \tfrac{1}{3}$$
$$= \tfrac{2}{9}\, h\, \Sigma I_{CL}$$

Example. The half ordinates of a waterplane 180 m long are as follows:

Section	AP	½	1	2	3	4	5	6	7	8	9	9½	FP
½ ord	0	5·0	8·0	10·5	12·5	13·5	13·5	12·5	11·0	7·5	3·0	1·0	0 m

Calculate the second moment of area of the waterplane about the centreline.

Section	½ ordinate	(½ ordinate)³	SM	2nd moment
AP	0	—	½	—
½	5·0	125·0	2	250·0
1	8·0	512·0	1½	768·0
2	10·5	1157·6	4	4630·4
3	12·5	1953·1	2	3906·2
4	13·5	2460·4	4	9841·6
5	13·5	2460·4	2	4920·8
6	12·5	1953·1	4	7812·4
7	11·0	1331·0	2	2662·0
8	7·5	421·9	4	1687·6
9	3·0	27·0	1½	40·5
9½	1·0	1·0	2	2·0
FP	0	—	½	—
				36521·5

Common interval = 18 m

Second moment of area of waterplane about centreline

$$= \frac{2}{9} \times 18 \times 36\ 521{\cdot}5$$
$$= 146\ 086\ \text{m}^4$$

It should be noted that the second moment of area about a transverse axis is considerably greater than the second moment about the centreline.

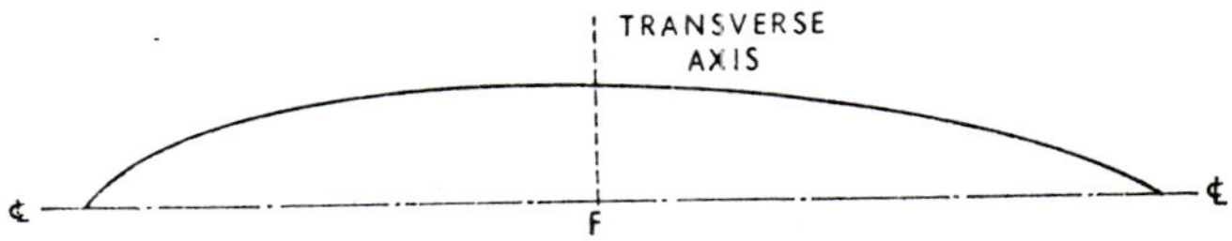

Fig. 27

Since each waterplane is symmetrical it is never necessary to calculate the first moment about the centreline. There are many occasions, however, on which the first moment of area of a tank surface must be calculated as shown by the following example.

Example. A double bottom tank extends from the centreline to the ship side. The widths of the tank surface, at regular intervals of h, are y_1, y_2, y_3, y_4 and y_5.

Calculate the second moment of area of the tank surface about a longitudinal axis through its centroid.

It is necessary in this calculation to determine the area, centroid from the centreline and the second moment of area.

Width	SM	Product for Area	(Width)²	SM	Product for 1st moment	(Width)³	SM	Product for 2nd moment
y_1	1	$1y_1$	y_1^2	1	$1y_1^2$	y_1^3	1	$1y_1^3$
y_2	4	$4y_2$	y_2^2	4	$4y_2^2$	y_2^3	4	$4y_2^3$
y_3	2	$2y_3$	y_3^2	2	$2y_3^2$	y_3^3	2	$2y_3^3$
y_4	4	$4y_4$	y_4^2	4	$4y_4^2$	y_4^3	4	$4y_4^3$
y_5	1	$1y_5$	y_5^2	1	$1y_5^2$	y_5^3	1	$1y_5^3$
		Σa			Σm			Σi

$$\text{Area } a = \frac{h}{3} \Sigma a$$

First moment of area about centreline

$$= \frac{h}{3} \times \Sigma m \times \tfrac{1}{2}$$

$$= \frac{h}{6} \Sigma m$$

Centroid from centreline $= \dfrac{\dfrac{h}{6} \Sigma m}{\dfrac{h}{3} \Sigma a}$

$$\bar{y} = \frac{\Sigma m}{2 \Sigma a}$$

Second moment of area about centreline

$$i_{CL} = \frac{h}{9} \Sigma i$$

Second moment of area about centroid

$$= i_{CL} - a \bar{y}^2$$

Example. A double bottom tank 21 m long has a watertight centre girder. The widths of the tank top measured from the centreline to the ship's side are 10·0, 9·5, 9·0, 8·0, 6·5, 4·0 and 1·0 m respectively. Calculate the second moment of area of the tank surface about a longitudinal axis through its centroid, for one side of the ship only.

Width	SM	Product for Area	$(\text{Width})^2$	SM	Product for 1st moment	$(\text{Width})^3$	SM	Product for 2nd moment
10·0	1	10·0	100·00	1	100·00	1000·0	1	1000·0
9·5	4	38·0	90·25	4	361·00	857·4	4	3429·6
9·0	2	18·0	81·00	2	162·00	729·0	2	1458·0
8·0	4	32·0	64·00	4	256·00	512·0	4	2048·0
6·5	2	13·0	42·25	2	84·50	274·6	2	549·2
4·0	4	16·0	16·00	4	64·00	64·0	4	256·0
1·0	1	1·0	1·00	1	1·00	1·0	1	1·0
		128·0			1028·50			8741·8

$$\text{Common interval} = \frac{21}{6} = 3\cdot5 \text{ m}$$

$$\text{Area of tank surface} = \frac{3\cdot5}{3} \times 128$$

$$= 149\cdot33 \text{ m}^2$$

$$\text{Centroid from centreline} = \frac{1028\cdot5}{2 \times 128}$$

$$= 4\cdot018 \text{ m}$$

Second moment of area about centreline

$$= \frac{3\cdot5}{9} \times 8741\cdot8$$

$$= 3400\cdot0 \text{ m}^4$$

Second moment of area about centroid

$$= 3400\cdot0 - 149\cdot33 \times 4\cdot018^2$$

$$= 3400\cdot0 - 2410\cdot8$$

$$= 989\cdot2 \text{ m}^4$$

A further application of first and second moments of area is the calculation of the load exerted by a liquid on a bulkhead and the position of the centre of pressure.

Let the widths of a bulkhead at intervals of h, commencing from the top, be y_0, y_1, y_2, . . . y_6 (Fig. 28)

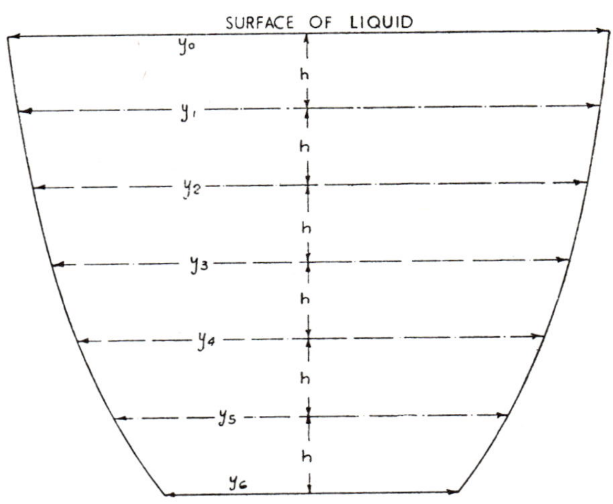

Fig. 28

Assume the bulkhead to be flooded to the top edge with liquid of density ρ on one side only

Width	SM	Product for Area	Lever	Product for 1st moment	Lever	Product for 2nd moment
y_0	1	$1y_0$	0	—	0	—
y_1	4	$4y_1$	1	$4y_1$	1	$4y_1$
y_2	2	$2y_2$	2	$4y_2$	2	$8y_2$
y_3	4	$4y_3$	3	$12y_3$	3	$36y_3$
y_4	2	$2y_4$	4	$8y_4$	4	$32y_4$
y_5	4	$4y_5$	5	$20y_5$	5	$100y_5$
y_6	1	$1y_6$	6	$6y_6$	6	$36y_6$
		Σa		Σm		Σi

$$\text{Area of bulkhead} = \frac{h}{3}\Sigma a$$

First moment of area of bulkhead about surface of liquid

$$= \frac{h^2}{3}\Sigma m$$

It was shown previously that:

$$\text{Load on bulkhead} = \rho g \times \text{first moment of area}$$

$$= \rho g \times \frac{h^2}{3}\Sigma m$$

Second moment of area of bulkhead about surface of liquid

$$= \frac{h^3}{3}\Sigma i$$

Centre of pressure from surface of liquid

$$= \frac{\text{second moment of area}}{\text{first moment of area}}$$

$$= \frac{h\Sigma i}{\Sigma m}$$

(Note: It is not necessary to calculate the area unless requested to do so).

Example. A fore peak bulkhead is 4·8 m deep and 5·5 m wide at the deck. At regular intervals of 1·2 m below the deck, the horizontal widths are 5·0, 4·0, 2·5 and 0·5 m respectively. The bulkhead is flooded to the top edge with sea water on one side only. Calculate:

(a) area of bulkhead

(b) load on bulkhead

(c) position of centre of pressure

Depth	Width	SM	Product for Area	Lever	Product for 1st moment	Lever	Product for 2nd moment
4·8	5·5	1	5·5	0	—	0	—
3·6	5·0	4	20·0	1	20·0	1	20·0
2·4	4·0	2	8·0	2	16·0	2	32·0
1·2	2·5	4	10·0	3	30·0	3	90·0
0	0·5	1	0·5	4	2·0	4	8·0
			44·0		68·0		150·0

Common interval = 1·2 m

(a) Area of bulkhead $= \frac{1 \cdot 2}{3} \times 44$

$= 17 \cdot 6 \text{ m}^2$

(b) Load on bulkhead $= 1 \cdot 025 \times 9 \cdot 81 \times \frac{1 \cdot 2^2}{3} \times 68 \cdot 0$

$= 328 \cdot 2 \text{ kN}$

(c) Centre of pressure from surface

$= 1 \cdot 2 \times \frac{150}{68}$

$= 2 \cdot 647 \text{ m}$

SIMPSON'S SECOND RULE

Simpson's Second Rule is based on the assumption that the curved portion of a plane forms part of a cubic expression ($y = ax^3 + bx^2 + cx + d$), and gives the area contained between *four* consecutive, equally spaced-ordinates.

$$\text{Area ABCD} = \tfrac{3}{8} h \,(1y_1 + 3y_2 + 3y_3 + 1y_4)$$

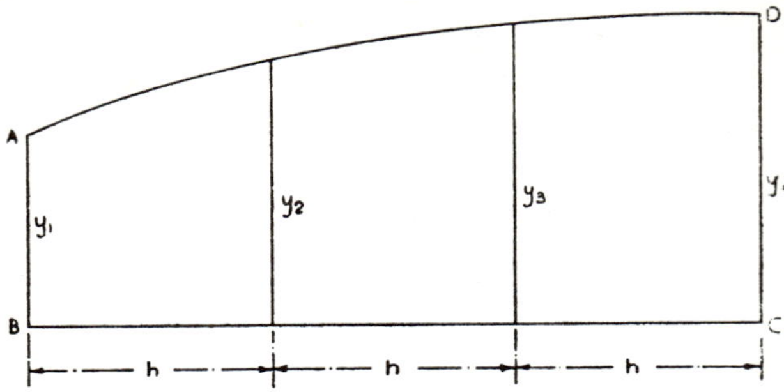

Fig. 29

This rule may be applied to successive areas in the same way as the first rule.

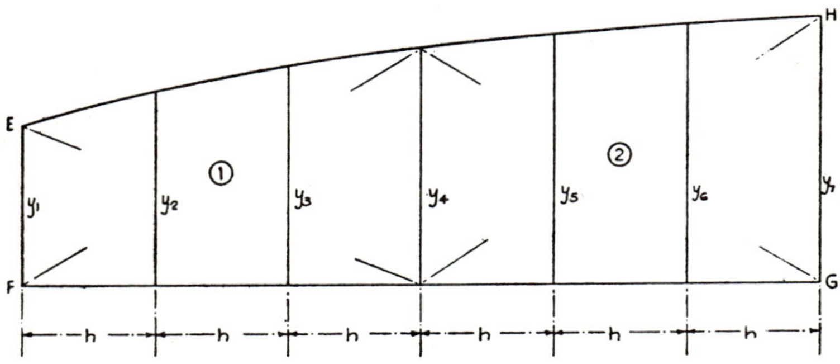

Fig. 30

$$\text{Area } 1 = \tfrac{3}{8}h(1y_1+3y_2+3y_3+1y_4)$$
$$\text{Area } 2 = \tfrac{3}{8}h(1y_4+3y_5+3y_6+1y_7)$$
$$\text{Area EFGH} = \tfrac{3}{8}h[(1y_1+3y_2+3y_3+1y_4)+(1y_4+3y_5+3y_6+1y_7)]$$
$$= \tfrac{3}{8}h[1y_1+3y_2+3y_3+2y_4+3y_5+3y_6+1y_7]$$

It may be shown that Simpson's Second Rule may be used to determine the area of a plane having a number of equal intervals divisible by *three*, i.e. 4, 7, 10, 13, 16 ordinates.

Example A waterplane 135 m long has equally-spaced half ordinates of 1·2, 4·4, 6·7, 7·8, 8·0, 8·0, 7·7, 6·1, 3·8 and 0 m respectively. Calculate the area of the waterplane.

½ ordinate	SM	Product for Area
1·2	1	1·2
4·4	3	13·2
6·7	3	20·1
7·8	2	15·6
8·0	3	24·0
8·0	3	24·0
7·7	2	15·4
6·1	3	18·3
3·8	3	11·4
0	1	0
		143·2

$$\text{Common interval} = \frac{135}{9} = 15 \text{ m}$$

$$\text{Area of waterplane} = \tfrac{3}{8} \times 15 \times 143{\cdot}2 \times 2$$

$$= 1611{\cdot}0 \text{ m}^2$$

TEST EXAMPLES 3

1. A ship 180 m long has ½ widths of waterplane of 1, 7·5, 12, 13·5, 14, 14, 14, 13·5, 12, 7 and 0 m respectively. Calculate:

(a) waterplane area
(b) TPC
(c) waterplane area coefficient.

2. The waterplane areas of a ship at 1·5 m intervals of draught, commencing at the keel, are 865, 1735, 1965, 2040, 2100, 2145 and 2215 m^2 respectively. Calculate the displacement at 9 m draught.

3. A ship 140 m long and 18 m beam floats at a draught of 9 m. The immersed cross-sectional areas at equal intervals are 5, 60, 116, 145, 152, 153, 153, 151, 142, 85 and 0 m^2 respectively. Calculate:

(a) displacement
(b) block coefficient
(c) midship section area coefficient
(d) prismatic coefficient.

4. The ½ ordinates of a waterplane 120 m long are as follows:

Section	AP	½	1	1½	2	3	4	5	6	7	8	8½	9	9½	FP
½ ord	1·2	3·5	5·3	6·8	8·0	8·3	8·5	8·5	8·5	8·4	8·2	7·9	6·2	3·5	0 m

Calculate:

(a) waterplane area
(b) distance of centroid from midships.

*f*5. The TPC values of a ship at 1·5 m intervals of draught, commencing at the keel, are 4·0, 6·1, 7·8, 9·1, 10·3, 11·4 and 12·0 respectively. Calculate at a draught of 9 m:

(a) displacement
(b) KB

*f*6. The ½ breadths of the load waterplane of a ship 150 m long, commencing from aft, are 0·3, 3·8, 6·0, 7·7, 8·3, 9·0, 8·4, 7·8, 6·9, 4·7 and 0 m respectively. Calculate:

(a) area of waterplane
(b) distance of centroid from midships
(c) second moment of area about a transverse axis through the centroid

*f*7. The displacement of a ship at draughts of 0, 1, 2, 3 and 4 m are 0, 189, 430, 692 and 977 tonne. Calculate the distance of the centre of buoyancy above the keel when floating at a draught of 4 m, given:

VCB below waterline

$$= \frac{\text{area between displacement curve and draught axis}}{\text{displacement}}$$

*f*8. The widths of a deep tank bulkhead at equal intervals of 1·2 m commencing at the top, are 8·0, 7·5, 6·5, 5·7, 4·7, 3·8 and 3·0 m. Calculate the load on the bulkhead and the position of the centre of pressure, if the bulkhead is flooded to the top edge with sea water on one side only.

*f*9. A forward deep tank 12 m long extends from a longitudinal bulkhead to the ship's side. The widths of the tank surface measured from the longitudinal bulkhead at regular intervals are 10, 9, 7, 4 and 1 m. Calculate the second moment of area of the tank surface about a longitudinal axis passing through its centroid.

*f*10. A ship 160 m long has ½ ordinates of waterplane of 1·6, 5·7, 8·8, 10·2, 10·5, 10·5, 10·5, 10·0, 8·0, 5·0 and 0 m respectively. Calculate the second moment of area of the waterplane about the centreline.

*f*11. The immersed cross-sectional areas of a ship 120 m long, commencing from aft, are 2, 40, 79, 100, 103, 104, 104, 103, 97, 58 and 0 m^2. Calculate:

(a) displacement

(b) longitudinal position of the centre of buoyancy.

12. The ½ ordinates of the waterplane of a ship 180 m long are 1·6, 6·0, 9·2, 10·5, 11, 11, 10·2, 8·3, 5·1 and 0 m respectively. Calculate the area of the waterplane.

CHAPTER 4

CENTRE OF GRAVITY

The centre of gravity of an object is the point at which the whole weight of the object may be regarded as acting. If the object is suspended from this point, then it will remain balanced and will not tilt.

The distance of the centre of gravity from any axis is the total moment of *force* about that axis divided by the total *force*. If a body is composed of a number of different types of material, the force may be represented by the *weights* of the individual parts.

$$\text{Centre of gravity from axis} = \frac{\text{moment of weight about axis}}{\text{total weight}}$$

At any point on the earth's surface, the value of g remains constant. Hence the weight may be represented by *mass*, and:

$$\text{Centre of gravity from axis} = \frac{\text{moment of mass about axis}}{\text{total mass}}$$

If the body is of the same material throughout, then the weight depends upon the *volume* and moments of *volume* may therefore be used.

$$\text{Centre of gravity from axis} = \frac{\text{moment of volume about axis}}{\text{total volume}}$$

The centre of gravity of a uniform lamina is midway through the thickness. Since both the thickness and the density are constant, moments of *area* may be used. This system may also be applied to determine the centre of gravity, or, more correctly, *centroid* of an area.

$$\text{Centroid from axis} = \frac{\text{moment of area about axis}}{\text{total area}}$$

The position of the centre of gravity of a ship may be found by taking moments of the individual masses. The actual calculation of the centre of gravity of a ship is a very lengthy process, and since many of the masses must be estimated, is not considered to be sufficiently accurate for stability calculations. Such a calculation is usually carried out for a passenger ship in the initial design stages, but the results are confirmed by an alternative method when the ship is completed. Once the position of the centre of gravity of an empty ship is known, however, the centre of gravity of the ship in any loaded condition may be found.

It is usual to measure the *vertical* position of the centre of gravity (VCG) of the ship above the keel and this distance is denoted by *KG*. The height of the centre of gravity of an item on the ship above the keel is denoted by *Kg*. The *longitudinal* position of the centre of gravity (LCG) is usually given as a distance forward or aft of midships. If the ship is upright, the *transverse* centre of gravity lies on the centreline of the ship and no calculation is necessary.

Example. A ship of 8500 tonne displacement is composed of masses of 2000, 3000, 1000, 2000, and 500 tonne at positions 2, 5, 8, 10 and 14 m above the keel. Determine the height of the centre of gravity of the ship above the keel.

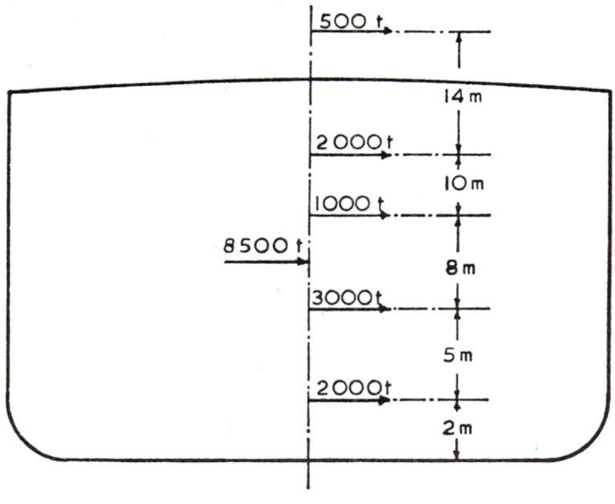

Fig. 31

This example is preferably answered in table form.

mass (tonne)	Kg (m)	Vertical moment (t m)
2000	2	4 000
3000	5	15 000
1000	8	8 000
2000	10	20 000
500	14	7 000
8500		54 000

$$KG = \frac{\text{total moment}}{\text{total displacement}}$$

$$= \frac{54\,000}{8500}$$

$$= 6{\cdot}353 \text{ m}$$

Example. A ship of 6000 tonne displacement is composed of masses of 300, 1200 and 2000 tonne at distances 60, 35 and 11 m aft of midships, and masses of 1000, 1000 and 500 tonne at distances 15, 30 and 50 m forward of midships. Calculate the distance of the centre of gravity of the ship from midships.

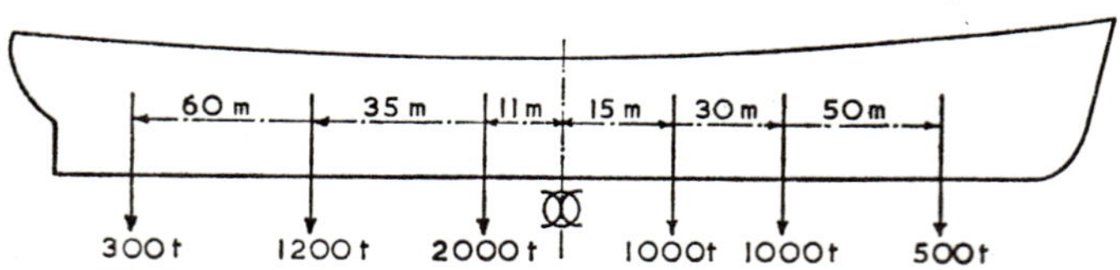

Fig. 32

A table is again preferred.

Mass (tonne)	Lcg from midships (m)	moment forward (t m)	moment aft (t m)
300	60 aft	—	18 000
1200	35 aft	—	42 000
2000	11 aft	—	22 000
1000	15 forward	15 000	—
1000	30 forward	30 000	—
500	50 forward	25 000	—
6000		70 000	82 000

The moment aft is greater than the moment forward and therefore the centre of gravity must lie aft of midships.

$$\text{Excess moment aft} = 82\,000 - 70\,000$$
$$= 12\,000 \text{ tonne m}$$

Centre of gravity aft of midships

$$= \frac{\text{excess moment}}{\text{total displacement}}$$

$$= \frac{12\,000}{6000}$$

$$= 2{\cdot}00 \text{ m}$$

SHIFT IN CENTRE OF GRAVITY DUE TO ADDITION OF MASS

When a mass is added to a ship, the centre of gravity of the ship moves towards the added mass. The distance moved by the ship's centre of gravity depends upon the magnitude of the added mass, the distance of the mass from the ship's centre of gravity and the displacement of the ship. If a mass is placed on the port side of the ship in the forecastle, the centre of gravity moves forward, upwards and to port. The actual distance and direction of this movement is seldom required but the separate components are most important, i.e. the longitudinal, vertical and transverse distances moved. When an item on a ship is removed, the centre of gravity moves away from the original position of that item.

Example. A ship of 4000 tonne displacement has its centre of gravity 1·5 m aft of midships and 4 m above the keel. 200 tonne of cargo are now added 45 m forward of midships and 12 m above the keel. Calculate the new position of the centre of gravity.

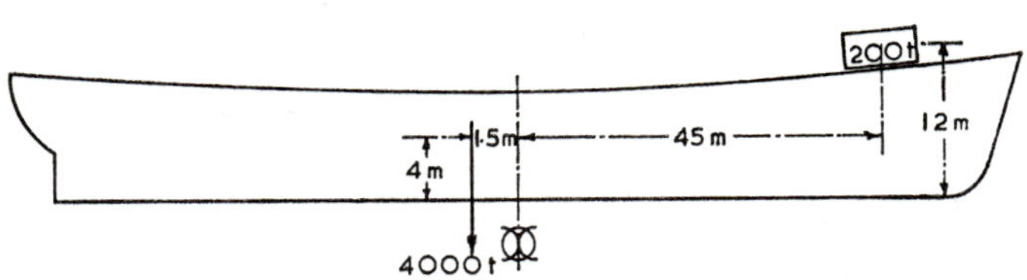

Fig. 33

Taking moments about midships:

$$\text{Moment aft of midships} = 4000 \times 1{\cdot}5 = 6000 \text{ t m}$$

$$\text{Moment forward of midships} = 200 \times 45 = 9000 \text{ t m}$$

$$\text{Excess moment forward} = 9000 - 6000 = 3000 \text{ t m} = 3000 \text{ t m}$$

$$\text{Centre of gravity from midships} = \frac{\text{excess moment}}{\text{total displacement}} = \frac{3000}{4000 + 200} = 0{\cdot}714 \text{ m forward}$$

Taking moments about the keel:

$$\text{Centre of gravity from keel} = \frac{4000 \times 4 + 200 \times 12}{4000 + 200} = \frac{16\,000 + 2400}{4200}$$

$$KG = 4{\cdot}381 \text{ m}$$

Thus the centre of gravity rises 0·381 m.

The same answer may be obtained by taking moments about the original centre of gravity, thus:

$$\text{Moment of ship about centre of gravity} = 4000 \times 0$$

$$\text{Moment of added mass about centre of gravity} = 200(12 - 4) = 1600 \text{ t m}$$

$$\text{Rise in centre of gravity} = \frac{\text{total moment}}{\text{total displacement}} = \frac{1600}{4200} = 0{\cdot}381 \text{ m}$$

If the actual distance moved by the centre of gravity is required, it may be found from the longitudinal and vertical movements.

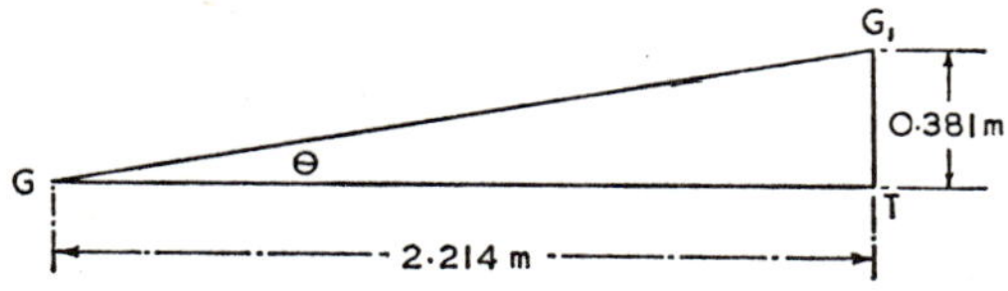

Fig. 34

Longitudinal shift in the centre of gravity

$$GT = 1{\cdot}5 + 0{\cdot}714$$
$$= 2{\cdot}214 \text{ m}$$
$$GG_1 = \sqrt{GT^2 + TG_1{}^2}$$
$$= \sqrt{2{\cdot}214^2 + 0{\cdot}381^2}$$
$$= 2{\cdot}247 \text{ m}$$

The angle θ which the centre of gravity moves relative to the horizontal may be found from Fig. 34.

$$\tan\theta = \frac{0{\cdot}381}{2{\cdot}214}$$
$$= 0{\cdot}712$$

from which $\theta = 9° \, 45'$

SHIFT IN CENTRE OF GRAVITY DUE TO MOVEMENT OF MASS

When a mass which is already on board a ship is moved in any direction, there is a corresponding movement in the centre of gravity of the ship in the same direction.

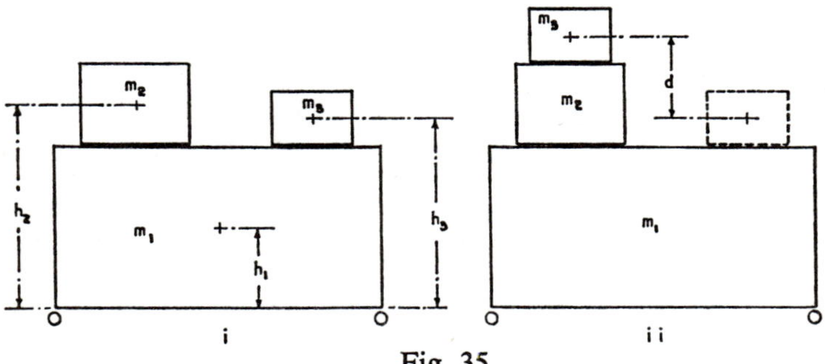

Fig. 35

Consider a system composed of masses of m_1, m_2 and m_3 as shown in Fig. 35 (i), the centre of gravity of each being h_1, h_2 and h_3 respectively from the base O—O. The distance of the centre of gravity of the system from the base may be determined by dividing the total moment of mass about O—O by the total mass.

$$\text{Centre of gravity from O—O} = \frac{\text{total moment of mass}}{\text{total mass}}$$

$$= \frac{m_1h_1 + m_2h_2 + m_3h_3}{m_1 + m_2 + m_3}$$

$$= y$$

If m_3 is now raised through a distance d to the position shown in Fig. 35 (ii), the centre of gravity of the system is also raised. New centre of gravity from O—O

$$= \frac{m_1h_1 + m_2h_2 + m_3(h_3 + d)}{m_1 + m_2 + m_3}$$

$$= \frac{m_1h_1 + m_2h_2 + m_3h_3}{m_1 + m_2 + m_3} + \frac{m_3d}{m_1 + m_2 + m_3}$$

$$= y + \frac{m_3d}{m_1 + m_2 + m_3}$$

Thus it may be seen that:

$$\text{Shift in centre of gravity} = \frac{m_3d}{m_1 + m_2 + m_3}$$

or,

$$\text{Shift in centre of gravity} = \frac{\text{mass moved} \times \text{distance moved}}{\text{TOTAL mass}}$$

This expression is most useful in ship calculations and is applied throughout stability and trim work. It should be noted that it is not necessary to know either the position of the centre of gravity of the ship, or the position of the mass relative to the centre of gravity of the ship. The rise in the centre of gravity is the same whether the mass is moved from the tank top to the deck or from the deck to the mast head as long as the distance moved is the same. The centre of gravity of the ship moves in the same direction as the centre of gravity of the mass. Thus if a mass is moved forward and down, the centre of gravity of the ship also moves forward and down.

Example. A ship of 5000 tonne displacement has a mass of 200 tonne on the fore deck 55 m forward of midships. Calculate the shift in the centre of gravity of the ship if the mass is moved to a position 8 m forward of midships.

$$\text{Shift in centre of gravity} = \frac{\text{mass moved} \times \text{distance moved}}{\text{displacement}}$$

$$= \frac{200 \times (55 - 8)}{5000}$$

$$= 1{\cdot}88 \text{ m}$$

EFFECT OF A SUSPENDED MASS

When a mass hangs freely from a point on a ship, its centre of gravity lies directly below that point. If the vessel now heels, the mass moves in the direction of the heel until it again lies vertically below the point of suspension, and no matter in which direction the vessel heels, the centre of gravity of the mass is always below this point. Thus it may be seen that the position of the centre of gravity of a hanging mass, relative to the ship, is at the point of suspension.

This principle proves to be very important when loading a ship by means of the ship's derricks. If, for example, a mass lying on the tank top is being discharged, then as soon as the mass is clear of the tank top its centre of gravity is virtually raised to the derrick head, causing a corresponding rise in the centre of gravity of the ship. If the mass is now raised to the derrick head there is no further change in the centre of gravity of the ship.

Ships which are equipped to load heavy cargoes by means of heavy lift derricks must have a standard of stability which will prevent excessive heel when the cargo is suspended from the derrick. A similar principle is involved in the design of ships which carry hanging cargo such as chilled meat. The meat is suspended by hangers from the underside of the deck and therefore the centre of gravity of the meat must be taken as the deck from which it hangs.

Example. A ship of 10 000 tonne displacement has a mass of 60 tonne lying on the deck. A derrick, whose head is 7·5 m above the centre of gravity of the mass, is used to place the mass on the tank top 10·5 m below the deck. Calculate the shift in the vessel's centre of gravity when the mass is:

(a) just clear of the deck
(b) at the derrick head
(c) in its final position.

(a) When the mass is just clear of the deck its centre of gravity is raised to the derrick head.

$$\text{Shift in centre of gravity} = \frac{\text{mass moved} \times \text{distance moved}}{\text{displacement}}$$

$$= \frac{60 \times 7{\cdot}5}{10\,000}$$

$$= 0{\cdot}045 \text{ m up}$$

(b) When the mass is at the derrick head there is no further movement of the centre of gravity of the ship.

Shift in centre of gravity $= 0{\cdot}045$ m up

$$\text{(c) Shift in centre of gravity} = \frac{60 \times 10{\cdot}5}{10\,000}$$

$$= 0{\cdot}063 \text{ m down}$$

TEST EXAMPLES 4

1. A ship of 4000 tonne displacement has its centre of gravity 6 m above the keel. Find the new displacement and position of the centre of gravity when masses of 1000, 200, 5000 and 3000 tonne are added at positions 0·8, 1·0, 5·0 and 9·5 m above the keel.

2. The centre of gravity of a ship of 5000 tonne displacement is 6 m above the keel and 1·5 m forward of midships. Calculate the new position of the centre of gravity if 500 tonne of cargo are placed in the 'tween decks 10 m above the keel and 36 m aft of midships.

3. A ship has 300 tonne of cargo in the hold, 24 m forward of midships. The displacement of the vessel is 6000 tonne and its centre of gravity is 1·2 m forward of midships.

Find the new position of the centre of gravity if this cargo is moved to an after hold, 40 m from midships.

4. An oil tanker of 17 000 tonne displacement has its centre of gravity 1 m aft of midships and has 250 tonne of oil fuel in its forward deep tank 75 m from midships.

This fuel is transferred to the after oil fuel bunker whose centre is 50 m from midships.

200 tonne of fuel from the after bunker is now burned.

Calculate the new position of the centre of gravity:

(a) after the oil has been transferred

(b) after the oil has been used.

5. A ship of 3000 tonne displacement has 500 tonne of cargo on board. This cargo is lowered 3 m and an additional 500 tonne of cargo is taken on board 3 m vertically above the original position of the centre of gravity. Determine the alteration in position of the centre of gravity.

6. A ship of 10 000 tonne displacement has its centre of gravity 3 m above the keel. Masses of 2000, 300 and 50 tonne are removed from positions 1·5, 4·5 and 6m above the keel. Find the new displacement and position of the centre of gravity.

f 7. A vessel of 8000 tonne displacement has 75 tonne of cargo on the deck. It is lifted by a derrick whose head is 10·5 m above the centre of gravity of the cargo, and placed in the lower hold 9 m below the deck and 14 m forward of its original position. Calculate the shift in the vessel's centre of gravity from its original position when the cargo is:

(a) just clear of the deck

(b) at the derrick head

(c) in its final position.

CHAPTER 5

STABILITY OF SHIPS

Statical stability is a measure of the tendency of a ship to return to the upright if inclined by an external force.

In theory it is possible to balance a pencil on its point on a flat surface. The pencil will be balanced if its centre of gravity is vertically above its point. In practice this is found to be impossible to achieve. It is, however, possible to balance the pencil on its flat end, since, if the pencil is very slightly inclined, the centre of gravity may still lie within the limits of the base and the pencil will tend to return to the upright. Fig. 36 is exaggerated to show this.

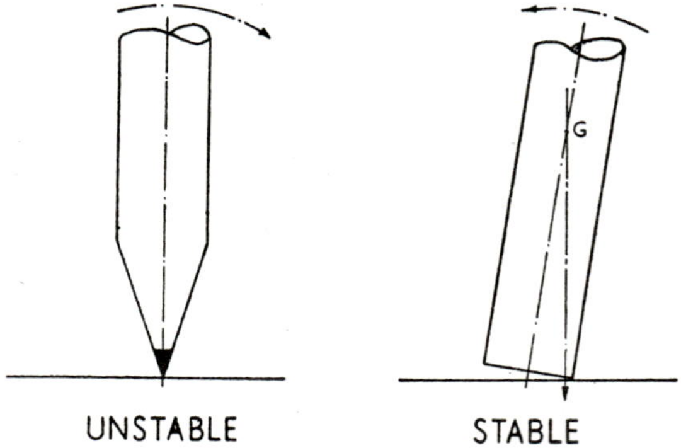

Fig. 36

The only times a ship may be assumed to be stationary and upright are before launching and when in dry dock. Thus it is essential to consider practical conditions and to assume that a ship is always moving. If the vessel is stated to be upright it should be regarded as rolling slightly about the upright position.

In the upright position (Fig. 37), the weight of the ship acts vertically down through the centre of gravity G, while the upthrust acts through the centre of buoyancy B. Since the weight is equal to the upthrust, and the centre of gravity and the centre of buoyancy are in the same vertical line, the ship is in equilibrium.

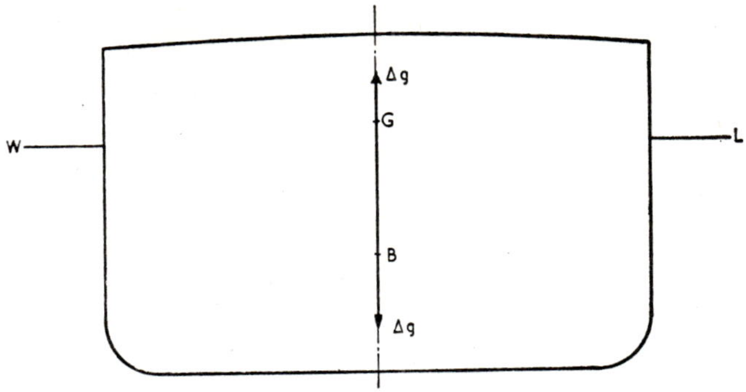

EQUILIBRIUM

Fig. 37

When the ship is inclined by an external force to an angle θ, the centre of gravity remains in the same position but the centre of buoyancy moves from B to B_1 (Fig. 38).

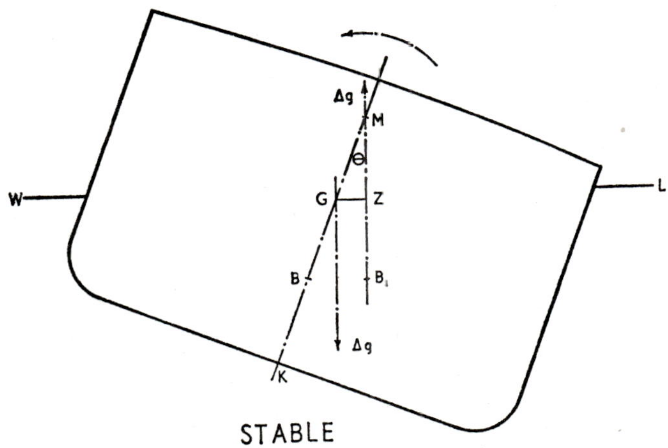

STABLE

Fig. 38

The buoyancy, therefore, acts up through B_1 while the weight still acts down through G, creating a moment of $\triangle g \times GZ$ which tends to return the ship to the upright. $\triangle g \times GZ$ is known as the *righting moment* and GZ the *righting lever*. Since this moment tends to right the ship the vessel is said to be *stable*.

For small angles of heel, up to about 10°, the vertical through the new centre of buoyancy B_1 intersects the centreline at M the *transverse metacentre*. It may be seen from Fig. 38 that:

$$GZ = GM \sin \theta$$

Thus for small angles of heel GZ is a function of GM, and since GM is independent of θ while GZ depends upon θ, it is useful to express the initial stability of a ship in terms of GM, the *metacentric height*. GM is said to be *positive* when G lies below M and the vessel is stable. A ship with a small metacentric height will have a small righting lever at any angle and will roll easily. The ship is then said to be *tender*. A ship with a large metacentric height will have a large righting lever at any angle and will have a considerable resistance to rolling. The ship is then said to be *stiff*. A stiff ship will be very uncomfortable, having a very small rolling period and in extreme cases may result in structural damage.

If the centre of gravity lies above the transverse metacentre (Fig. 39), the moment acts in the opposite direction, increasing the angle of heel. The vessel is then *unstable* and will not return to the upright, the metacentric height being regarded as *negative*.

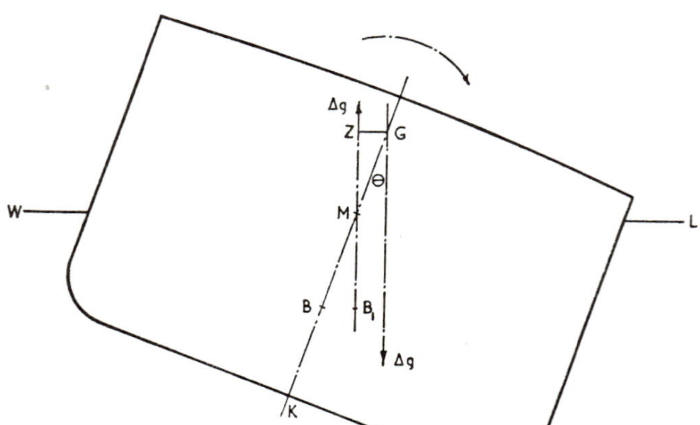

UNSTABLE

Fig. 39

When the centre of gravity and transverse metacentric coincide (Fig. 40), there is no moment acting on the ship which will therefore remain inclined to angle θ. The vessel is then said to be in *neutral equilibrium*.

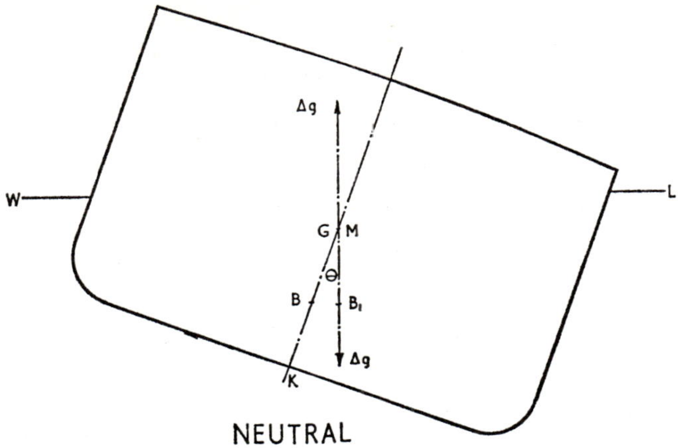

NEUTRAL
EQUILIBRIUM
Fig. 40

Since any reduction in the height of G will make the ship stable, and any rise in G will make the ship unstable, this condition is regarded as the point at which a ship *becomes* either stable or unstable.

TO FIND THE POSITION OF M

The distance of the transverse metacentre above the keel (KM) is given by $KM = KB + BM$.

KB is the distance of the centre of buoyancy above the keel and may be found by one of the methods shown previously.

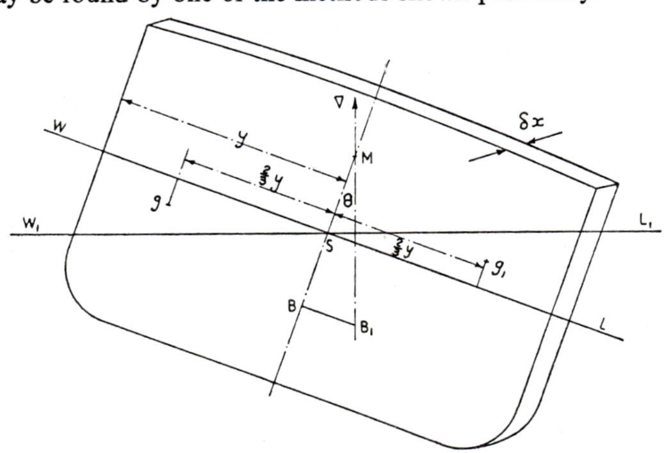

Fig. 41

BM may be found as follows:

Consider a ship whose volume of displacement is ∇, lying upright at waterline WL, the centre of buoyancy being on the centreline of the ship. If the ship is now inclined to a small angle θ, it will lie at waterline W_1L_1 which intersects the original waterline at S (Fig. 41). Since θ is small it may be assumed that S is on the centreline.

A wedge of buoyancy WSW_1 has been moved across the ship to L_1SL causing the centre of buoyancy to move from B to B_1.

Let v = volume of wedge

gg_1 = transverse shift in centre of gravity of wedge

$$\text{Then } BB_1 = \frac{v \times gg_1}{\nabla}$$

$$\text{But } BB_1 = BM \tan\theta$$

$$\therefore BM \tan\theta = \frac{v \times gg_1}{\nabla}$$

$$BM = \frac{v \times gg_1}{\nabla \tan\theta}$$

To determine the value of $v \times gg_1$, divide the ship into thin transverse strips of length δx, and let the half width of waterplane in way of one such strip be y.

$$\begin{aligned}
\text{Volume of strip of wedge} &= \tfrac{1}{2}y \times y \tan\theta\, \delta x \\
\text{Moment of shift of strip of wedge} &= \tfrac{4}{3}y \times \tfrac{1}{2}y^2 \tan\theta\, \delta x \\
&= \tfrac{2}{3}y^3 \tan\theta\, \delta x \\
\text{Total moment of shift of wedge} &= v \times gg_1 \\
&= \Sigma\, \tfrac{2}{3}y^3 \tan\theta\, \delta x \\
&= \tan\theta\, \tfrac{2}{3}\, \Sigma y^3\, \delta x
\end{aligned}$$

But $\frac{2}{3}\Sigma y^3 \delta x$ = second moment of area of waterplane about the centreline of the ship

$= I$

$$\therefore v \times gg_1 = I \tan\theta$$

$$BM = \frac{I \tan\theta}{\nabla \tan\theta}$$

$$BM = \frac{I}{\nabla}$$

Example. A box barge of length L and breadth B floats at a level keel draught d. Calculate the height of the transverse metacentre above the keel.

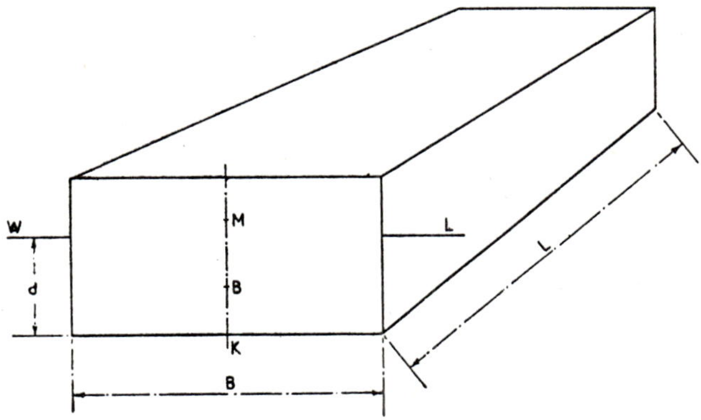

Fig. 42

$$KM = KB + BM$$

$$KB = \frac{d}{2}$$

$$BM = \frac{I}{\nabla}$$

$$I = \tfrac{1}{12} LB^3$$

$$\nabla = L.B.d$$

$$BM = \frac{LB^3}{12.L.B.d}$$

$$= \frac{B^2}{12d}$$

$$\therefore KM = \frac{d}{2} + \frac{B^2}{12d}$$

It should be noted that while the above expression is applicable only to a box barge, similar expressions may be derived for vessels of constant triangular or circular cross sections. The waterplane in each case is in the form of a *rectangle*, the second moment of which is $\frac{1}{12}$ × length × breadth³. As long as the length of a vessel having constant cross-section exceeds the breadth, the length does not affect the transverse stability of the ship.

Example. A vessel of constant triangular cross-section has a depth of 12 m and a breadth at the deck of 15 m.

Calculate the draught at which the vessel will become unstable if the centre of gravity is 6·675 m above the keel.

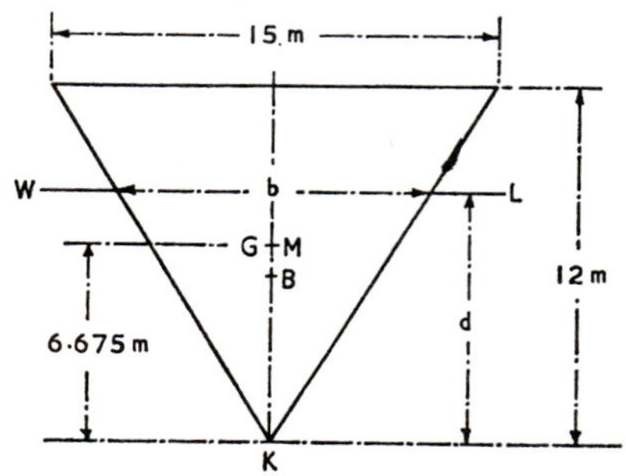

Fig. 43

Let d = draught
b = breadth at waterline

By similar triangles $$\frac{b}{d} = \frac{B}{D}$$

$$\therefore b = \frac{15}{12} d$$

$$= \frac{5}{4} d$$

$$KB = \tfrac{2}{3} d$$

$$\nabla = \tfrac{1}{2} L\, b\, d$$

$$I = \tfrac{1}{12} L\, b^3$$

(Note that b is the breadth at the waterline).

$$BM = \frac{I}{\nabla}$$

$$= \tfrac{1}{12} L\, b^3 \div \tfrac{1}{2} L\, b\, d$$

$$= \frac{b^2}{6d}$$

$$= \frac{1}{6d}\left(\frac{5}{4}d\right)^2$$

$$= \frac{25}{96}\, d$$

The vessel becomes unstable when G and M coincide.

Thus
$$KM = KG$$
$$= 6{\cdot}675 \text{ m}$$
$$6{\cdot}675 = \tfrac{2}{3}\, d + \tfrac{25}{96} d$$
$$= \frac{89}{96}\, d$$
$$d = 6{\cdot}675 \times \frac{96}{89}$$
$$\text{Draught } d = 7{\cdot}2 \text{ m}$$

METACENTRIC DIAGRAM

Since both KB and BM depend upon draught, their values for any ship may be calculated for a number of different draughts, and

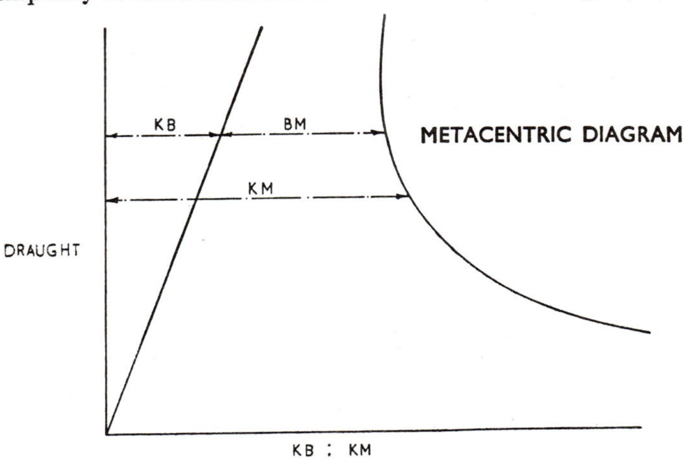

Fig. 44

plotted to form the *metacentric diagram* for the ship. The height of the transverse metacentre above the keel may then be found at any intermediate draught.

The metacentric diagram for a box barge is similar to that for a ship (Fig. 44), while the diagram for a vessel of constant triangular cross-section is formed by two straight lines starting from the origin (Fig. 45).

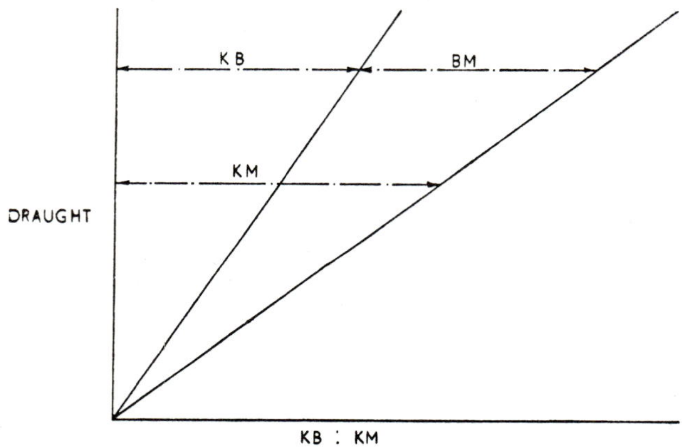

METACENTRIC DIAGRAM FOR VESSEL OF CONSTANT TRIANGULAR CROSS-SECTION

Fig. 45

Example. A vessel of constant rectangular cross-section is 12 m wide. Draw the metacentric diagram using 1·2 m intervals of draught up to the 7·2 m waterline.

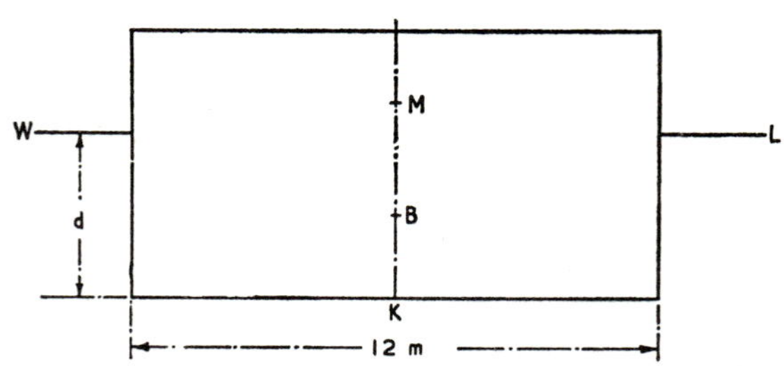

Fig. 46

It is useful in an example of this type to derive an expression for *KB* and *KM* in terms of the only variable-draught-and substitute the different draught values in tabular form.

$$KB = \frac{d}{2}$$

$$BM = \frac{B^2}{12d}$$

$$= \frac{12^2}{12d}$$

$$= \frac{12}{d}$$

$$KM = KB + BM$$

d	KB	BM	KM
0	0	∞	∞
1·2	0·6	10·00	10·60
2·4	1·2	5·00	6·20
3·6	1·8	3·33	5·13
4·8	2·4	2·50	4·90
6·0	3·0	2·00	5·00
7·2	3·6	1·67	5·27

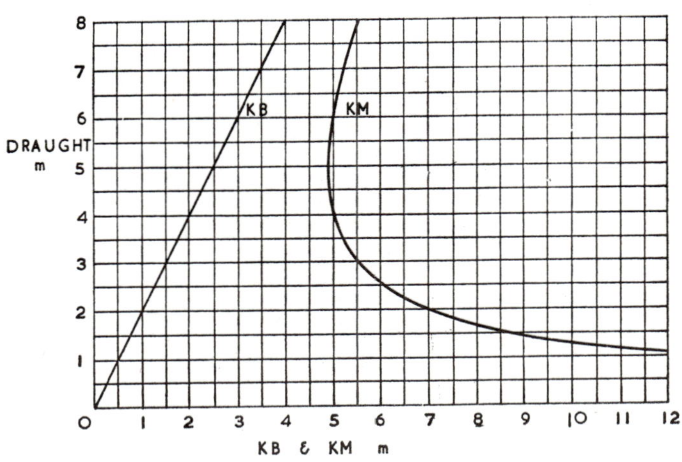

Fig. 47

INCLINING EXPERIMENT

This is a simple experiment which is carried out on the completed ship to determine the metacentric height, and hence the height of the centre of gravity of the ship. If the height of the centre of gravity of the empty ship is known, it is possible to calculate its position for any given condition of loading. It is therefore necessary to carry out the inclining experiment on the empty ship (or as near to empty as possible).

The experiment is commenced with the ship upright.

A small mass *m* is moved across the ship through a distance *d*. This causes the centre of gravity to move from its original position *G* on the centreline to G_1. (Fig. 48).

$$\text{If} \quad \triangle = \text{displacement of ship}$$

$$\text{Then} \quad GG_1 = \frac{m \times d}{\triangle}$$

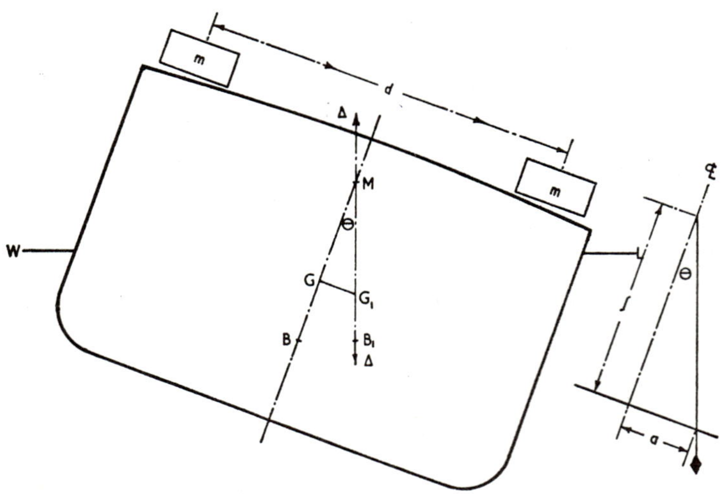

Fig. 48

The ship then heels to angle θ, when the centre of buoyancy moves from B to B_1, in the same vertical line as G_1. But the vertical through B_1 intersects the centreline at M, the transverse metacentre.

$$GG_1 = GM \tan \theta$$

$$\therefore \quad GM \tan \theta = \frac{m \times d}{\triangle}$$

$$GM = \frac{m \times d}{\triangle \tan \theta}$$

To determine the angle of heel it is necessary to suspend a pendulum from, say, the underside of a hatch. The deflection *a* of the pendulum may be measured when the mass is moved across the deck.

Thus if l = length of pendulum

$$\tan \theta = \frac{a}{l}$$

and $$GM = \frac{m \times d \times l}{\triangle \times a}$$

The height of the transverse metacentre above the keel may be found from the metacentric diagram and hence the height of the centre of gravity of the ship may be determined.

$$KG = KM - GM$$

Example. A mass of 6 tonne is moved transversely through a distance of 14 m on a ship of 4300 tonne displacement, when the deflection of an 11 m pendulum is found to be 120 mm. The transverse metacentre is 7·25 m above the keel.

Determine the height of the centre of gravity above the keel.

$$GM = \frac{m \times d}{\triangle \tan \theta}$$

$$= \frac{6 \times 14 \times 11}{4300 \times 120 \times 10^{-3}}$$

$$= 1{\cdot}79\text{m}$$

$$\begin{aligned} KG &= KM - GM \\ &= 7{\cdot}25 - 1{\cdot}79 \\ &= 5{\cdot}46 \text{ m} \end{aligned}$$

CONDUCT OF EXPERIMENT

The experiment must be carried out very carefully to ensure accurate results. At least two pendulums are used, one forward and one aft. They are made as long as possible and are suspended from some convenient point, e.g. the underside of the hatch. A stool is arranged in way of each pendulum on which the deflections are recorded. The pendulum bobs are immersed in water or light oil to dampen the swing.

Four masses A, B, C and D are placed on the deck, two on each side of the ship near midships, their centres being as far as possible from the centreline.

The mooring ropes are slackened and the ship-to-shore gangway removed. The draughts and density of water are read as accurately as possible.

The inclining masses are then moved, one at a time, across the ship until all four are on one side, then all four on the other side and finally two on each side. The deflections of the pendulums are recorded for each movement of mass. An average of these deflections is used to determine the metacentric height. Thus if there are eight movements of mass, and the recorded deflections of pendulum are a_1, a_2, a_3, a_4.....................a_8, then

$$\text{average deflection} = \frac{a_1 + a_2 + a_3 + a_4 + a_5 + a_6 + a_7 + a_8}{8}$$

The ship should be in a sheltered position, e.g. graving dock, and the experiment should be carried out in calm weather. Only those men required for the experiment should be allowed on board. Any

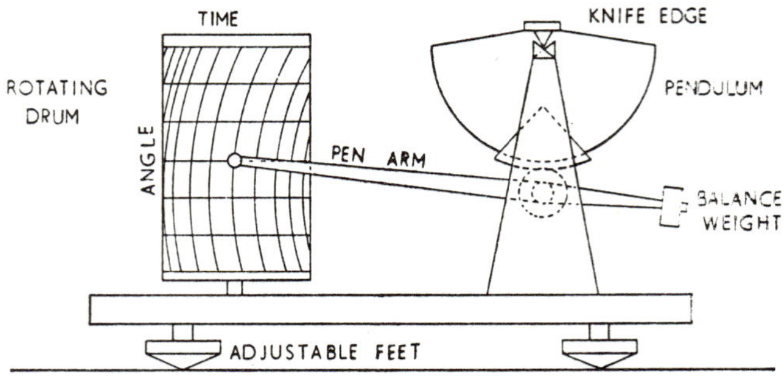

STABILOGRAPH

Fig. 49

movement of liquid affects the results and therefore all tanks should be empty or pressed up tight. The magnitude and position of any mass which is not included in the lightweight of the ship should be noted and it is therefore necessary to sound all tanks and inspect the whole ship. Corrections are made to the centre of gravity for any such masses.

An instrument for recording inclination is in use by many shipyards. It consists of a heavy metal pendulum balanced on knife edges, geared to a pen arm which records the angle of heel on a rotating drum. The advantages of using this instrument, known as a *Stabilograph*, are that a permanent record is obtained and the movement of the ship may be seen as the experiment is in progress. If, for instance, the mooring ropes are restricting the heel, the irregular movement will be seen on the drum.

FREE SURFACE EFFECT

When a tank on board a ship is not completely full of liquid, and the vessel heels, the liquid moves across the tank in the same direction as the heel. The centre of gravity of the ship moves away from the centreline, reducing the righting lever and increasing the angle of heel (Fig. 50).

The movement of the centre of gravity from G to G_1 has been caused by the transfer of a wedge of liquid across the tank. Thus if

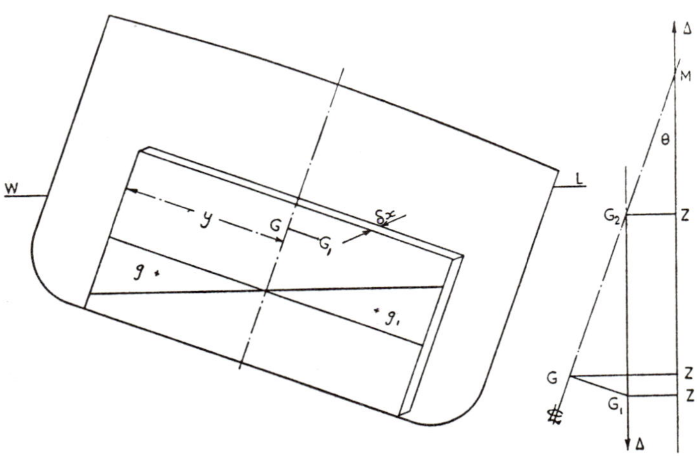

Fig. 50

m is the mass of the wedge and gg_1 the distance moved by its centre, then

$$GG_1 = \frac{m \times gg_1}{\triangle}$$

$$\begin{aligned} \text{But } m &= v \times \rho \\ \text{where } v &= \text{volume of wedge} \\ \rho &= \text{density of liquid} \\ \text{and } \triangle &= \nabla \times \rho_1 \\ \text{where } \nabla &= \text{volume of displacement} \\ \rho_1 &= \text{density of water} \end{aligned}$$

$$\therefore \qquad GG_1 = \frac{v \times \rho \times gg_1}{\nabla \times \rho_1}$$

Divide the tank into thin, transverse strips of length δx and let one such strip have a half width of free surface of y

$$\begin{aligned} \text{Volume of strip of wedge} &= \tfrac{1}{2}\, y \times y \tan\theta\, \delta x \\ &= \tfrac{1}{2} y^2 \tan\theta\, \delta x \\ \text{Mass of strip of wedge} &= \rho \times \tfrac{1}{2} y^2 \tan\theta\, \delta x \end{aligned}$$

Moment of transfer of strip of wedge

$$\begin{aligned} &= \tfrac{4}{3} y \times \rho \times \tfrac{1}{2} y^2 \tan\theta\, \delta x \\ &= \rho \times \tfrac{2}{3} y^3 \tan\theta\, \delta x \end{aligned}$$

Total moment of transfer of wedge

$$\begin{aligned} &= v\, \rho\, gg_1 \\ &= \rho \tan\theta\, \Sigma\, \tfrac{2}{3} y^3 \delta x \end{aligned}$$

But $\Sigma\, \tfrac{2}{3} y^3 \delta x$ = 2nd moment of area of free surface about the centreline of the tank

$= i$

$$\therefore GG_1 = \frac{\rho i \tan\theta}{\rho_1 \nabla}$$

The righting lever has therefore been reduced from GZ to G_1Z. But the righting lever is the perpendicular distance between the verticals through the centre of buoyancy and the centre of gravity, and this distance may be measured at any point. The vertical through G_1 intersects the centreline at G_2, and

$$G_2Z = G_1Z$$

$$\text{also } G_2Z = G_2M \sin\theta$$

but G_1Z does not equal $G_1M \sin\theta$

Since the initial stability of a ship is usually measured in terms of metacentric height, it is useful to assume that the effect of a free surface of liquid is to raise the centre of gravity from G to G_2, thus reducing the metacentric height of the vessel.

GG_2 is termed the *virtual reduction in metacentric height due to free surface* or, more commonly, *the free surface effect.*

$$\text{Now } GG_1 = GG_2 \tan\theta$$

$$\therefore GG_2 = \frac{\rho\, i \tan\theta}{\rho_1 \nabla \tan\theta}$$

Free surface effect $\qquad GG_2 = \dfrac{\rho\, i}{\rho_1 \nabla} \qquad \text{or} = \dfrac{\rho\, i}{\Delta}$

Example. A ship of 5000 tonne displacement has a rectangular tank 6m long and 10 m wide. Calculate the virtual reduction in metacentric height if this tank is partly full of oil (rd 0·8).

$$\rho = 1000 \times 0{\cdot}8 \ \text{kg/m}^3$$
$$i = \tfrac{1}{12}\, 6 \times 10^3 \ \text{m}^4$$
$$\rho_1 = 1025 \ \text{kg/m}^3$$
$$\nabla = \frac{5000}{1{\cdot}025} \ \text{m}^3$$
$$\therefore GG_2 = \frac{1000 \times 0{\cdot}8 \times 6 \times 10^3 \times 1{\cdot}025}{1025 \times 5000 \times 12}$$
$$= 0{\cdot}08 \ \text{m}$$

THE EFFECT OF TANK DIVISIONS ON FREE SURFACE

Consider a rectangular tank of length l and breadth b partly full of sea water.

(a) WITH NO DIVISIONS

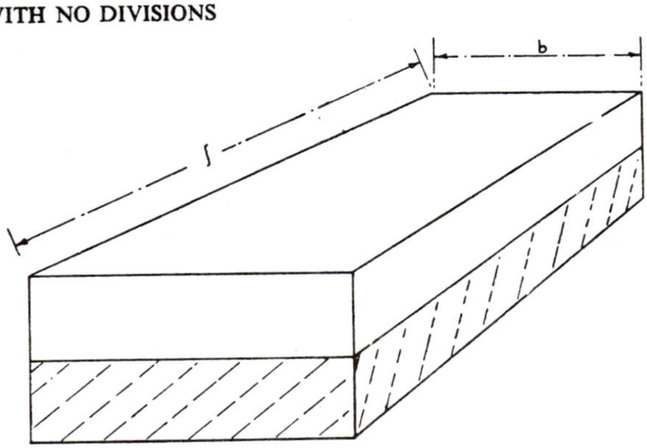

Fig. 51

$$GG_2 = \frac{\rho\, i}{\rho_1 \nabla}$$

$$= \frac{i}{\nabla} \text{ since } \rho = \rho_1$$

$$i = \tfrac{1}{12}\, l\, b^3$$

$$\therefore GG_2 = \frac{l\, b^3}{12 \nabla}$$

(b) WITH A MID-LENGTH, TRANSVERSE DIVISION

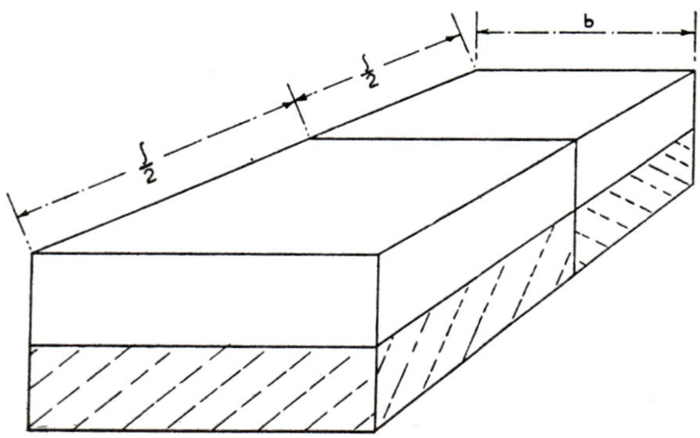

Fig. 52

For one tank $i = \tfrac{1}{12} \frac{l}{2} b^3$

For two tanks $i = \tfrac{1}{12} \frac{l}{2} b^3 \times 2$

$$= \tfrac{1}{12}\, l\, b^3$$

$$\therefore GG_2 = \frac{l\, b^3}{12 \nabla}$$

Thus as long as there is a free surface of liquid in both tanks there is no reduction in free surface effect. It would, however, be possible to fill one tank completely and have a free surface effect in only one tank.

(c) WITH A LONGITUDINAL, CENTRELINE DIVISION

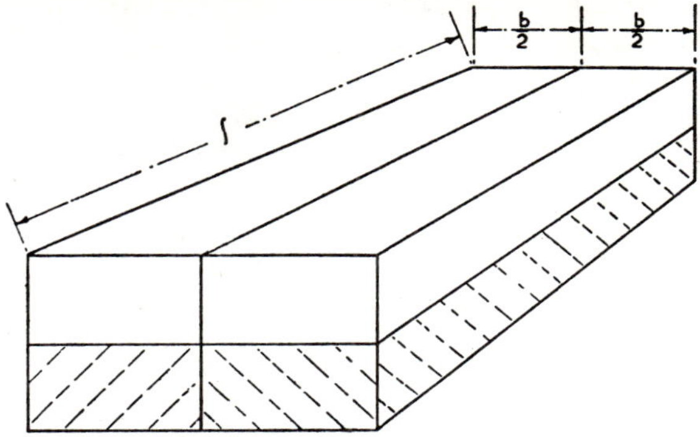

Fig. 53

$$\text{For one tank } i = \tfrac{1}{12}\, l \left(\frac{b}{2}\right)^3$$

$$\text{For two tanks } i = \tfrac{1}{12}\, l \left(\frac{b}{2}\right)^3 \times 2$$

$$= \tfrac{1}{4} \times \tfrac{1}{12}\, l\, b^3$$

$$\therefore \quad GG_2 = \frac{1}{4} \times \frac{l\, b^3}{12\, \nabla}$$

Thus the free surface effect is reduced to one quarter of the original by introducing a longitudinal division.

(d) WITH TWO LONGITUDINAL DIVISIONS FORMING THREE EQUAL TANKS

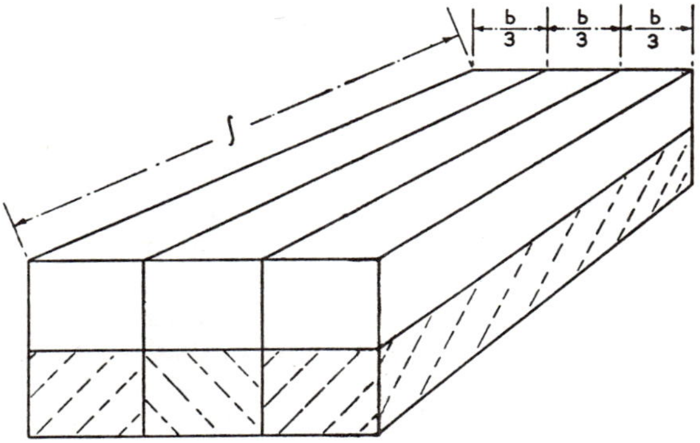

Fig. 54

$$\text{For one tank } i = \tfrac{1}{12}\, l \left(\frac{b}{3}\right)^3$$

$$\text{For three tanks } i = \tfrac{1}{12}\, l \left(\frac{b}{3}\right)^3 \times 3$$

$$= \tfrac{1}{9} \times \tfrac{1}{12}\, l\, b^3$$

$$\therefore GG_2 = \frac{1}{9} \times \frac{l\, b^3}{12\, \nabla}$$

It may be seen that the free surface effect is still further reduced by the introduction of longitudinal divisions.

If a tank is sub-divided by n longitudinal divisions forming *equal* tanks, then

$$GG_2 = \frac{1}{(n+1)^2}\, \frac{l\, b^3}{12\, \nabla}$$

PRACTICAL CONSIDERATIONS

The effect of a free surface of liquid may be most dangerous in a vessel with a small metacentric height and may even cause the vessel to become unstable. In such a ship, tanks which are required to carry liquid should be pressed up tight. If the ship is initially unstable and heeling to port, then any attempt to introduce water ballast will reduce the stability. Before ballasting, therefore, an attempt should be made to lower the centre of gravity of the ship by pressing up existing tanks and lowering masses in the ship. If water is introduced into a double bottom tank on the starboard side the vessel will flop to starboard and may possibly capsize. A small tank on the port side should therefore be filled completely before filling on the starboard side. The angle of heel will increase due to free surface and the effect of the added mass but there will be no sudden movement of the ship.

A particularly dangerous condition may occur when a fire breaks out in the upper 'tween decks of a ship or in the accommodation of a passenger ship. If water is pumped into the space, the stability of the ship will be reduced both by the added mass of water and by the free surface effect. Any accumulation of water should be avoided. Circumstances will dictate the method used to remove the water, and will vary with the ship, cargo and position of fire. It may be possible to discharge the water using a portable pump. In calm weather or in port a hole may be drilled in the side of the ship. A hole in the deck would allow the water to work its way into the bilges from where it may be pumped overboard, but it is doubtful if such a method would be possible except in rare circumstances. It may, however, be possible to remove the 'tween deck hatch covers thus restricting the height of water to about 150 mm. The cargo in the lower hold would be damaged by such a method but this would be preferable to losing the ship.

It is important to note that the free surface effect depends upon the displacement of the ship and the shape and dimensions of the *free surface*. It is independent of the total mass of liquid in the tank and of the position of the tank in the ship.

The ship with the greatest free surface effect is, of course, the oil tanker, since space must be left in the tanks for expansion of oil. Originally tankers were built with centreline bulkhead and expansion trunks. Twin longitudinal bulkheads were then introduced without expansion trunks and were found to be successful, since the loss in metacentric height due to free surface was designed for. It is not possible to design dry cargo vessels in the same way, since the position of the centre of gravity of the ship varies considerably with the nature and disposition of the cargo. Thus while the free surface effect in a tanker is greater than in a dry cargo ship, it is of more importance in the latter.

The effect of a suspended mass on the stability of a ship may be treated in the same way as a free surface. It may be shown, as stated in Chapter 4, that the centre of gravity of the mass may be taken as acting at the point of suspension.

*f*STABILITY AT LARGE ANGLES OF HEEL

When a ship heels to an angle greater than about 10°, the principles on which the initial stability were based are no longer true. The proof of the formula for *BM* was based on the assumption that the two waterplanes intersect at the centreline and that the wedges are right angled triangles. Neither of these assumptions may be made for large angles of heel, and the stability of the ship must be determined from first principles.

The righting lever is the perpendicular distance from a vertical axis through the centre of gravity *G* to the centre of buoyancy B_1. This distance may be found by dividing the moment of buoyancy about this axis by the buoyancy. In practice recourse is made to an instrument known as an integrator which may be used to determine the area of any plane and the moment of the plane about a given axis. The method used is as follows.

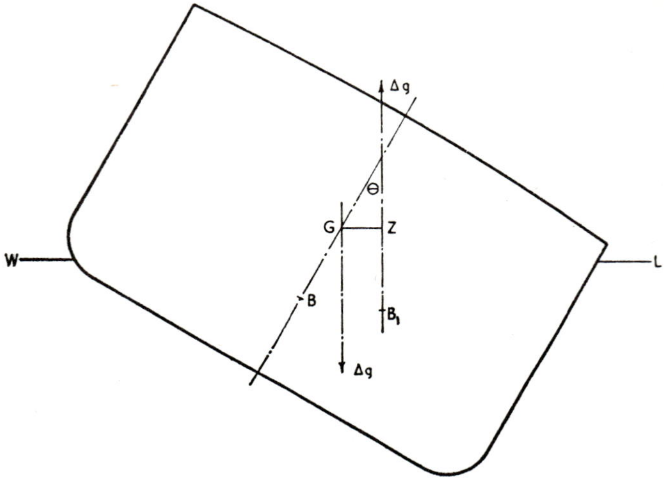

Fig. 55

The position of the centre of gravity G must be assumed at some convenient position above the keel, since the actual position is not known. Sections through the ship are drawn at intervals along the ship's length. These sections are inclined to an angle of, say 15°. The integrator is set with its axis in the vertical through G. The

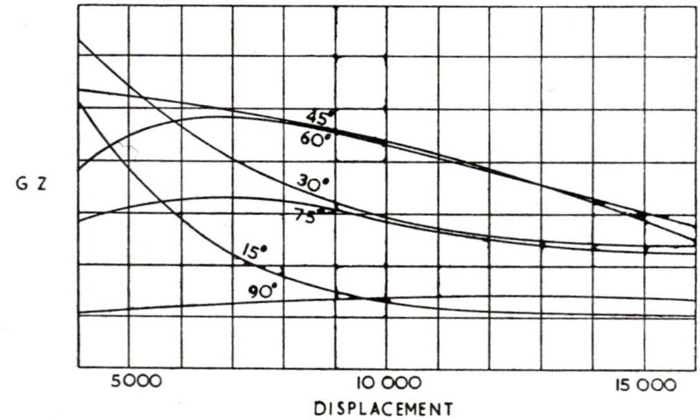

CROSS CURVES OF STABILITY

Fig. 56

outline of each section is traced by the integrator up to a given waterline and the displacement and righting lever obtained. This is repeated for different waterlines and for angles of 30°, 45°, 60°, 75° and 90°. The GZ values at each angle are plotted on a base of displacement to form the *cross curves of stability* for the ship.

The displacement, height of centre of gravity and metacentric height of a vessel may be calculated for any loaded condition. At this displacement the righting levers may be obtained at the respective angles for the assumed position of the centre of gravity. These values must be amended to suit the actual height of the centre of gravity.

Let G = assumed position of centre of gravity
G_1 = actual position of centre of gravity

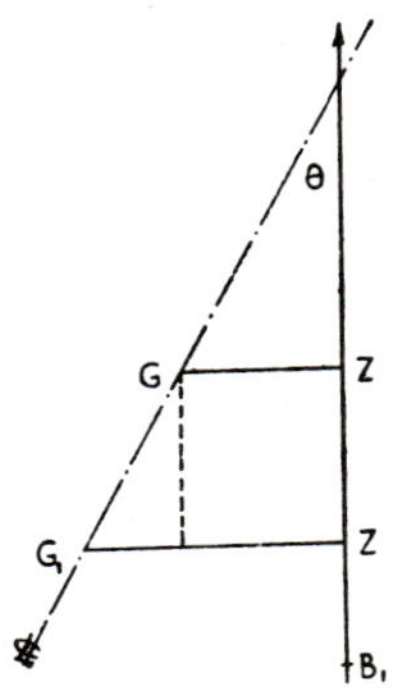

Fig. 57

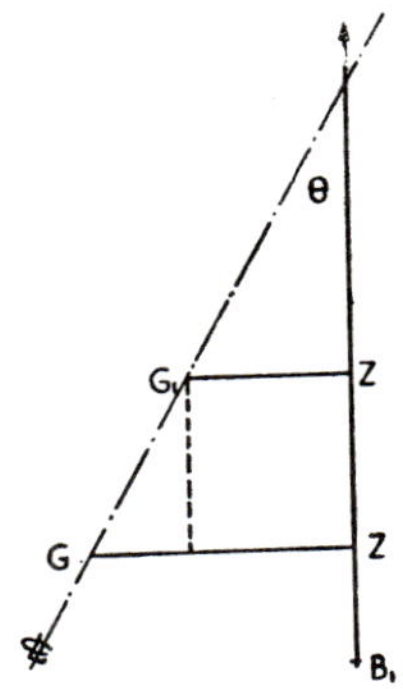

Fig. 58

If G_1 lies *below* G (Fig. 57), then the ship is *more* stable and

$$G_1Z = GZ + GG_1 \sin \theta$$

If G_1 lies *above* G (Fig. 58), then the ship is *less* stable and

$$G_1Z = GZ - GG_1 \sin \theta$$

The amended righting levers are plotted on a base of angle of heel to form the *Curve of Statical Stability* for the ship in this condition of loading. The initial slope of the curve lies along a line drawn from the origin to GM plotted vertically at one radian (57·3°).

The area under this curve to any given angle, multiplied by the gravitational weight of the ship, is the work done in heeling the ship to that angle and is known as the *Dynamical Stability*.

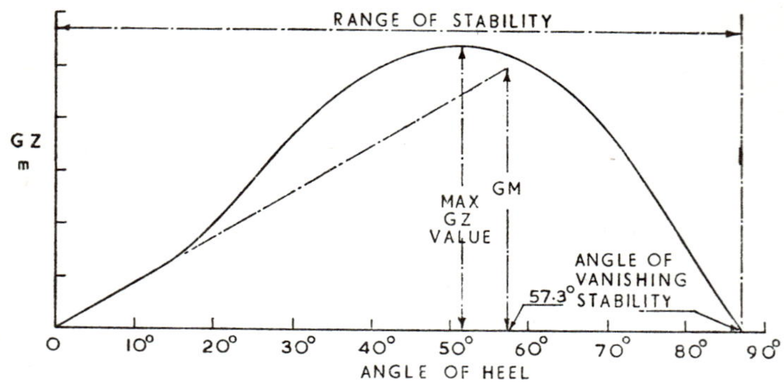

CURVE OF STATICAL STABILITY

Fig. 59

Example. A vessel has the following righting levers at a particular draught, based on an assumed KG of 7·2 m

θ	0°	15°	30°	45°	60°	75°	90°
GZ	0	0·43	0·93	1·21	1·15	0·85	0·42 m

The vessel is loaded to this draught but the actual KG is found to be 7·8 m and the GM 1-0 m.

Draw the amended statical stability curve.

$$GG_1 = 0{\cdot}6 \text{ m}$$
$$G_1Z = GZ - GG_1 \sin\theta$$

(i.e. the vessel is *less* stable than suggested by the original values).

Angle θ	$\sin\theta$	$GG_1 \sin\theta$	GZ	G_1Z
0	0	—	0	0
15°	0·259	0·15	0·43	0·28
30°	0·500	0·30	0·93	0·63
45°	0·707	0·42	1·21	0·79
60°	0·866	0·52	1·15	0·63
75°	0·966	0·58	0·85	0·27
90°	1·000	0·60	0·42	−0·18

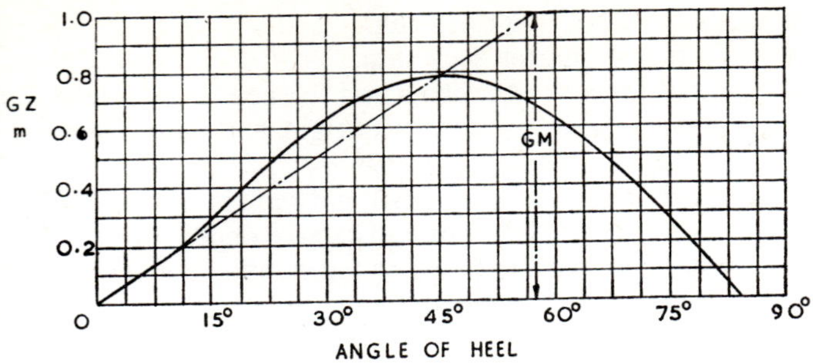

Fig. 60

The shape of the stability curve of a ship depends largely on the metacentric height and the freeboard. A tremendous change takes place in this curve when the weather deck edge becomes immersed. Thus a ship with a large freeboard will normally have large range of stability while a vessel with a small freeboard will have a much smaller range. Fig. 61 shows the effect of freeboard on two ships with the same metacentric height.

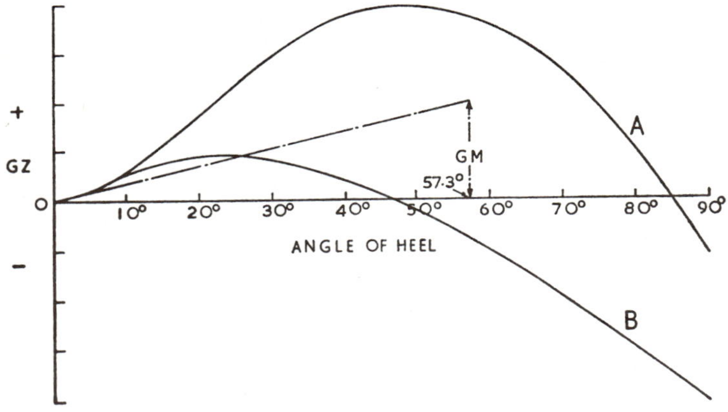

Fig. 61

Vessel A is a closed shelter deck ship.
Vessel B is a raised quarter deck ship.

It is essential for a vessel with small freeboard, such as an oil tanker, to have a large metacentric height and thus extend the range of stability.

If a vessel is initially unstable it will not remain upright but will either heel to the *Angle of Loll* or will capsize depending upon the degree of instability and the shape of the stability curve (Fig. 62).

Vessel A will heel to an angle of loll of about 8° but still remains a fairly stable ship, and while this heel would be very inconvenient, the vessel would not be in a dangerous condition.

If vessel B is unstable it will capsize since at all angles the righting lever is negative.

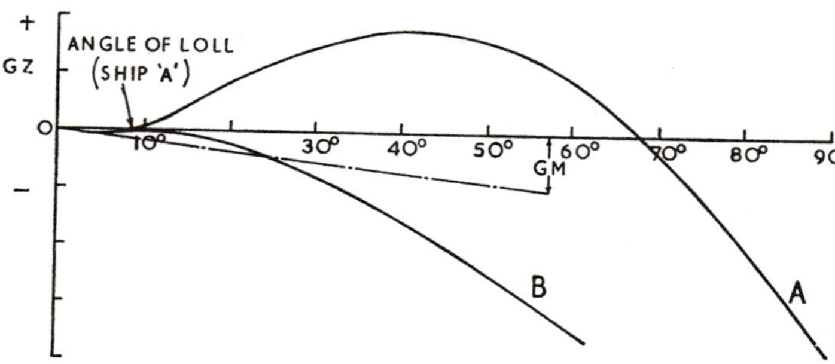

Fig. 62

Example. The righting levers of a ship of 15 000 tonne displacement at angles of heel of 15°, 30°, 45° and 60° are 0·29, 0·70, 0·93 and 0·90 m respectively. Calculate the dynamical stability of the ship at 60° heel.

Angle	GZ	SM	Product for area
0	0	1	0
15°	0·29	4	1·16
30°	0·70	2	1·40
45°	0·93	4	3·72
60°	0·90	1	0·90
			7·18

Note: The common interval must be expressed in *radians*.

$$h = \frac{15}{57{\cdot}3}$$

$$\text{Area under curve} = \tfrac{1}{3} \cdot \frac{15}{57{\cdot}3} \cdot 7{\cdot}13$$

$$\begin{aligned} &= 0{\cdot}6012 \\ \text{Dynamical stability} &= 15\,000 \times 9{\cdot}81 \times 0{\cdot}6012 \\ &= 88{\cdot}47 \times 10^3 \text{ kJ} \\ &= 88{\cdot}47 \text{ MJ} \end{aligned}$$

STABILITY OF A WALL SIDED-SHIP

If a vessel is assumed to be wall-sided in the vicinity of the waterplane, the righting lever may be estimated from the expression.

$$GZ = \sin\theta\,(GM + \tfrac{1}{2}\,BM\,tan^2\,\theta)$$

This formula may be regarded as reasonably accurate for vessels which are deeply loaded, up to the point at which the deck edge enters the water.

If a vessel is initially unstable it will either capsize or heel to the angle of loll. At this angle of loll θ the vessel does not tend to return to the upright or incline to a greater angle. The righting lever is therefore zero. Hence if the vessel is assumed to be wall-sided:

$$0 = \sin\theta\,(GM + \tfrac{1}{2}\,BM\,tan^2\theta)$$

and since sin θ cannot be zero unless θ is zero:

$$\begin{aligned} 0 &= GM + \tfrac{1}{2}BM\,tan^2\theta \\ \tfrac{1}{2}BMtan^2\theta &= -GM \end{aligned}$$

$$\tan^2\theta = -\,2\,\frac{GM}{BM}$$

From which $$\tan\theta = \pm\sqrt{-\,2\,\frac{GM}{BM}}$$

Since GM must be negative for this condition, $-\,2\,\dfrac{GM}{BM}$ must be *positive.*

Thus the angle of loll may be determined for any given unstable condition.

Example. A ship of 12 000 tonne displacement has a second moment of area about the centreline of $72 \times 10^3\ \text{m}^4$. If the metacentric height is $-0{\cdot}05$ m, calculate the angle of loll.

$$\nabla = \frac{12\,000}{1{\cdot}025}\ \text{m}^3$$

$$BM = \frac{I}{\nabla}$$

$$= \frac{72 \times 10^3 \times 1{\cdot}025}{12\,000}\ \text{m}$$

$$= 6{\cdot}150\ \text{m}$$

$$\tan\theta = \pm\sqrt{-2\,\frac{(-0{\cdot}05)}{6{\cdot}15}}$$

$$= \pm\,0{\cdot}1275$$

From which $\theta = \pm\,7°\,16'$

i.e. the vessel may heel 7° 16′ either to port or to starboard.

A more practical application of this expression may be found when the vessel is listing at sea. It is necessary first to bring the ship upright and then provide sufficient stability for the remainder of the voyage. Thus it is essential to estimate the negative metacentric height causing the angle of loll in order to ensure that these conditions are realised.

Example. At one point during a voyage the above vessel is found to have an angle of loll of 13°. Calculate the initial metacentric height.

From above $BM = 6{\cdot}150$ m

$$\tan\theta = \pm\sqrt{\frac{-2GM}{BM}}$$

$$0{\cdot}2309 = \pm\sqrt{\frac{-2GM}{6{\cdot}15}}$$

$$GM = -\,\frac{0{\cdot}2309^2 \times 6{\cdot}15}{2}$$

Initial metacentric height $= -\,0{\cdot}164$ m.

TEST EXAMPLES 5

1. A ship displaces 12 000 tonne, its centre of gravity is 6·50 m above the keel and its centre of buoyancy is 3·60 m above the keel If the second moment of area of the waterplane about the centreline is $42 \cdot 5 \times 10^3$ m^4 find the metacentric height.

2. A vessel of 10 000 tonne displacement has a second moment of area of waterplane about the centreline of 60×10^3 m^4. The centre of buoyancy is 2·75 m above the keel. The following are the disposition of the masses on board the ship.

4000 tonne 6·30 m above the keel

2000 tonne 7·50 m above the keel

4000 tonne 9·15 m above the keel

Calculate the metacentric height.

3. A vessel of constant rectangular cross-section has a breadth of 12 m and metacentric height of one quarter of the draught. The vertical centre of gravity lies on the waterline. Calculate the draught.

*f*4. A raft is made from two cylinders each 1·5 m diameter and 6 m long. The distance between the centres of the cylinders is 3 m. If the draught is 0·75 m, calculate the transverse *BM*.

5. A vessel of constant rectangular cross-section is 7·2 m wide.

(a) Draw the metacentric diagram using 0·5 m intervals of draught up to the 4·0 m waterline.

(b) If the centre of gravity is 3 m above the keel, determine from the metacentric diagram the limits of draught between which the vessel will be unstable.

6. A vessel of constant triangular cross-section is 9 m wide at the deck and has a depth to deck of 7·5 m. Draw the metacentric diagram using 0·5 m intervals of draught up to the 3·0 m waterline.

7. An inclining experiment was carried out on a ship of 8000 tonne displacement. A mass of 10 tonne was moved 14 m across the deck causing a pendulum 8·5 m long to deflect 110 mm. The transverse metacentre was 7·15 m above the keel. Calculate the metacentric height and the height of the centre of gravity above the keel.

8. An inclining experiment was carried out on a ship of 4000 tonne displacement, when masses of 6 tonne were moved transversely through 13·5 m. The deflections of a 7·5 m pendulum were 81, 78, 85, 83, 79, 82, 84, and 80 mm respectively.
Calculate the metacentric height.

*f*9. A ship of 5000 tonne displacement has a rectangular double bottom tank 9 m wide and 12 m long, half full of sea water. Calculate the virtual reduction in metacentric height due to free surface.

*f*10. A ship of 8000 tonne displacement has its centre of gravity 4·5 m above the keel and transverse metacentre 5·0 m above the keel when a rectangular tank 7·5 m long and 15 m wide contains sea water. A mass of 10 tonne is moved 12 m across the deck. Calculate the angle of heel.:
(a) if there is no free surface of water,
(b) if the water does not completely fill the tank.

*f*11. A ship of 6000 tonne displacement has its centre of gravity 5·9 m above the keel and transverse metacentre 6·8 m above the keel. A rectangular double bottom tank 10·5 m long, 12 m wide and 1·2 m deep is now half-filled with sea water. Calculate the metacentric height.

*f*12. An oil tanker 24 m wide displaces 25 000 tonne when loaded in nine equal tanks, each 10 m long, with oil rd 0·8. Calculate the total free surface effect with:
(a) no longitudinal divisions,
(b) a longitudinal centreline bulkhead,
(c) twin longitudinal bulkheads forming three equal tanks.
(d) twin longitudinal bulkheads, the centre compartment having a width of 12 m.

*f*13. A ship of 12 500 tonne displacement and 15 m beam has a metacentric height of 1·10 m. A mass of 80 tonne is lifted from its position in the centre of the lower hold by one of the ship's derricks, and placed on the quay 2 m from the ship's side. The ship heels to a maximum angle of 3·5° when the mass is being moved.
(a) Does the *GM* alter during the operation?
(b) Calculate the height of the derrick head above the original centre of gravity of the mass.

*f*14. The righting levers of a ship, for an assumed KG of 3·5 m, are 0, 0·25, 0·46, 0·51, 0·39, 0·10 and −0·38 m at angles of heel of 0, 15°, 30°, 45°, 60°, 75° and 90° respectively.

When the ship is loaded to the same displacement the centre of gravity is 3·0 m above the keel and the metacentric height 1·25 m. Draw the amended curve of statical stability.

*f*15. The righting moments of a ship at angles of heel of 0, 15°, 30°, 45° and 60° are 0, 1690, 5430, 9360 and 9140 kN m respectively. Calculate the dynamical stability at 60°.

*f*16. A ship of 18 000 tonne displacement has KB 5·25 m, KG 9·24 m and second moment of area about the centreline of 82×10^3 m^4.

Using the wall-sided formula calculate the righting levers at intervals of 5° heel up to 20° and sketch the stability curve up to this angle.

*f*17. A ship of 7200 tonne displacement has KG 5.20 m, KB 3.12 m and KM 5.35 m. 300 tonne of fuel at Kg 0.6 m are now used. Ignoring free surface effect and assuming the KM remains constant, calculate the angle to which the vessel will heel.

CHAPTER 6

TRIM

f CHANGE IN DRAUGHTS DUE TO ADDED MASSES

TRIM is the difference between the draughts forward and aft. Thus if a ship floats at draughts of 6 m forward and 7 m aft, it is said to trim 1 m by the *stern*. If the draught forward is greater than the draught aft the vessel is said to trim by the *head*.

CENTRE OF FLOTATION (LCF) is the centroid of the waterplane and is the axis about which a ship changes trim when a mass is added, removed or moved longitudinally.

If a small mass m is added to a ship at the centre of flotation, there is an increase in mean draught but no change in trim, since the centre of gravity of the added mass is at the same position as the centre of the added layer of buoyancy. A large mass (e.g. one exceeding, say, one twentieth of the displacement) will cause a considerable increase in draught and hence a change in waterplane area and centre of flotation.

MEAN DRAUGHT

The mean draught of a vessel is the draught at which the vessel would lie in level keel conditions. Since the vessel changes trim about the LCF, the draught at this point remains constant for any given displacement whether the vessel is at level keel or trimmed. Hence the mean draught may be taken as the draught at the LCF.

The mean of the end draughts may be compared with the actual draught amidships to determine whether the vessel is hogging or sagging, but is of little relevance in hydrostatic calculations.

EFFECT OF ADDING SMALL MASSES

It is useful to assume that when a small mass is added to the ship it is first placed at the centre of flotation and then moved forward or aft to its final position. Thus the effect of an added mass on the draughts may be divided into:

(a) a bodily increase in draught

(b) a change in trim due to the movement of the mass from the centre of flotation to its final position.

The bodily increase in draught may be found by dividing the mass by the TPC.

The change in trim due to any longitudinal movement of mass may be found by considering its effect on the centre of gravity of the ship.

Consider a ship of displacement $\triangle$ and length L, lying at waterline WL and having a mass m on the deck (Fig. 63). The centre of gravity G and the centre of buoyancy B lie in the same vertical line.

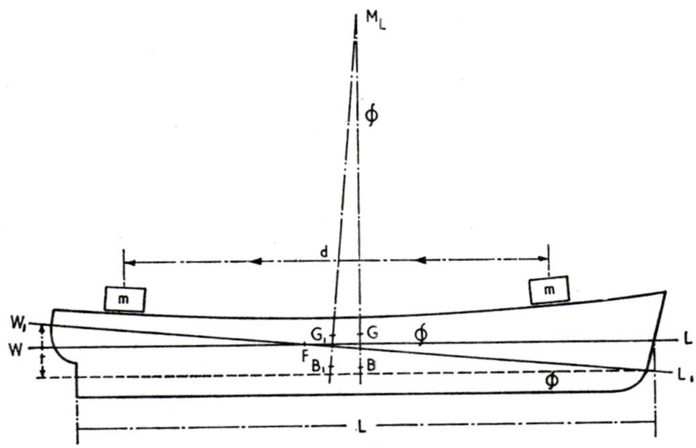

Fig. 63

If the mass is moved a distance d aft, the centre of gravity moves aft from G to G_1, and

$$GG_1 = \frac{m \times d}{\triangle}$$

The ship then changes trim through the centre of flotation F until it lies at waterline W_1L_1. This change in trim causes the centre of buoyancy to move aft from B to B_1, in the same vertical line as G_1. The vertical through B_1 intersects the original vertical through B at M_L, the *longitudinal metacentre*. GM_L is known as the *longitudinal metacentric height*,

$$GM_L = KB + BM_L - KG$$

$$BM_L = \frac{I_F}{\nabla}$$

Where I_F = second moment of area of the waterplane about a transverse axis through the centre of flotation F.

If the vessel trims through an angle ϕ, then

$$GG_1 = GM_L \tan \phi$$

and

$$GM_L \tan \phi = \frac{m \times d}{\triangle}$$

$$\tan \phi = \frac{m \times d}{\triangle \times GM_L}$$

Draw RL_1 parallel to WL.

$$\begin{aligned}\text{Change in trim} &= W_1W + LL_1 \\ &= W_1 R \\ &= \frac{t}{100} \text{ m}\end{aligned}$$

Where t = change in trim in cm over length L m.

But

$$\tan \phi = \frac{t}{100L}$$

$\therefore$

$$\frac{t}{100L} = \frac{m \times d}{\triangle \times GM_L}$$

$$t = \frac{m \times d \times 100L}{\triangle \times GM_L} \text{ cm}$$

The change in trim may therefore be calculated from this expression. $m \times d$ is known as the trimming moment.

It is useful to know the moment which will cause a change in trim of one cm.

$$m \times d = \frac{t \times \triangle \times GM_L}{100\ L} \text{ tonne m}$$

Let

$$t = 1 \text{ cm}$$

Then moment to change trim one cm

$$\text{MCTI cm} = \frac{\triangle \times GM_L}{100L} \text{ tonne m}$$

Change in trim $$t = \frac{\text{trimming moment}}{\text{MCTIcm}} \text{ cm}$$

$$= \frac{m \times d}{\text{MCTIcm}} \text{ cm by the stern}$$

It is now possible to determine the effect of this change in trim on the end draughts. Since the vessel changes trim by the stern, the forward draught will be reduced while the after draught will be increased.

By similar triangles.

$$\frac{t}{L} = \frac{LL_1}{FL} = \frac{W_1W}{WF}$$

t, LL_1 and W_1W may be expressed in cm while L, FL and WF are expressed in m.

Change in draught forward $LL_1 = -\frac{t}{L} \times \text{FL cm}$

Change in draught aft $W_1W = +\frac{t}{L} \times \text{WF cm}$

Example. A ship of 5000 tonne displacement, 96 m long, floats at draughts of 5·60 m forward and 6·30 m aft. The TPC is 11·5, GM_L 105 m and centre of flotation 2·4 m aft of midships.

Calculate:

(a) the MCTIcm

(b) the new end draughts when 88 tonne are added 31 m forward of midships.

(a) $$\text{MCTI cm} = \frac{\triangle \times GM_L}{100L}$$

$$= \frac{5000 \times 105}{100 \times 96}$$

$$= 54{\cdot}69 \text{ tonne m}$$

(b)

$$\text{Bodily sinkage} = \frac{88}{11{\cdot}5}$$
$$= 7{\cdot}65 \text{ cm}$$
$$d = 31 + 2{\cdot}4$$
$$= 33{\cdot}4 \text{ m from } F$$
$$\text{Trimming moment} = 88 \times 33{\cdot}4 \text{ tonne m}$$
$$\text{Change in trim} = \frac{88 \times 33{\cdot}4}{54{\cdot}69}$$
$$= 53{\cdot}74 \text{ cm by the head}$$

$$\text{Distance from } F \text{ to fore end} = \frac{96}{2} + 2{\cdot}4$$
$$= 50{\cdot}4 \text{ m}$$
$$\text{Distance from } F \text{ to after end} = \frac{96}{2} - 2{\cdot}4$$
$$= 45{\cdot}6 \text{ m}$$
$$\text{Change in trim forward} = +\frac{53{\cdot}74}{96} \times 50{\cdot}4$$
$$= +28{\cdot}22 \text{ cm}$$
$$\text{Change in trim aft} = -\frac{53{\cdot}74}{96} \times 45{\cdot}6$$
$$= -25{\cdot}52 \text{ cm}$$
$$\text{New draught forward} = 5{\cdot}60 + 0{\cdot}076 + 0{\cdot}282$$
$$= 5{\cdot}958 \text{ m}$$
$$\text{New draught aft} = 6{\cdot}30 + 0{\cdot}076 - 0{\cdot}255$$
$$= 6{\cdot}121 \text{ m}$$

If a number of items are added to the ship at different positions along its length, the total mass and nett trimming moment may be used to determine the final draughts.

Example. A ship 150 m long has draughts of 7·70 m forward and 8·25 m aft, MCT I cm 250 tonne m, TPC 26 and LCF 1·8 m forward of midships. Calculate the new draughts after the following masses have been added:

50 tonne, 70 m aft of midships
170 tonne, 36 m aft of midships
100 tonne, 5 m aft of midships
130 tonne, 4 m forward of midships
40 tonne, 63 m forward of midships

Mass (tonne)	Distance from F (m)	moment forward (tonne m)	moment aft (tonne m)
50	71·8A	–	3590
170	37·8A	–	6426
100	6·8A	–	680
130	2·2F	286	–
40	61·2F	2448	–
Total 490		2734	10 696

$$\text{Excess moment aft} = 10\,696 - 2734$$
$$= 7962 \text{ tonne m}$$

$$\text{Change in trim} = \frac{7962}{250}$$
$$= 31{\cdot}85 \text{ cm by the stern}$$

$$\text{Change in trim forward} = -\frac{31{\cdot}85}{150}\left(\frac{150}{2} - 1{\cdot}8\right)$$
$$= -15{\cdot}54 \text{ cm}$$

$$\text{Change in trim aft} = +\frac{31{\cdot}85}{150}\left(\frac{150}{2} + 1{\cdot}8\right)$$
$$= +16{\cdot}31 \text{ cm}$$

$$\text{Bodily sinkage} = \frac{490}{26}$$
$$= 18{\cdot}85 \text{ cm}$$

$$\text{New draught forward} = 7{\cdot}70 + 0{\cdot}189 - 0{\cdot}155$$
$$= 7{\cdot}734 \text{ m}$$

$$\text{New draught aft} = 8{\cdot}25 + 0{\cdot}189 + 0{\cdot}163$$
$$= 8{\cdot}602 \text{ m}$$

DETERMINATION OF DRAUGHTS AFTER THE ADDITION OF LARGE MASSES

When a large mass is added to a ship the resultant increase in draught is sufficient to cause changes in all the hydrostatic details. It then becomes necessary to calculate the final draughts from first principles. Such a problem exists every time a ship loads or discharges the major part of its deadweight.

The underlying principle is that after loading or discharging the vessel is in equilibrium and hence the final centre of gravity is in the same vertical line as the final centre of buoyancy.

For any given condition of loading it is possible to calculate the displacement $\triangle$ and the longitudinal position of the centre of gravity G relative to midships.

From the hydrostatic curves or data, the mean draught may be obtained at this displacement, and hence the value of MCTI cm and the distance of the LCB and LCF from midships. These values are calculated for the level keel condition and it is unlikely that the LCB will be in the same vertical line as G. Thus a trimming moment acts on the ship. This trimming moment is the displacement multiplied by the longitudinal distance between B and G, known as the *trimming lever*.

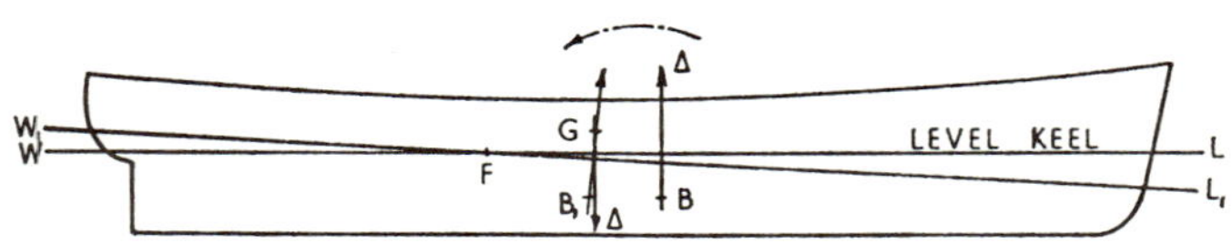

Fig. 64

The trimming moment, divided by the MCTI cm, gives the change in trim from the level keel condition, i.e. the total trim of the vessel. The vessel changes trim about the LCF and hence it is possible to calculate the end draughts. When the vessel has changed trim in this manner, the new centre of buoyancy B_1 lies in the same vertical line as G.

Example. A ship 125 m long has a light displacement of 4000 tonne with LCG 1·60 m aft of midships. The following items are now added:

Cargo 8500 tonne Lcg 3·9 m forward of midships
Fuel 1200 tonne Lcg 3·1 m aft of midships
Water 200 tonne Lcg 7·6 m aft of midships
Stores 100 tonne Lcg 30·5 m forward of midships.

At 14 000 tonne displacement the mean draught is 7·80 m, MCTI cm 160 tonne m, LCB 2·00 m forward of midships and LCF 1·5 m aft of midships.

Calculate the final draughts.

Item	mass(t)	Lcg(m)	moment forward	moment aft
Cargo	8500	3·9F	33 150	—
Fuel	1200	3·1A	—	3720
Water	200	7·6A	—	1520
Stores	100	30·5F	3050	—
Lightweight	4000	1·6A	—	6400
Displacement	14 000		36 200	11 640

$$\text{Excess moment forward} = 36\,200 - 11\,640$$
$$= 24\,560 \text{ tonne m}$$

$$\text{LCG from midships} = \frac{24\,560}{14\,000}$$

$$= 1{\cdot}754 \text{ m forward}$$
$$\text{LCB from midships} = 2{\cdot}000 \text{ m forward}$$
$$\text{trimming lever} = 1{\cdot}754 - 2{\cdot}000$$
$$= 0{\cdot}245 \text{ m aft}$$
$$\text{trimming moment} = 14\,000 \times 0{\cdot}246 \text{ tonne m}$$

$$\text{trim} = \frac{14\,000 \times 0{\cdot}246}{160}$$

$$= 21{\cdot}5 \text{ cm by the stern}$$

$$\text{Change in draught forward} = -\frac{21{\cdot}5}{125}\left(\frac{125}{2} + 1{\cdot}5\right)$$

$$= -11{\cdot}0 \text{ cm}$$

$$\text{Change in draught aft} = +\frac{21{\cdot}5}{125}\left(\frac{125}{2} - 1{\cdot}5\right)$$

$$= +10{\cdot}5 \text{ cm}$$
$$\text{Draught forward} = 7{\cdot}80 - 0{\cdot}110$$
$$= 7{\cdot}690 \text{ m}$$
$$\text{Draught aft} = 7{\cdot}80 + 0{\cdot}105$$
$$= 7{\cdot}905 \text{ m}$$

CHANGE IN MEAN DRAUGHT DUE TO CHANGE IN DENSITY

The displacement of a ship floating freely at rest is equal to the mass of the volume of water which it displaces. For any given displacement, the volume of water displaced must depend upon the density of the water. When a ship moves from sea water into river water without change in displacement, there is a slight increase in draught.

Consider a ship of displacement $\triangle$ tonne, waterplane area A_w m^2, which moves from sea water of ρ_S t/m^3 into river water of ρ_R t/m^3 without change in displacement.

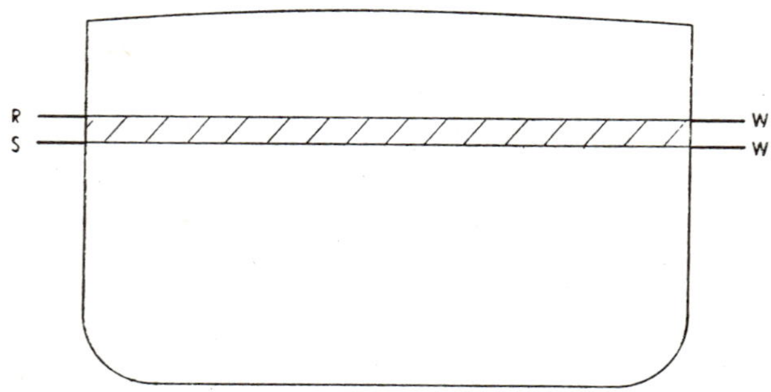

Fig. 65

Volume of displacement in sea water

$$\nabla_S = \frac{\triangle}{\rho_S} \text{ m}^3$$

Volume of displacement in river water

$$\nabla_R = \frac{\triangle}{\rho_R} \text{ m}^3$$

Change in volume of displacement

$$v = \nabla_R - \nabla_S$$

$$= \frac{\triangle}{\rho_R} - \frac{\triangle}{\rho_S}$$

$$= \triangle \left(\frac{1}{\rho_R} - \frac{1}{\rho_S}\right) \text{ m}^3$$

This change in volume causes an increase in draught. Since the increase is small, the waterplane area may be assumed to remain constant and the increase in mean draught may therefore be found by dividing the change in volume by the waterplane area.

$$\text{Increase in draught} = \frac{\triangle}{A_w}\left(\frac{1}{\rho_R} - \frac{1}{\rho_S}\right) \text{ m}$$

$$= \frac{100\,\triangle}{A_w}\left(\frac{\rho_S - \rho_R}{\rho_R \times \rho_S}\right) \text{ cm}$$

The tonne per cm immersion for sea water is given by

$$\text{TPC} = \frac{A_w}{100} \times \rho_S$$

$$\therefore \quad A_w = \frac{100\ \text{TPC}}{\rho_S} \text{ m}^2$$

Substituting for A_w in the formula for increase in draught:

$$\text{Increase in draught} = \frac{100\,\triangle\,\rho_S}{100\ \text{TPC}}\left(\frac{\rho_S - \rho_R}{\rho_R \times \rho_S}\right)$$

$$= \frac{\triangle}{\text{TPC}}\left(\frac{\rho_S - \rho_R}{\rho_R}\right) \text{ cm}$$

A particular case occurs when a ship moves from sea water of 1·025 t/m³ into fresh water of 1·000 t/m³, the TPC being given in the sea water.

$$\text{Increase in draught} = \frac{\triangle}{\text{TPC}}\left(\frac{1{\cdot}025 - 1{\cdot}000}{1{\cdot}000}\right)$$

$$= \frac{\triangle}{40\ \text{TPC}} \text{ cm}$$

This is known as the *fresh water allowance*, used when computing the freeboard of a ship and is the difference between the S line and the F line on the freeboard markings.

Example. A ship of 10 000 tonne displacement has a waterplane area of 1300 m². The ship loads in water of 1·010 t/m³ and moves into water of 1·026 t/m³. Find the change in mean draught.

Since the vessel moves into water of a greater density there will be a reduction in mean draught.

$$\text{Reduction in mean draught} = \frac{100\,\triangle}{A_w}\left(\frac{\rho_S - \rho_R}{\rho_R \times \rho_S}\right) \text{ cm}$$

$$= \frac{100 \times 10\,000}{1300}\left(\frac{1{\cdot}026 - 1{\cdot}010}{1{\cdot}010 \times 1{\cdot}026}\right)$$

$$= 11{\cdot}88 \text{ cm}$$

When a vessel moves from water of one density to water of a different density, there may be a change in displacement due to the consumption of fuel and stores, causing an additional change in mean draught. If the vessel moves from sea water into river water, it is possible in certain circumstances for the increase in draught due to change in density to be equal to the reduction in draught due to the removed mass. In such a case there will be no change in mean draught.

Example. 215 tonne of oil fuel and stores are used in a ship while passing from sea water of 1·026 t/m³ into river water of 1·002 t/m³. If the mean draught remains unchanged, calculate the displacement in the river water.

Let $\triangle$ = displacement in river water

Then $\triangle + 215$ = displacement in sea water

Since the draught remains unaltered, the volume of displacement in the river water must be equal to the volume of displacement in the sea water.

$$\nabla_R = \frac{\triangle}{\rho_R}$$

$$= \frac{\triangle}{1{\cdot}002} \text{ m}^3$$

$$\nabla_S = \frac{\triangle + 215}{\rho_S}$$

$$= \frac{\triangle + 215}{1{\cdot}026} \text{ m}^3$$

Hence $$\nabla_R = \nabla_S$$

$$\frac{\triangle}{1{\cdot}002} = \frac{\triangle + 215}{1{\cdot}026}$$

$$1{\cdot}026\,\triangle = 1{\cdot}002\,\triangle + 1{\cdot}002 \times 215$$
$$0{\cdot}024\,\triangle = 1{\cdot}002 \times 215$$

$$\triangle = \frac{1{\cdot}002 \times 215}{0{\cdot}024}$$

$$= 8976 \text{ tonne}$$

f CHANGE IN TRIM DUE TO CHANGE IN DENSITY

When a ship passes from sea water into river water, or vice versa, without change in displacement, there is a change in trim in addition to the change in mean draught. This change in trim is always small.

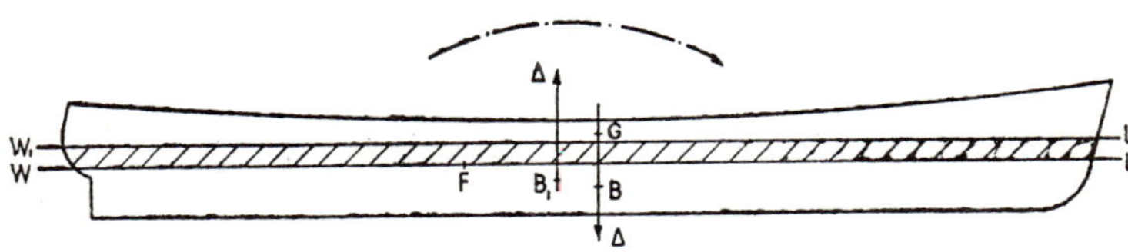

Fig. 66

Consider a ship of displacement $\triangle$ lying at waterline WL in sea water of density ρ_S t/m³. The centre of gravity *G* and the centre of buoyancy *B* are in the same vertical line.

If the vessel now moves into river water of ρ_R t/m³, there is a bodily increase in draught and the vessel lies at waterline W_1L_1. The volume of displacement has been increased by a layer of volume *v* whose centre of gravity is at the centre of flotation *F*. This causes the centre of buoyancy to move from *B* to B_1, the centre of gravity remaining at *G*.

Volume of displacement in sea water

$$\nabla_S = \frac{\triangle}{\rho_S}\text{ m}^3$$

Volume of displacement in river water

$$\nabla_R = \frac{\Delta}{\rho_R} \text{ m}^3$$

Change in volume of displacement

$$\begin{aligned} v &= \nabla_R - \nabla_S \\ &= \Delta \left(\frac{1}{\rho_R} - \frac{1}{\rho_S} \right) \\ &= \Delta \left(\frac{\rho_S - \rho_R}{\rho_R \times \rho_S} \right) \text{ m}^3 \end{aligned}$$

Shift in centre of buoyancy

$$\begin{aligned} BB_1 &= \frac{v \times FB}{\nabla_R} \\ &= \Delta \left(\frac{\rho_S - \rho_R}{\rho_R \times \rho_S} \right) FB \times \frac{\rho_R}{\Delta} \\ &= FB \left(\frac{\rho_S - \rho_R}{\rho_S} \right) \text{ m} \end{aligned}$$

Since B_1 is no longer in line with G, a moment of $\Delta \times BB_1$ acts on the ship causing a change in trim by the head.

$$\text{Change in trim} = \frac{\Delta \times BB_1}{\text{MCTI cm}} \text{ cm}$$

$$= \frac{\Delta\, FB}{\text{MCTI cm}} \left(\frac{\rho_S - \rho_R}{\rho_S} \right) \text{ cm by the head}$$

Note: If the ship moves from the river water into sea water, it will change trim by the stern, and:

$$\text{Change in trim} = \frac{\Delta\, FB}{\text{MCTI cm}} \left(\frac{\rho_S - \rho_R}{\rho_R} \right) \text{ cm by the stern}$$

Example. A ship 120 m long and 9100 tonne displacement floats at a level keel draught of 6·50 m in fresh water of 1·000 t/m³. MCTI cm 130 tonne m, TPC in sea water 16·5, LCB 2·30 m forward of midships. LCF 0·6 m aft of midships.

Calculate the new draughts if the vessel moves into sea water of 1·024 t/m³ without change in displacement.

$$\text{Reduction in mean draught} = \frac{\triangle}{\text{TPC}}\left(\frac{\rho S - \rho R}{\rho R}\right)$$

$$= \frac{9100}{16\cdot5}\left(\frac{1\cdot024 - 1\cdot000}{1\cdot000}\right)$$

$$= 13\cdot24 \text{ cm}$$

$$\text{Change in trim} = \frac{\triangle\ FB}{\text{MCTI cm}}\left(\frac{\rho_S - \rho_R}{\rho_R}\right)$$

$$= \frac{9100 \times (2\cdot30 + 0\cdot60)}{130}\left(\frac{1\cdot024 - 1\cdot000}{1\cdot000}\right)$$

$$= 4\cdot87 \text{ cm by the stern}$$

$$\text{Change forward} = -\frac{4\cdot87}{120}\left(\frac{120}{2} + 0\cdot6\right)$$

$$= -\ 2\cdot46 \text{ cm}$$

$$\text{Change aft} = +\frac{4\cdot87}{120}\left(\frac{120}{2} - 0\cdot6\right)$$

$$= +\ 2\cdot41 \text{ cm}$$

$$\text{New draught forward} = 6\cdot50 - 0\cdot132 - 0\cdot025$$

$$= 6\cdot343 \text{ m}$$

$$\text{New draught aft} = 6\cdot50 - 0\cdot132 + 0\cdot024$$

$$= 6\cdot392 \text{ m}$$

CHANGE IN MEAN DRAUGHT DUE TO BILGING

BUOYANCY is the upthrust exerted by the water on the ship and depends upon the volume of water displaced by the ship up to the waterline.

RESERVE BUOYANCY is the potential buoyancy of a ship and depends upon the intact, watertight volume above the waterline. When a mass is added to a ship, or buoyancy is lost due to bilging, the reserve buoyancy is converted into buoyancy by increasing the draught. If the loss in buoyancy exceeds the reserve buoyancy the vessel will sink.

PERMEABILITY μ is the volume of a compartment into which water may flow if the compartment is laid open to the sea, expressed as a ratio or percentage of the total volume of the compartment. Thus if a compartment is completely empty, the permeability is 100 per cent. The permeability of a machinery space is about 85 per cent and accommodation about 95 per cent. The permeability of a cargo hold varies considerably with the type of cargo, but an average value may be taken as 60 per cent.

The effects of bilging a mid-length compartment may be shown most simply by considering a box barge of length L, breadth B and draught d having a mid-length compartment of length l, permeability μ.

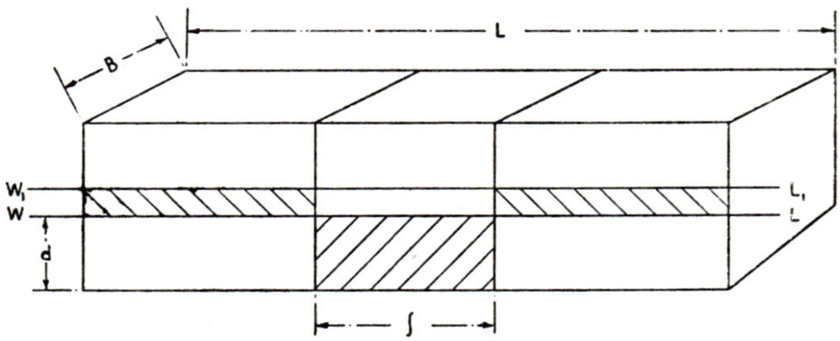

Fig. 67

If this compartment is bilged, buoyancy is lost and must be replaced by increasing the draught. The volume of buoyancy lost is the volume of the compartment up to waterline WL, less the volume of water excluded by the cargo in the compartment.

$$\text{Volume of lost buoyancy} = \mu\, lBd$$

This is replaced by the increase in draught multiplied by the area of the intact part of the waterplane, i.e. the area of waterplane on each side of the bilged compartment plus the area of cargo which projects through the waterplane in the bilged compartment.

$$\begin{aligned}\text{Area of intact waterplane} &= (L-l)B + lB(1-\mu)\\ &= LB - lB + lB - \mu lB\\ &= (L - \mu l)B\end{aligned}$$

$$\text{Increase in draught} = \frac{\text{volume of lost buoyancy}}{\text{area of intact waterplane}}$$

$$= \frac{\mu\, lBd}{(L-\mu l)B}$$

$$= \frac{\mu l d}{L-\mu l}$$

μl may be regarded as the *effective length* of the bilged compartment.

Example. A box barge 30 m long and 8 m beam floats at a level keel draught of 3 m and has a mid-length compartment 6 m long. Calculate the new draught if this compartment is bilged:

(a) with $\mu = 100\%$
(b) with $\mu = 75\%$.

(a) Volume of lost buoyancy $= 6 \times 8 \times 3 \text{ m}^3$
Area of intact waterplane $= (30-6) \times 8 \text{ m}^2$

$$\text{Increase in draught} = \frac{6 \times 8 \times 3}{24 \times 8}$$

$= 0{\cdot}75$ m
new draught $= 3 + 0{\cdot}75$
$= 3{\cdot}75$ m

(b) Volume of lost buoyancy $= 0{\cdot}75 \times 6 \times 8 \times 3 \text{ m}^3$
Area of intact waterplane $= (30 - 0{\cdot}75 \times 6) \times 8 \text{ m}^2$

$$\text{Increase in draught} = \frac{0{\cdot}75 \times 6 \times 8 \times 3}{25{\cdot}5 \times 8}$$

$= 0{\cdot}529$ m
New draught $= 3 + 0{\cdot}529$
$= 3{\cdot}529$ m

f CHANGE IN DRAUGHTS DUE TO BILGING AN END COMPARTMENT

If a bilged compartment does not lie at the mid-length, then there is a change in trim in addition to the change in mean draught.

Consider a box barge of length L, breadth B and draught d having an empty compartment of length l at the extreme fore end.

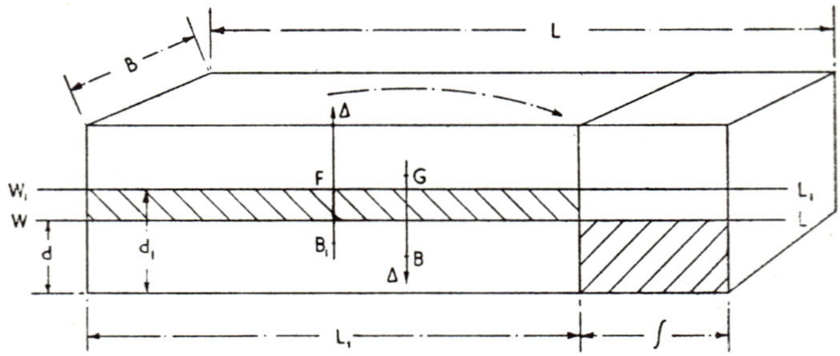

Fig. 68

Before bilging the vessel lies at waterline WL, the centre of gravity G and the centre of buoyancy B lying in the same vertical line.

After bilging the end compartment, the vessel lies initially at waterline W_1L_1. The new mean draught d_1 may be calculated as shown previously assuming that the compartment is amidships.

The volume of lost buoyancy has been replaced by a layer whose centre is at the middle of the length L_1. This causes the centre of buoyancy to move aft from B to B_1, a distance of $\frac{1}{2}l$. Thus a moment of $\triangle \times BB_1$ acts on the ship causing a *considerable* change in trim by the head. The vessel changes trim about the centre of flotation F which is the centroid of the *intact* waterplane i.e. the mid-point of L_1.

$$\text{Trimming moment} = \triangle \times BB_1$$

$$\text{Change in trim} = \frac{\triangle \times BB_1}{\text{MCTI cm}} \text{ cm by the head}$$

$$\text{MCTI cm} = \frac{\triangle \times GM_L}{100\,L} \text{ tonne m}$$

GM_L must be calculated for the *intact* waterplane

$$KB_1 = \frac{d_1}{2}$$

$$B_1M_L = \frac{L_1^3\,B}{12\,\nabla}$$

$$\text{where } \nabla = L\,B\,d$$
$$= L_1\,Bd_1$$
$$GM_L = KB_1 + B_1M_L - KG$$

$$\text{Change in trim} = \frac{\triangle \times \frac{1}{2}\, l}{\triangle \times GM_L} \times 100\, L$$

$$= \frac{50\, L\, l}{GM_L} \text{ cm by the head.}$$

Example. A box barge 120 m long and 8 m beam floats at an even keel draught of 3 m and has an empty compartment 6 m long at the extreme fore end. The centre of gravity is 2·8 m above the keel. Calculate the final draughts if this compartment is bilged.

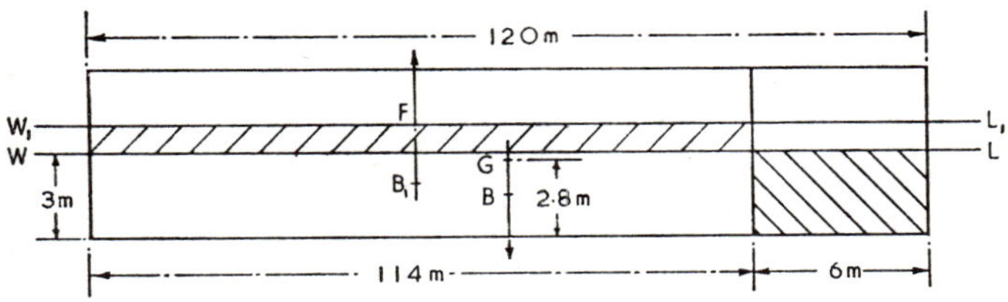

Fig. 69

$$\text{Increase in mean draught} = \frac{6 \times 8 \times 3}{(120 - 6) \times 8}$$

$$= 0{\cdot}158 \text{ m}$$

$$\text{New draught } d_1 = 3{\cdot}158 \text{ m}$$

$$KB_1 = \frac{d_1}{2}$$

$$= 1{\cdot}579 \text{ m}$$

$$B_1M_L = \frac{114^3 \times 8}{12 \times 120 \times 8 \times 3}$$

$$= 342{\cdot}94 \text{ m}$$
$$GM_L = 1{\cdot}58 + 342{\cdot}94 - 2{\cdot}80$$
$$= 341{\cdot}72 \text{ m}$$

$$\text{Change in trim} = \frac{50 \times 120 \times 6}{341\cdot72}$$

$$= 105\cdot3 \text{ cm by the head}$$

$$\text{Change forward} = + \frac{105\cdot3}{120} \times \left(\frac{120}{2} + 3\right)$$

$$= + 55\cdot3 \text{ cm}$$

$$\text{Change aft} = - \frac{105\cdot3}{120} \times \left(\frac{120}{2} - 3\right)$$

$$= - 50\cdot0 \text{ cm}$$

$$\text{New draught forward} = 3\cdot158 + 0\cdot553$$
$$= 3\cdot711 \text{ m}$$

$$\text{New draught aft} = 3\cdot158 - 0\cdot500$$
$$= 2\cdot658 \text{ m}$$

TEST EXAMPLES 6

*f*1. A ship 125 m long displaces 12 000 tonne. When a mass of 100 tonne is moved 75 m from forward to aft there is a change in trim of 65 cm by the stern. Calculate:

(a) MCTI cm
(b) the longitudinal metacentric height
(c) the distance moved by the centre of gravity of the ship.

*f*2. A ship 120 m long floats at draughts of 5·50 m forward and 5·80 m aft; MCTI cm 80 tonne m, TPC 13, LCF 2·5 m forward of midships. Calculate the new draughts when a mass of 110 tonne is added 24 m aft of midships.

*f*3. A ship 130 m long displaces 14 000 tonne when floating at draughts of 7·50 m forward and 8·10 m aft. GM_L 125 m, TPC 18, LCF 3 m aft of midships.

Calculate the final draughts when a mass of 180 tonne lying 40 m aft of midships is removed from the ship.

*f*4. The draughts of a ship 90 m long are 5·80 m forward and 6·40 m aft. MCTI cm 50 tonne m; TPC 11 and LCF 2 m aft of midships. Determine the point at which a mass of 180 tonne should be placed so that the after draught remains unaltered, and calculate the final draught forward.

*f*5. A ship 150 m long floats at draughts of 8·20 m forward and 8·90 m aft. MCTI cm 260 tonne m, TPC 28 and LCF 1·5 m aft of midships. It is necessary to bring the vessel to an even keel and a double bottom tank 60 m forward of midships is available.

Calculate the mass of water required and the final draught.

*f*6. A ship whose length is 110 m has MCTI cm 55 tonne m; TPC 9; LCF 1·5 m forward of midships and floats at draughts of 4·20 m forward and 4·45 m aft.

Calculate the new draughts after the following masses have been added:

20 tonne 40 m aft of midships
50 tonne 23 m aft of midships
30 tonne 2 m aft of midships
70 tonne 6 m forward of midships
15 tonne 30 m forward of midships.

*f*7. The draughts of a ship 170 m long are 6·85 m forward and 7·50 m aft. MCTI cm 300 tonne m; TPC 28; LCF 3·5 m forward of midships.

Calculate the new draughts after the following changes in loading have taken place;

160 tonne added 63 m aft of midships
200 tonne added 27 m forward of midships
120 tonne removed 75 m aft of midships
70 tonne removed 16 m aft of midships.

*f*8. A ship 80 m long has a light displacement of 1050 tonne and LCG 4·64 m aft of midships.

The following items are then added:

Cargo 2150 tonne, Lcg 4·71 m forward of midships
Fuel 80 tonne, Lcg 32·55 m aft of midships
Water 15 tonne, Lcg 32·90 m aft of midships
Stores 5 tonne, Lcg 33·60 m forward of midships.

The following hydrostatic particulars are available.

Draught m	Displacement tonne	MCTI cm tonne m	LCB from midships m	LCF from midships m
5·00	3533	43·10	1·00F	1·27A
4·50	3172	41·26	1·24F	0·84A

Calculate the final draughts of the loaded vessel.

9. A ship of 15 000 tonne displacement has a waterplane area of 1950 m². It is loaded in river water of 1·005 t/m³ and proceeds to sea where the density is 1·022 t/m³.

Calculate the change in mean draught.

10. A ship of 7000 tonne displacement has a waterplane area of 1500 m². In passing from sea water into river water of 1005 kg/m³ there is an increase in draught of 10 cm. Find the density of the sea water.

11. The ½ ordinates of the waterplane of a ship of 8200 tonne displacement, 90 m long, are 0, 2·61, 3·68, 4·74, 5·84, 7·00, 7·30, 6·47, 5·35, 4·26, 3·16, 1·88 and 0 m respectively. It floats in sea water of 1·024 t/m³. Calculate:

(a) TPC
(b) mass necessary to increase the mean draught by 12 cm
(c) change in mean draught when moving into water of 1·005 t/m³.

12. A ship consumes 360 tonne of fuel, stores and water when moving from sea water of 1·025 t/m^3 into fresh water of 1·000 t/m^3 and on arrival it is found that the draught has remained constant.

Calculate the displacement in the sea water.

*f*13. A ship 90 m long displaces 5200 tonne and floats at draughts of 4·95 m forward and 5·35 m aft when in sea water of 1023 kg/m^3. The waterplane area is 1100 m^2, GM_L 95 m, LCB 0·6 m forward of midships and LCF 2·2 m aft of midships.

Calculate the new draughts when the vessel moves into fresh water of 1002 kg/m^3.

*f*14. A ship of 22 000 tonne displacement is 160 m long, MCTI cm 280 tonne m, waterplane area 3060 m^2, centre of buoyancy 1 m aft of midships and centre of flotation 4 m aft of midships. It floats in water of 1·007 t/m^3 at draughts of 8·15 m forward and 8·75 m aft.

Calculate the new draughts if the vessel moves into sea water of 1·026 t/m^3.

15. A box barge 60 m long and 10 m wide floats at an even keel draught of 4 m. It has a compartment amidships 12 m long.

Calculate the new draught if this compartment is laid open to the sea when:

(a) μ is 100%
(b) μ is 85%
(c) μ is 60%.

16. A box barge 50 m long and 8 m wide floats at a draught of 3 m and has a mid-length compartment 9 m long containing coal (rd 1·28) which stows at 1·22 m^3/t.

Calculate the new draught if this compartment is bilged.

17. A vessel of constant rectangular cross-section is 60 m long, 12 m beam and floats at a draught of 4·5 m. It has a mid-length compartment 9 m long which extends right across the ship and up to the deck, but is sub-divided by a horizontal watertight flat 3 m above the keel.

Find the new draught if this compartment is bilged:

(a) below the flat
(b) above the flat.

18. A box barge 25 m long and 4 m wide floats in fresh water at a draught of 1·2 m and has an empty mid-length compartment 5 m long. The bottom of the barge is lined with teak (rd 0·805) 120 mm thick. After grounding all the teak is torn off and the centre compartment laid open to the sea. Calculate the final draught.

*f*19. A box barge 100 m long, 12 m beam and 4 m draught has a compartment at the extreme fore end 8 m long, sub-divided by a horizontal watertight flat 2 m above the keel. The centre of gravity is 3 m above the keel.

Calculate the end draughts if the compartment is bilged.

(a) at the flat, water flowing into both compartments
(b) below the flat
(c) above the flat.

CHAPTER 7

RESISTANCE

When a ship moves through the water at any speed, a force or resistance is exerted by the water on the ship. The ship must therefore exert an equal thrust to overcome the resistance and travel at that speed. If, for example, the resistance of the water on the ship at 17 knots is 800 kN, and the ship provides a thrust of 800 kN, then the vessel will travel at 17 knots.

The total resistance or tow-rope resistance R_t of a ship may be divided into two main sections:

(a) frictional resistance R_f

(b) residuary resistances R_r

Hence $$R_t = R_f + R_r$$

FRICTIONAL RESISTANCE R_f

As the ship moves through the water, friction between the hull and the water causes a belt of eddying water adjacent to the hull to be drawn along with the ship, although at a reduced speed. The belt moves aft and new particles of water are continually set in motion, the force required to produce this motion being provided by the ship.

The frictional resistance of a ship depends upon:

(i) the speed of the ship
(ii) the wetted surface area
(iii) the length of the ship
(iv) the roughness of the hull
(v) the density of the water.

Wm Froude developed the formula:

$$R_f = f\,S\,V^n \text{ N}$$

where f is a coefficient which depends upon the length of the ship L, the roughness of the hull and the density of the water.
S is the wetted surface area in m^2
V is the ship speed in knots.
n is an index of about 1·825

The value of f for a mild steel hull in sea water is given by :

$$f = 0{\cdot}417 + \frac{0{\cdot}773}{L + 2{\cdot}862}$$

Thus f is reduced as the length of the ship is increased.

In a slow or medium-speed ship the frictional resistance forms the major part of the total resistance, and may be as much as 75% of R_t. The importance of surface roughness may be seen when a ship is badly fouled with marine growth or heavily corroded, when the speed of the ship may be considerably reduced.

1 knot = 1·852 km/h

Example. A ship whose wetted surface area is 5150 m^2 travels at 15 knots. Calculate the frictional resistance and the power required to overcome this resistance.

$f = 0{\cdot}422$, $n = 1{\cdot}825$

$$R_f = f\,S\,V^n$$
$$= 0{\cdot}422 \times 5150 \times 15^{18\cdot25}$$
$$= 303\ 700\ N$$
$$\text{Power} = R_f\,(\text{N}) \times v\,(\text{m/s})$$
$$= 303\ 700 \times 15 \times \frac{1852}{3600}\ \text{W}$$
$$= 2344\ \text{kW}$$

Example. A plate drawn through fresh water at 3m/s has a frictional resistance of 12 N/m^2.
Estimate the power required to overcome the frictional resistance of a ship at 12 knots if the wetted surface area is 3300 m^2 and the index of speed is 1-9.

$$12\ \text{knots} = 12 \times \frac{1852}{3600}$$
$$= 6{\cdot}175\ \text{m/s}$$

At 3 m/s

$$R_f = 12 \times 3300$$
$$= 39\ 600\ \text{N in FW}$$
$$= 39\ 600 \times 1{\cdot}025\ \text{N in SW}$$

At 12 knots.

$$R_f = 39\ 600 \times 1{\cdot}025 \times \left(\frac{6{\cdot}175}{3}\right)^{1\cdot9}$$
$$= 160\ 000\ \text{N}$$
$$\text{Power} = 160\ 000 \times 6{\cdot}175$$
$$= 988\ 000\ \text{W}$$
$$= 988{\cdot}0\ \text{kW}$$

RESIDUARY RESISTANCES R_r

The residuary resistances of a ship may be divided into:

i) Resistance caused by the formation of streamlines round the ship, i.e. due to the change in the direction of the water. If the water changes direction abruptly, such as round a box barge, the resistance may be considerable, but in modern, well-designed ships should be very small.

ii) Eddy resistance caused by sudden changes in form. This resistance will be small in a ship where careful attention is paid to detail. The eddy resistance due to fitting rectangular sternframe and single plate rudder may be as much as 5% of the total resistance of the ship. By streamlining the sternframe and fitting a double plate rudder, eddy resistance is practically negligible.

iii) Resistance caused by the formation of waves as the ship passes through the water. In slow or medium-speed ships the wavemaking resistance is small compared with the frictional resistance. At high speeds, however, the wave making resistance is considerably increased and may be 50% or 60% of the total resistance.

Several attempts have been made to reduce the wave making resistance of ships, with varying degrees of success. One method which has proved to be successful is the use of the bulbous bow. The wave produced by the bulb interferes with the wave produced by the stem, resulting in a reduced height of bow wave and consequent reduction in the energy required to produce the wave.

The relation between the frictional resistance and the residuary resistances is shown in Fig. 70.

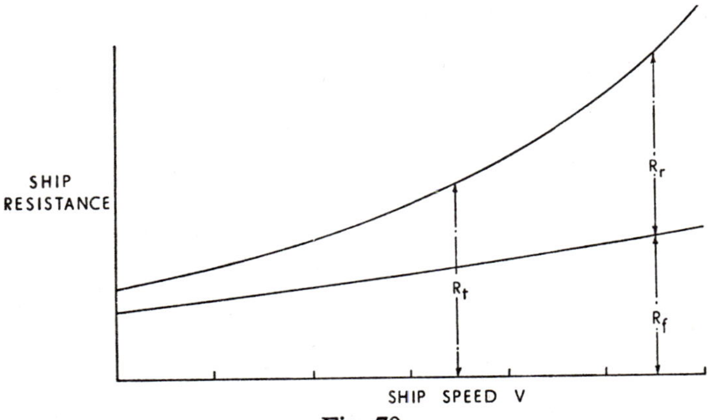

Fig. 70

Residuary Resistances follow Froude's Law of Comparison:

The residuary resistances of similar ships are in the ratio of the cube of their linear dimensions if their speeds are in the ratio of the square root of their linear dimensions.

Thus $$\frac{R_{r1}}{R_{r2}} = \left(\frac{L_1}{L_2}\right)^3 \quad \text{if} \quad \frac{V_1}{V_2} = \sqrt{\frac{L_1}{L_2}}$$

or $$\frac{R_{r1}}{R_{r2}} = \frac{\triangle_1}{\triangle_2} \quad \text{if} \quad \frac{V_1}{V_2} = \left(\frac{\triangle_1}{\triangle_2}\right)^{\frac{1}{6}}$$

V_1 and V_2 are termed *corresponding speeds.*

Thus at corresponding speeds:

$$\frac{V_1}{\sqrt{L_1}} = \frac{V_2}{\sqrt{L_2}}$$

$\frac{V}{\sqrt{L}}$ is known as the *speed-length ratio.*

It may therefore be seen that at corresponding speeds the wave making characteristics of similar ships are the same. At high speeds the speed-length ratio is high and the wave making resistance is large. To give the same wave making characteristics, the corresponding speed of a much smaller, similar ship will be greatly reduced and may not be what is popularly regarded to be a high speed. A ship is therefore considered slow or fast in relation to its speed-length ratio.

If $\frac{V}{\sqrt{L}}$ is below 1·0 the ship is said to be slow (V in knots: L in m)

If $\frac{V}{\sqrt{L}}$ is above 1·5 the ship is said to be fast.

Thus a speed of 15 knots would be regarded as slow for a ship 225m long, but fast for a ship 100 m long.

Example. The residuary resistance of a model 7 m long is 20N when towed at $3\frac{1}{2}$ knots.

Calculate the power required to overcome the residuary resistance of a similar ship 140 m long at its corresponding speed.

$$V_2 = V_1 \sqrt{\frac{L_2}{L_1}}$$

$$= 3{\cdot}5 \sqrt{\frac{140}{7}}$$

$$= 15{\cdot}65 \text{ knots.}$$

$$R_{r2} = R_{r1} \left(\frac{L_2}{L_1}\right)^3$$

$$= 20 \left(\frac{140}{7}\right)^3$$

$$= 160\,000 \text{ N}$$

$$\text{Power} = R_r \times v$$

$$160\,000 \times 15{\cdot}65 \times \frac{1852}{3600}$$

$$= 1288 \text{ kW}$$

The calculation of residuary resistance is usually based on the results of model experiments. A wax model of the ship is towed at its corresponding speed in a towing tank and the total resistance of the model measured. The frictional resistance of the model is calculated and subtracted from the total resistance, leaving the residuary resistance. The residuary resistance of the model is then used to determine the residuary resistance of the ship.

Once the total resistance of the ship is known it is possible to determine the power required to overcome this resistance. This is known as the *effective power* (ep) of the ship.

The model is tested without appendages such as rudder and bilge keels. An allowance must therefore be made for these appendages and also the general disturbance of the water at sea compared with tank conditions, This allowance is known as the *ship correlation factor* (SCF).

The power obtained directly from the model tests is known as the *effective power* (*naked*) (ep_n). The true effective power is the ep_n multiplied by the ship correlation factor.

Example. A 6 m model of a ship has a wetted surface area of 8 m². When towed at a speed of 3 knots in fresh water the total resistance is found to be 38N.

If the ship is 130 m long, calculate the effective power at the corresponding speed.

Take $n = 1{\cdot}825$ and calculate f from the formula. SCF 1·15

Model

$$R_t = 38 \text{ N in fresh water}$$
$$= 38 \times 1{\cdot}025$$
$$= 38{\cdot}95 \text{ N in sea water}$$

$$f = 0{\cdot}417 + \frac{0{\cdot}773}{L + 2{\cdot}862}$$

$$= 0{\cdot}417 + \frac{0{\cdot}773}{8{\cdot}862}$$

$$= 0{\cdot}504$$
$$R_f = 0{\cdot}504 \times 8 \times 3^{1{\cdot}825}$$
$$= 29{\cdot}94 \text{ N}$$
$$R_r = R_t - R_f$$
$$= 38{\cdot}95 - 29{\cdot}94$$
$$= 9{\cdot}01 \text{ N}$$

Ship

$$R_r \propto L^3$$

$$\therefore \quad R_r = 9{\cdot}01 \times \left(\frac{130}{6}\right)^3$$

$$= 91\,600 \text{ N}$$
$$S \propto L^2$$

$$\therefore \quad S = 8 \times \left(\frac{130}{6}\right)^2$$

$$= 3755 \text{ m}^2$$

$$V \propto \sqrt{L}$$

$$\therefore \quad V = 3\sqrt{\frac{130}{6}}$$

$$= 13{\cdot}96 \text{ knots}$$

$$f = 0{\cdot}417 + \frac{0{\cdot}773}{132{\cdot}862}$$

$$= 0{\cdot}4228$$
$$R_f = 0{\cdot}4228 \times 3755 \times 13{\cdot}96^{1{\cdot}825}$$
$$= 195\,000 \text{ N}$$
$$R_t = 195\,000 + 91\,600$$
$$= 286\,600 \text{ N}$$

$$\text{ep}_n = 286\,600 \times 13{\cdot}96 \times \frac{1852}{3600}$$

$$= 2059 \text{ kW}$$

Effective power $\quad \text{ep} = 2059 \times 1{\cdot}15$

$$= 2368 \text{ kW}$$

ADMIRALTY COEFFICIENT

It is sometimes necessary to obtain an approximation to the power of a ship without resorting to model experiments, and several methods are available. One system which has been in use for several years is the Admiralty Coefficient method. This is based on the assumption that for small variations in speed the total resistance may be expressed in the form:

$$Rt \propto \rho S V^n$$

It was seen earlier that

$$S \propto \triangle^{\frac{2}{3}}$$

Hence with constant density

$$R_t \propto \triangle^{\frac{2}{3}} V^n$$

But $\quad \text{power} \propto R_t \times V$

$$\propto \triangle^{\frac{2}{3}} V^{n+1}$$

or $\quad \text{power} = \dfrac{\triangle^{\frac{2}{3}} V^{n+1}}{\text{a coefficient}}$

The coefficient is known as the Admiralty Coefficient.

Originally this method was used to determine the power supplied by the engine. Since types of machinery vary considerably it is now considered that the relation between displacement, speed and shaft power (sp) is of more practical value.

Most merchant ships may be classed as slow or medium-speed, and for such vessels the index n may be taken as 2. Thus

$$\text{Admiralty Coefficient C} = \frac{\Delta^{\frac{2}{3}} V^3}{\text{sp}}$$

where Δ = displacement in tonne
V = ship speed in knots
sp = shaft power in kW

The Admiralty Coefficient may be regarded as constant for similar ships at their corresponding speeds. Values of **C** vary between about 350 and 600 for different ships, the higher values indicating more efficient ships.

For small changes in speed, the value of **C** may be regarded as constant for any ship at constant displacement.

At corresponding speeds $V \propto \Delta^{\frac{1}{6}}$

$$\therefore \quad V^3 \propto \Delta^{\frac{3}{6}}$$

$$\text{sp} \propto \Delta^{\frac{2}{3}} V^3$$

$$\propto \Delta^{\frac{2}{3}} \times \Delta^{\frac{3}{6}}$$

$$\propto \Delta^{\frac{7}{6}}$$

ie

$$\frac{\text{sp}_1}{\text{sp}_2} = \left(\frac{\Delta_1}{\Delta_2}\right)^{\frac{7}{6}}$$

Thus if the shaft power of one ship is known, the shaft power for a similar ship may be obtained at the corresponding speed.

Example. A ship of 14 000 tonne displacement has an Admiralty Coefficient of 450.
Calculate the shaft power required at 16 knots.

$$\text{sp} = \frac{\Delta^{\frac{2}{3}} V^3}{\text{C}}$$

$$= \frac{14\,000^{\frac{2}{3}} \times 16^3}{450}$$

$$= 5286 \text{ kW}$$

Example. A ship of 15 000 tonne displacement requires 3500 kW at a particular speed.
Calculate the shaft power required by a similar ship of 18 000 tonne displacement at its corresponding speed.

$$\text{sp} \propto \Delta^{7/6}$$

$$\therefore \qquad \text{sp} = 3500 \times \left(\frac{18\,000}{15\,000}\right)^{7/6}$$

$$= 4330 \text{ kW}$$

The index of speed *n* for high speed ships may be considerably more than 2 and thus the shaft power may vary as the speed to some index greater than 3 (e.g. 4). This higher index, however, is only applicable within the high speed range.

$$\text{i.e.} \qquad \frac{\text{sp}_1}{\text{sp}_2} = \left(\frac{V_1}{V_2}\right)^4$$

where both V_1 and V_2 are within the high speed range.

Example. A ship travelling at 20 knots requires 12 000 kW shaft power.
Calculate the shaft power at 22 knots if, within this speed range, the index of speed is 4.

$$\frac{\text{sp}_1}{\text{sp}_2} = \left(\frac{V_1}{V_2}\right)^4$$

$$\text{sp}_2 = 12\,000 \times \left(\frac{22}{20}\right)^4$$

$$= 17\,570 \text{ kW}$$

FUEL COEFFICIENT AND FUEL CONSUMPTION

The fuel consumption of a ship depends upon the power developed, indeed the overall efficiency of power plant is often measured in terms of the *specific fuel consumption* which is the consumption per unit of power, expressed in kg/h. Efficient diesel engines may have a specific fuel consumption of about 0·20 kg/kW h, while that for a steam turbine may be about 0·30 kg/kW h. The specific fuel consumption of a ship at different speeds follows the form shown in Fig. 71.

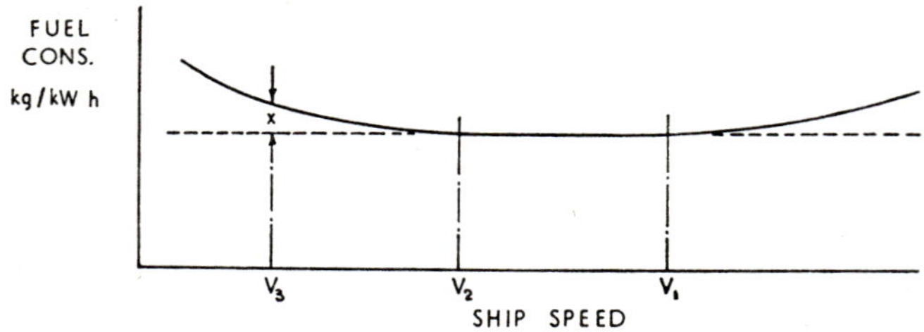

Fig. 71

Between V_1 and V_2 the specific consumption may be regarded as constant for practical purposes, and if the ship speed varies only between these limits, then:

$$\text{Fuel consumption / unit time} \propto \text{power developed}$$
$$\propto \text{sp}$$
and since
$$\text{sp} \propto \Delta^{\frac{2}{3}} V^3$$
$$\text{Fuel consumption/unit time} \propto \Delta^{\frac{2}{3}} V^3$$

or
$$\text{Fuel consumption / day} = \frac{\Delta^{\frac{2}{3}} V^3}{\text{fuel coefficient}} \text{ tonne}$$

Values of fuel coefficient vary between about 40 000 and 120 000, the higher values indicating more efficient ships.

Example. The fuel coefficient of a ship of 14 000 tonne displacement is 75 000. Calculate the fuel consumption per day if the vessel travels at 12½ knots.

$$\text{Fuel consumption per day} = \frac{14\,000^{2/3} \times 12{\cdot}5^3}{75\,000}$$
$$= 15{\cdot}12 \text{ tonne}$$

If the displacement and fuel coefficient remain constant, i.e. between V_1 and V_2 Fig. 71:
Fuel consumption / unit time $\propto$ speed^3

Hence
$$\frac{\text{cons}_1}{\text{cons}_2} = \left(\frac{V_1}{V_2}\right)^3$$

Example. A ship uses 20 tonne of fuel per day at 13 knots. Calculate the daily consumption at 11 knots.

$$\text{New daily consumption} = 20 \times \left(\frac{11}{13}\right)^3$$

$$= 12{\cdot}11 \text{ tonne.}$$

The total fuel consumption for any voyage may be found by multiplying the daily consumption by the number of days required to complete the voyage.

If D is the distance travelled at V knots, then:

$$\text{Number of days} \propto \frac{D}{V}$$

$$\text{But daily consumption} \propto V^3$$

$$\therefore \text{total voyage consumption} \propto V^3 \times \frac{D}{V}$$

$$\propto V^2 D$$

i.e. $$\frac{\text{voy. cons}_1}{\text{voy. cons}_2} = \left(\frac{V_1}{V_2}\right)^2 \times \frac{D_1}{D_2}$$

Hence for any given distance travelled the voyage consumption varies as the speed *squared.*

Example. A vessel uses 125 tonne of fuel on a voyage when travelling at 16 knots. Calculate the mass of fuel saved if, on the return voyage, the speed is reduced to 15 knots, the displacement of the ship remaining constant.

$$\text{New voyage consumption} = 125 \left(\frac{15}{16}\right)^2$$

$$= 110 \text{ tonne}$$

$$\therefore \text{Saving in fuel} = 125 - 110$$
$$= 15 \text{ tonne}$$

A general expression for voyage consumption is:

$$\frac{\text{new voy. cons.}}{\text{old voy. cons.}} = \left(\frac{\text{new displ.}}{\text{old displ.}}\right)^{\frac{2}{3}} \times \left(\frac{\text{new speed}}{\text{old speed}}\right)^{2} \times \frac{\text{new dist.}}{\text{old dist.}}$$

All of the above calculations are based on the assumption that the ship speed lies between V_1 and V_2 Fig. 71. If the speed is reduced to V_3, however, the specific consumption may be increased by $x\%$. In this case the daily consumption and voyage consumption are also increased by $x\%$.

Example. A ship has a daily fuel consumption of 30 tonne at 15 knots. The speed is reduced to 12 knots and at this speed the consumption per unit power is 8% more than at 15 knots. Calculate the new consumption per day.

$$\text{New daily consumption} = 1{\cdot}08 \times 30 \times \left(\frac{12}{15}\right)^{3}$$

$$= 16{\cdot}6 \text{ tonne}$$

It should be noted that if a formula for fuel consumption is given in any question, the formula must be used for the *complete question.*

TEST EXAMPLES 7

1. A ship has a wetted surface area of 3200 m^2. Calculate the power required to overcome frictional resistance at 17 knots if $n = 1{\cdot}825$ and $f = 0{\cdot}424$.

2. A plate towed edgewise in sea water has a resistance of 13 N/m^2 at 3 m/s.
A ship travels at 15 knots and has a wetted surface area of 3800 m^2. If the frictional resistance varies as speed $^{1\cdot97}$ calculate the power required to overcome frictional resistance.

3. The frictional resistance per square metre of a ship is 12 N at 180 m/min. The ship has a wetted surface area of 4000 m^2 and travels at 14 knots. Frictional resistance varies as speed $^{1\cdot9}$. If frictional resistance is 70% of the total resistance, calculate the effective power.

4. A ship is 125 m long, 15 m beam and floats at a draught of 7·8 m. Its block coefficient is 0·72. Calculate the power required to overcome frictional resistance at 17·5 knots if $n = 1{\cdot}825$ and $f = 0{\cdot}423$. Use Taylor's formula for wetted surface, with $c = 2{\cdot}55$.

5. The residuary resistance of a one-twentieth scale model of a ship in sea water is 36 N when towed at 3 knots. Calculate the residuary resistance of the ship at its corresponding speed and the power required to overcome residuary resistance at this speed.

6. A ship of 14 000 tonne displacement has a residuary resistance of 113 kN at 16 knots. Calculate the corresponding speed of a similar ship of 24 000 tonne displacement and the residuary resistance at this speed.

*f*7. The frictional resistance of a ship in fresh water at 3 m/s is 11 N/m^2. The ship has a wetted surface area of 2500 m^2 and the frictional resistance is 72% of the total resistance and varies as speed $^{1\cdot92}$. If the effective power is 1100 kW, calculate the speed of the ship.

*f*8. A 6 m model of a ship has a wetted surface area of 7 m^2, and when towed in fresh water at 3 knots, has a total resistance of 35 N. Calculate the effective power of the ship, 120 m long, at its corresponding speed.
$n = 1{\cdot}825$: f from formula: SCF $= 1{\cdot}15$

9. A ship of 12 000 tonne displacement has an Admiralty Coefficient of 550. Calculate the shaft power at 16 knots.

10. A ship requires a shaft power of 2800 kW at 14 knots, and the Admiralty Coefficient is 520. Calculate:

(a) the displacement

(b) the shaft power if the speed is reduced by 15%.

11. A ship of 8000 tonne displacement has an Admiralty Coefficient of 470. Calculate its speed if the shaft power provided is 2100 kW.

*f*12. A ship 150 m long and 19 m beam floats at a draught of 8 m and has a block coefficient of 0·68.

(a) If the Admiralty Coefficient is 600, calculate the shaft power required at 18 knots.

(b) If the speed is now increased to 21 knots, and within this speed range resistance varies as speed^3, find the new shaft power.

13. A ship of 15 000 tonne displacement has a fuel coefficient of 62 500. Calculate the fuel consumption per day at 14½ knots.

14. A ship of 9000 tonne displacement has a fuel coefficient of 53 500. Calculate the speed at which it must travel to use 25 tonne of fuel per day.

15. A ship travels 2000 nautical miles at 16 knots and returns with the same displacement at 14 knots. Find the saving in fuel on the return voyage if the consumption per day at 16 knots is 28 tonne.

16. The daily fuel consumption of a ship at 15 knots is 40 tonne. 1100 nautical miles from port it is found that the bunkers are reduced to 115 tonne. If the ship reaches port with 20 tonne of fuel on board, calculate the reduced speed and the time taken in hours to complete the voyage.

17. A ship uses 23 tonne of fuel per day at 14 knots. Calculate the speed if the consumption per day is:

(a) increased by 15%

(b) reduced by 12%

(c) reduced to 18 tonne.

18. The normal speed of a ship is 14 knots and the fuel consumption per hour is given by $0{\cdot}12 + 0{\cdot}001\ V^3$ tonne, with V in knots, Calculate:

(a) the total fuel consumption over a voyage of 1700 nautical miles

(b) the speed at which the vessel must travel to save 10 tonne of fuel per day.

19. A ship's speed is increased by 20% above normal for 8 hours, reduced by 10% below normal for 10 hours and for the remaining 6 hours of the day the speed is normal. Calculate the precentage variation in fuel consumption in that day from normal.

20. A ship's speed was 18 knots. A reduction of 3·5 knots gave a saving in fuel consumption of 22 tonne per day. Calculate the consumption per day at 18 knots..

*f*21. The daily fuel consumption of a ship at 17 knots is 42 tonne. Calculate the speed of the ship if the consumption is reduced to 28 tonne per day, and the specific consumption at the reduced speed is 18% more than at 17 knots.

CHAPTER 8

PROPELLERS

The after side of a marine propeller is the driving face and is in the form of a helical screw. This screw is formed by a number of blades, from 3 to 7, set at an angle to the plane of rotation.

DIAMETER D The diameter of the propeller is the diameter of the circle or disc cut out by the blade tips.

PITCH P If the propeller is assumed to work in an unyielding fluid, then in one revolution of the shaft the propeller will move forward a distance which is known as the pitch.

PITCH RATIO p, or face pitch ratio is the face pitch divided by the diameter. Thus

$$p = \frac{P}{D}$$

THEORETICAL SPEED V_T is the distance the propeller would advance in unit time if working in an unyielding fluid. Thus if the propeller turns at N rev/min,

$$V_T = P \times N \text{ m/min}$$
$$= \frac{P \times N \times 60}{1852} \text{ knots.}$$

APPARENT SLIP Since the propeller works in water, the ship speed V will normally be less than the theoretical speed. The difference between the two speeds is known as the apparent slip and is usually expressed as a ratio or percentage of the theoretical speed.

$$\text{Apparent slip speed} = V_T - V \text{ knots}$$

$$\text{Apparent slip} = \frac{V_T - V}{V_T} \times 100\%.$$

If the ship speed is measured relative to the surrounding water i.e. by means of a log line, the theoretical speed will invariably exceed the ship speed, giving a *positive* apparent slip. If, however the ship speed is measured relative to the land, then any movement of water will affect the apparent slip, and should the vessel be travelling in a following current the ship speed may exceed the theoretical speed, resulting in a *negative* apparent slip.

WAKE In its passage through the water the ship sets in motion particles of water in its neighbourhood, caused, as mentioned earlier, by friction between the hull and the water. This moving water is known as the wake and is important in propeller calculations since the propeller works in wake water. The speed of the ship relative to the wake is termed the SPEED OF ADVANCE V_a. The wake speed is often expressed as a fraction of the ship speed.

$$\text{Wake fraction } w = \frac{V - V_a}{V}$$

The wake fraction may be obtained approximately from the expression

$$w = 0{\cdot}5C_b - 0{\cdot}05$$

where C_b is the block coefficient.

REAL SLIP or TRUE SLIP is the difference between the theoretical speed and the speed of advance, expressed as a ratio or percentage of the theoretical speed.

$$\text{Real slip speed} = V_T - V_a \text{ knots}$$

$$\text{Real slip} = \frac{V_T - V_a}{V_T} \times 100\%$$

The real slip is always positive and is independent of current.

Example. A propeller of 4·5 m pitch turns at 120 rev/min and drives the ship at 15·5 knots. If the wake fraction is 0·30 calculate the apparent slip and the real slip.

$$\text{Theoretical speed } V_T = \frac{4{\cdot}5 \times 120 \times 60}{1852}$$
$$= 17{\cdot}49 \text{ knots.}$$

$$\text{Apparent slip} = \frac{17{\cdot}49 - 15{\cdot}5}{17{\cdot}49} \times 100$$
$$= 11{\cdot}38\%$$

$$\text{Wake fraction } w = \frac{V - V_a}{V}$$

$$\therefore \quad wV = V - V_a$$
$$V_a = V - wV$$
$$= V(1 - w)$$
$$= 15{\cdot}5 \times 0{\cdot}7$$
$$= 10{\cdot}85 \text{ knots}$$

$$\text{Real ship} = \frac{17{\cdot}49 - 10{\cdot}85}{17{\cdot}49} \times 100$$
$$= 37{\cdot}96\%$$

The relation between the different speeds may be shown clearly by the following line diagram.

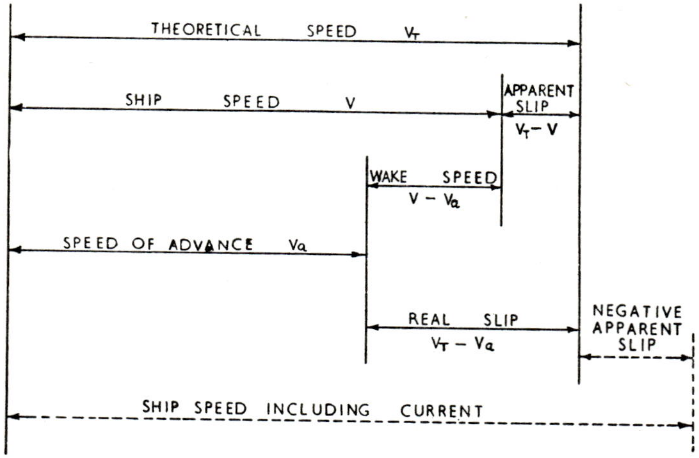

Fig. 72

PROJECTED AREA A_p is the sum of the blade areas projected onto a plane which is perpendicular to the axis of the screw.

DEVELOPED AREA is the actual area of the driving faces
(a) clear of the boss A_d
(b) including the boss area A_b

BLADE AREA RATIO BAR is the developed area excluding boss divided by the area of the circle cut out by the blade tips

$$\text{BAR} = \frac{A_d}{\frac{\pi}{4} D^2}$$

DISC AREA RATIO D.A.R. is the developed area including boss divided by the area of the circle cut out by the blade tips.

$$\text{D.A.R.} = \frac{A_b}{\frac{\pi}{4} D^2}$$

THRUST

The thrust exerted by a propeller may be calculated approximately by regarding the propeller as a reaction machine. Water is received into the propeller disc at the speed of advance and projected aft at the theoretical speed.

Consider a time interval of one second.

Let A = effective disc area in m^2
= disc area—boss area
ρ = density of water in kg/m^3
P = pitch of propeller in m
n = rev/s
v_a = speed of advance in m/s

Mass of water passing through disc in one second.

$$M = \rho APn \text{ kg}$$

$$\text{Change in velocity} = Pn - v_a \quad \text{m/s}$$

Since this change occurs in one second,

$$\text{Acceleration } a = Pn - v_a \quad \text{m/s}^2$$

But
$$\text{real slip } s = \frac{Pn - v_a}{Pn}$$

$$\therefore \quad Pn - v_a = sPn$$
and
$$a = sPn$$

$$\text{Force} = \text{mass} \times \text{acceleration}$$

$$\therefore \quad \text{Thrust T} = M \times a = \rho APn \times sPn = \rho AP^2n^2 \text{ N}$$

It is interesting to note that increased slip leads to increased thrust and that the propeller will not exert a thrust with zero slip. The power produced by the propeller is known as the *thrust power* tp.

$$\text{tp} = \text{thrust } (N) \times \text{speed of advance (m/s) W}$$

$$= T \times v_a \text{ W}$$

Hence
$$\frac{\text{tp}_1}{\text{tp}_2} = \frac{T_1 \, v_{a1}}{T_2 \, v_{a2}}$$

If the power remains constant, but the external conditions vary, then

$$T_1 \, v_{a1} = T_2 \, v_{a2}$$

and since the speed of advance depends upon rev/min,

$$T_1 \, N_1 = T_2 \, N_2$$

Now the thrust is absorbed by the thrust collars and hence the thrust varies directly as the pressure t on the thrust collars.

$$\therefore \quad t_1 \, N_1 = t_2 \, N_2$$

This indicates that if, with constant power, the ship meets a head wind, the speed will reduce but the pressure on the thrust collars will increase.

Example. The tp of a ship is 2000 kW and the pressure on the thrust 20 bar at 120 rev/min.

Calculate the pressure on the thrust when the tp is 1800 kW at 95 rev/min.

$$\frac{20 \times 120}{2000} = \frac{t_2 \times 95}{1800}$$

$$t_2 = \frac{1800 \times 20 \times 120}{2000 \times 95}$$

$$= 22{\cdot}74 \text{ bar}$$

RELATION BETWEEN POWERS

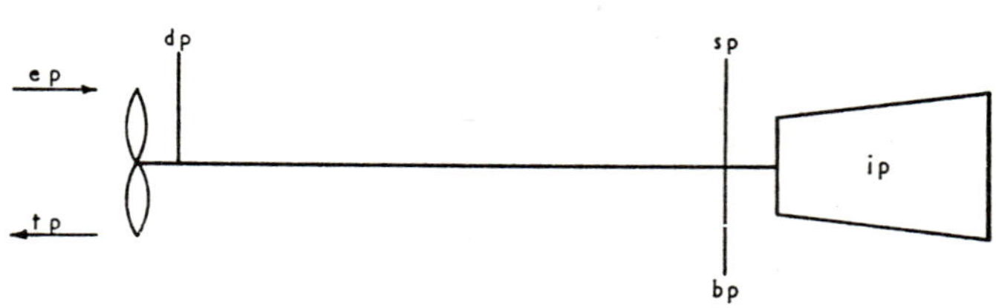

Fig. 73

The power produced by the engine is the indicated power ip. The mechanical efficiency of the engine is usually between about 80% and 90% and therefore only this percentage of the ip is transmitted to the shaft, giving the shaft power sp or brake power bp.

$$\text{sp or bp} = \text{ip} \times \text{mechanical efficiency}$$

Shaft losses vary between about 3% and 5% and therefore the power delivered to the propeller, the delivered power dp, is almost 95% of the sp.

$$\text{dp} = \text{sp} \times \text{transmission efficiency}$$

The delivered power may be calculated from the torque on the shaft

$$\text{dp} = \text{torque} \times 2\pi n$$

The propeller has an efficiency of 60% to 70% and hence the thrust power tp is given by:

$$\text{tp} = \text{dp} \times \text{propeller efficiency}$$

The action of the propeller in accelerating the water creates a suction on the after end of the ship. The thrust exerted by the propeller must exceed the total resistance by this amount. The relation between thrust and resistance may be expressed in the form

$$R_t = T(1 - t)$$

where t is the thrust deduction factor.

The thrust power will therefore differ from the effective power. The ratio of ep to tp is known as the hull efficiency which is a little more than unity for single screw ships and about unity for twin screw ships.

$$\text{ep} = \text{tp} \times \text{hull efficiency}$$

In an attempt to estimate the power required by the machinery from the calculation of ep, a quasi propulsive coefficient QPC is introduced. This is the ratio of ep to dp and obviates the use of hull efficiency and propeller efficiency. The prefix quasi is used to show that the mechanical efficiency of the machinery and the transmission losses have not been taken into account.

$$\text{ep} = \text{dp} \times \text{QPC}$$

The true propulsive coefficient is the relation between the ep and the ip, although in many cases sp is used in place of ip

i.e. $\quad \text{ep} = \text{ip} \times \text{propulsive coefficient}$

or $\quad \text{ep} = \text{sp} \times \text{propulsive coefficient}$

Example. The total resistance of a ship at 13 knots is 180 kN, the QPC is 0·70, shaft losses 5% and the mechanical efficiency of the machinery 87%.

Calculate the indicated power.

$$\text{ep} = R_t \times v$$

$$= 180 \times 10^3 \times 13 \times \frac{1852}{3600} \text{ W}$$

$$= 1204 \text{ kW}$$

$$\text{dp} = \frac{\text{ep}}{\text{QPC}}$$

$$= \frac{1204}{0 \cdot 7}$$

$$= 1720 \text{ kW}$$

$$\text{sp} = \frac{\text{dp}}{\text{transmission efficiency}}$$

$$= \frac{1720}{0 \cdot 95}$$

$$= 1810 \text{ kW}$$

$$\text{ip} = \frac{\text{sp}}{\text{mechanical efficiency}}$$

$$= \frac{1810}{0 \cdot 87}$$

$$\text{indicated power} = 2080 \text{ kW}$$

f RELATION BETWEEN MEAN PRESSURE AND SPEED

The indicated power of a steam reciprocating engine is given by:

$$\text{ip} \propto \text{pm ALN}$$

where
p_m = mean effective pressure (m.e.p.)
A = area of piston
L = length of stroke
N = number of strokes per minute or revolutions per minute

Since A and L remain constant,

$$\text{power} \propto p_m \times \text{N}$$

But it was shown earlier, that with constant displacement

$$\text{power} \propto \text{speed}^3$$

and since

$$\text{speed} \propto \text{propeller pitch} \times \text{revs/min.}$$
$$\propto P \times N$$
$$\text{power} \propto P^3 N^3$$

$\therefore$

$$p_m \times N = P^3 N^3$$
$$p_m \propto P^3 N^2$$

or

$$\frac{p_{m1}}{p_{m2}} = \left(\frac{P_1}{P_2}\right)^3 \times \left(\frac{N_1}{N_2}\right)^2$$

If the pitch of the propeller remains constant, then

$$\frac{p_{m1}}{p_{m1}} = \left(\frac{N_1}{N_2}\right)^2$$

Example. The mean effective pressure of an engine is 5·5 bar at 115 rev/min. Calculate the mep if the revs are reduced to 95 per minute.

$$p_{m2} = p_{m1}\left(\frac{N_2}{N_1}\right)^2$$

$$= 5{\cdot}5\left(\frac{95}{115}\right)^2$$

$$= 3{\cdot}753 \text{ bar}$$

MEASUREMENT OF PITCH

If the propeller is assumed to have no forward motion, then a point on the blade, distance R from the centre of boss will move a distance of $2\pi R$ in one revolution. If the propeller is now assumed to work in an unyielding fluid, then in one revolution it will advance a distance of P, the pitch. The pitch angle θ may be defined as

$$\tan\theta = \frac{P}{2\pi R}$$

$$\therefore \qquad \text{Pitch} = \tan\theta \times 2\pi R$$

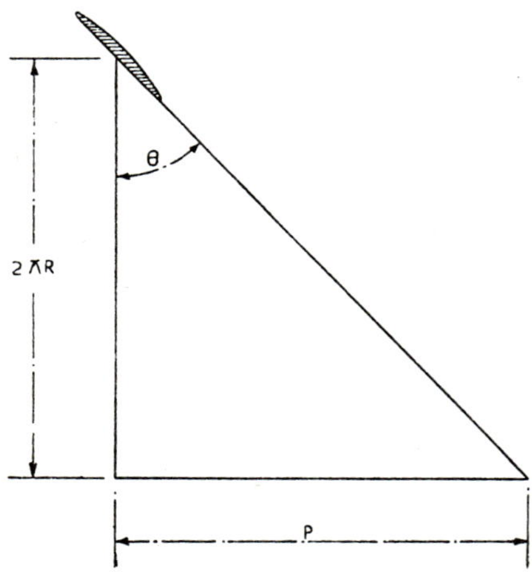

Fig. 74

The pitch of a propeller may be measured without removing the propeller from the ship, by means of a simple instrument known as a pitchometer. One form of this instrument consists of a protractor with an adjustable arm. The face of the boss is used as a datum, and a spirit level is set horizontal when the pitchometer is set on the datum. The instrument is then set on the propeller blade at the required distance from the boss and the arm containing the level moved until it is horizontal, a reading of pitch angle or pitch may then be read from the protractor at the required radius (Fig 75).

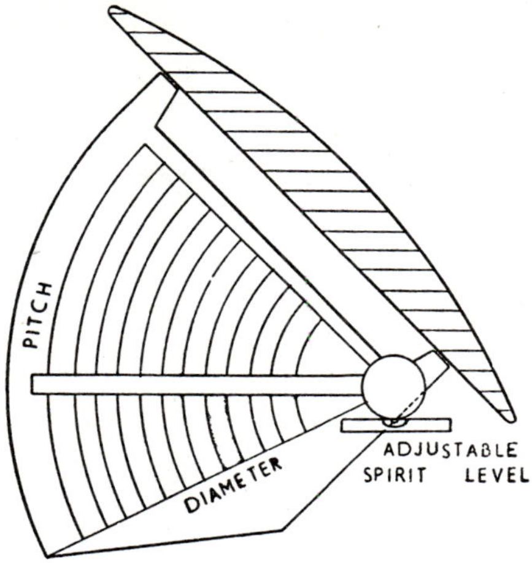

Fig. 75

An alternative method is to turn the propeller until one blade is horizontal. A weighted cord is draped over the blade at any given radius as shown in Fig. 76.

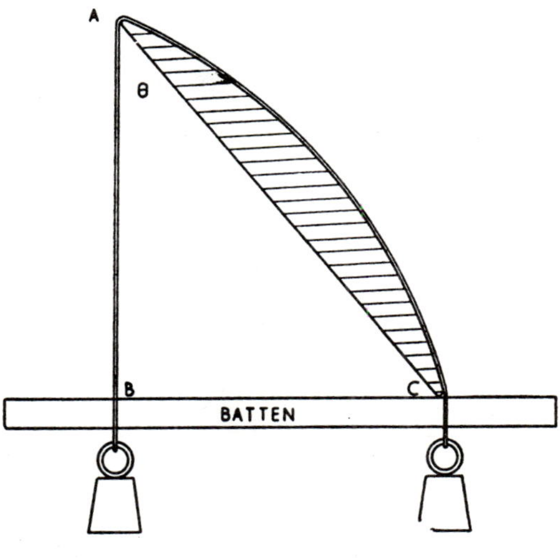

Fig. 76

A batten is placed horizontally at the lower edge of the blade with the aid of a spirit level. The distances AB and BC are then measured. θ is the pitch angle, and

$$\tan \theta = \frac{BC}{AB}$$

But

$$\text{Pitch} = \tan \theta \times 2\pi R$$

$\therefore$

$$\text{Pitch} = \frac{BC}{AB} \times 2\pi R$$

CAVITATION

The thrust of a propeller varies approximately as the square of the revolutions. Thus as the speed of rotation is increased there is a considerable increase in thrust. The distribution of pressure due to thrust over the blade section is approximately as shown in Fig. 77.

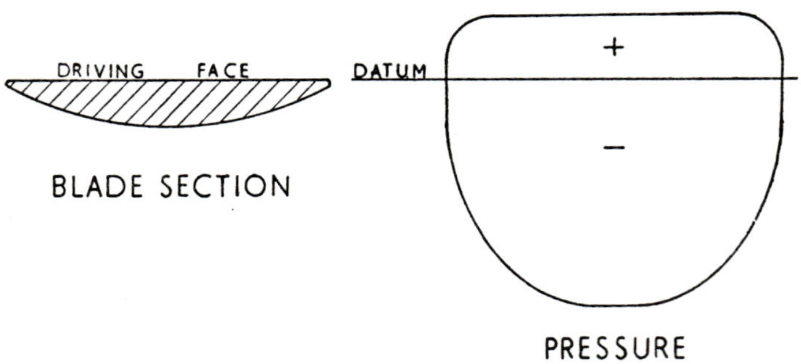

Fig. 77

The nett pressure at any point on the back of the blade is the algebraic sum of the atmospheric pressure, water pressure and negative pressure or suction caused by the thrust. When this suction is high at any point, the nett pressure may fall below the vapour pressure of the water at water temperature, causing a cavity or bubble to form on the blade. This cavity is filled with water vapour and with air which disassociates from the sea water. As the blade turns, the bubble moves across the blade to a point where the nett pressure is higher, causing the cavity to collapse. The forming and collapsing of these cavities is known as *cavitation*.

When the cavity collapses, the water pounds the blade material, and since the breakdown occurs at the same position each time, causes severe erosion of the blades and may produce holes in the blade material several inches deep. Cavitation also causes reduction in thrust and efficiency, vibration and noise. It may be reduced or avoided by reducing the revolutions and by increasing the blade area for constant thrust, thus reducing the negative pressure. Since cavitation is affected by pressure and temperature, it is more likely to occur in propellers operating near the surface than in those deeply submerged, and will occur more readily in the tropics than in cold regions.

BUILT AND SOLID PROPELLERS

Propellers may be either cast in a solid unit or built up from a boss to which are bolted the separate blades. Built propellers have the advantage that if a blade is damaged or lost it may be replaced cheaply without removing the propeller. The bolt holes in the boss are slotted to allow alterations in pitch. Unfortunately the diameter of boss is increased and leads to a reduction in efficiency. Blades have been known to slacken and it is therefore necessary to check the propeller both for slackness and pitch when the ship is in dry dock. Variations in pitch from blade to blade lead to hull vibration.

Solid propellers are fitted to the majority of modern ships. The blades and boss form an integral unit which may be manufactured to within very fine limits. The pitch is checked accurately on all blades and the propeller balanced at the factory to reduce the possibility of torsional vibration. The cost of repair is higher than with built propeller, in the event of damage, but the overall maintenance costs are reduced.

TEST EXAMPLES 8

1. A ship travels at 14 knots when the propeller, 5 m pitch, turns at 105 rev/min. If the wake fraction is 0·35, calculate the apparent and real slip.

2. A propeller of 5·5 m diameter has a pitch ratio of 0·8. When turning at 120 rev/min, the wake fraction is found to be 0·32 and the real slip 35%

Calculate the ship speed, speed of advance and apparent slip.

3. A ship of 12 400 tonne displacement is 120 m long, 17·5 m beam and floats at a draught of 7·5 m. The propeller has a face pitch ratio of 0·75 and, when turning at 100 rev/min, produces a ship speed of 12 knots with a real slip of 30%. Calculate the apparent slip, pitch and diameter of the propeller. The wake fraction w may be found from the expression:

$$w = 0{\cdot}5C_b - 0{\cdot}05$$

*f*4. When a propeller of 4·8 m pitch turns at 110 rev/min, the apparent slip is found to be $-s\%$ and the real slip $+1{\cdot}5s\%$. If the wake speed is 25% of the ship speed, calculate the ship speed, the apparent slip and the real slip.

*f*5. A propeller 4·6 m diameter has a pitch of 4·3 m and boss diameter of 0·75 m. The real slip is 28% at 95 rev/min.

Calculate the speed of advance, thrust and thrust power.

6. The pressure exerted on the thrust is 17·5 b at 115 rev/min. Calculate the thrust pressure at 90 rev/min.

7. The power required to drive a ship at a given speed was 3400 kW and the pressure on the thrust 19·5 b.

Calculate the new thrust pressure if the speed is reduced by 12% and the corresponding power is 2900 kW.

*f*8. A ship of 15 000 tonne displacement has an Admiralty Coefficient, based on shaft power, of 420. The mechanical efficiency of the machinery is 83%, shaft losses 6%, propeller efficiency 65% and QPC 0·71. At a particular speed the thrust power is 2550 kW. Calculate: (a) indicated power,
(b) effective power,
(c) ship speed.

*f*9. A propeller of 4 m pitch has an efficiency of 67%. When turning at 125 rev/min the real slip is 36% and the delivered power 2800 kW.

Calculate the thrust of the propeller.

*f*10. The mean referred pressure was found to be 4·5 b at 130 rev/min. Calculate the pressure at 105 rev/min.

*f*11. At 120 rev/min the m e p is 5·5 b.

Calculate the m e p and the rev/min if the power is reduced by 20%

12 The pitch angle, measured at a distance of 2 m from the centre of the boss, was found to be 21·5°.

Calculate the pitch of the propeller.

13. The pitch of a propeller is measured by means of a batten and cord. The horizontal ordinate is found to be 40 cm while the vertical ordinate is 1·15 m at a distance of 2·6 m from the centre of the boss. Calculate the pitch of the propeller and the blade width at that point.

CHAPTER 9

ƒRUDDER THEORY

When a rudder is turned from the centreline plane to any angle, the water flows round the rudder and creates an additional resistance on that side of the centreline. The force F which acts on the rudder parallel to the centreline has two components:

(a) the force created by the formation of streamlines round the rudder, i.e. due to the change in direction of the water.

(b) the suction on the after side of the rudder caused by eddying.

This force F follows the laws of fluid friction and may be determined from the expression.

$$F = k\,A\,v^2 \quad \text{N}$$

where k = a coefficient which depends upon the shape of the rudder, the rudder angle and the density of the water. When the ship speed is expressed in m/s, average values of k for sea water vary between about 570 and 610.

A = rudder area

v = ship speed.

The area of rudder is not specified by Classification Societies, but experience has shown that the area should be related to the area of the middle-line plane (i.e. length of ship × draught), and values of one sixtieth for fast ships and one seventieth for slow ships have been found successful.

$$\text{i.e. Area of rudder} = \frac{L \times d}{60} \text{ for fast ships}$$

$$= \frac{L \times d}{70} \text{ for slow ships.}$$

If the rudder is turned to an angle α, then the component of force acting normal to the plane of the rudder F_n is given by:

$$F_n = F \sin \alpha$$
$$= k\,A\,v^2 \sin \alpha$$

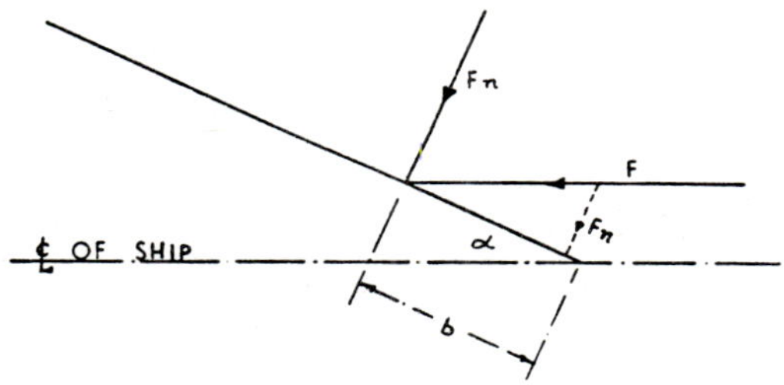

Fig. 78

This force F_n acts at the centre of effort of the rudder. The position of the centre of effort varies with the shape of the rudder and the rudder angle. For rectangular rudders the centre of effort is between 20% and 38% of the width of the rudder from the leading edge. The effect of the normal force is to tend to push the rudder back to its centreline position. Such movement is resisted by the rudder stock and the steering gear. It is therefore possible to calculate the turning moment or torque on the rudder stock.

If the centre of effort is b m from the centre of the rudder stock, then at any angle α

$$\text{Torque on stock } T = F_n \times b$$
$$= k\,A\,v^2\,b \sin \alpha \quad \text{N m}$$

From the basic torsion equation the diameter of the stock may be found for any given allowable stress.

$$\frac{T}{J} = \frac{q}{r}$$

where q = allowable stress in N/m²
r = radius of stock in m
J = second moment of area about a polar axis in m^4

$$= \frac{\pi d^4}{32}$$

$$= \frac{\pi r^4}{2}$$

For any rudder, at constant ship speed, values of torque may be plotted on a base of rudder angle. The area under this curve up to any angle is the work done in turning the rudder to this angle, and may be found by the use of Simpson's Rule. Care must be taken to express the common interval in radians, not degrees.

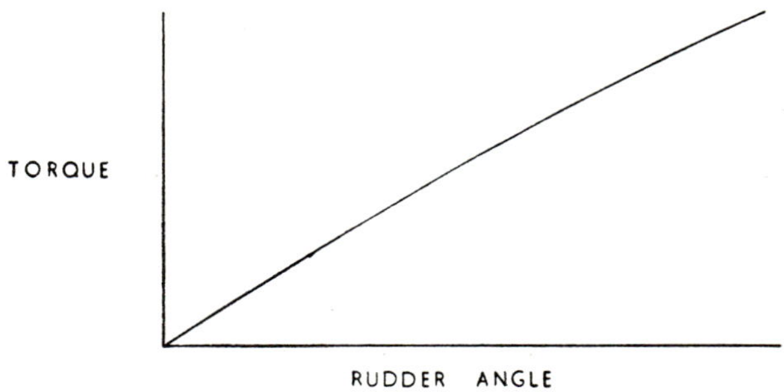

Fig. 79

If the centre of the rudder stock is between 20% and 38% of the width of the rudder from the leading edge, then at a given angle the centre of stock will coincide with the centre of effort and thus there will be no torque. The rudder is then said to be *balanced.* At any other rudder angle the centres of stock and effort will not coincide and there will be a torque of reduced magnitude. Thus it may be seen that the diameter of stock and power of the steering gear may be reduced if a balanced rudder is fitted.

It is usual to limit the rudder angle to 35° on each side of the centreline, since, if this angle is exceeded, the diameter of the turning circle is increased.

Example. A rudder has an area of 15 m^2 with its centre of effort 0·9 m from the centre of stock. The maximum rudder angle is 35° and it is designed for a service speed of 15 knots. Calculate the diameter of the rudder stock if the maximum allowable stress in the stock is 55 MN/m^2 and the rudder force parallel to the centre-line of the ship is given by:

$$F = 580\ Av^2 \text{ N with } v \text{ in m/s}$$

$$\text{Ship speed } v = 15 \times \frac{1852}{3600}$$

$$= 7{\cdot}717 \text{ m/s}$$

$$F = 580 \times 15 \times 7{\cdot}717^2$$
$$= 518\ 060 \text{ N}$$

$$\text{Torque } T = F_n\, b$$
$$= F \sin\alpha \times b$$
$$= 518\ 060 \times 0{\cdot}5736 \times 0{\cdot}9$$
$$= 267\ 440 \text{ N m}$$

$$\frac{T}{J} = \frac{q}{r}$$

$$J = \frac{Tr}{q}$$

$$\therefore \quad \frac{\pi r^4}{2} = \frac{Tr}{q}$$

$$r^3 = \frac{2T}{\pi q}$$

$$= \frac{2 \times 267\ 440}{3{\cdot}412 \times 55 \times 10^6}$$

$$= 0{\cdot}003\ 095 \text{ m}^3$$

$$r = 0{\cdot}145 \text{ m}$$

$$\text{Diameter of stock} = 0{\cdot}29 \text{ m}$$

ANGLE OF HEEL DUE TO FORCE ON RUDDER

When the rudder is turned from its central position, a transverse component of the normal rudder force acts on the rudder.

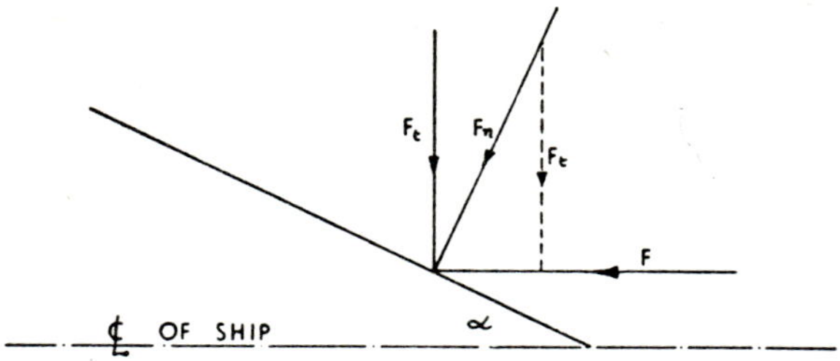

Fig. 80

Let F_n = rudder force normal to the plane of the rudder
F_t = transverse rudder force
α = rudder angle

$$\begin{aligned} \text{Then } F_t &= F_n \cos \alpha \\ &= F \sin \alpha \cos \alpha \\ &= k\, A\, v^2 \sin \alpha \cos \alpha \end{aligned}$$

This transverse force acts at the centre of the rudder N, and tends to push the ship sideways. A resistance R is exerted by the water on the ship, and acts at the *centre of lateral resistance L* which is the centroid of the projected, immersed plane of the ship (sometimes taken as the centre of buoyancy). This resistance is increased as the ship moves, until it reaches its maximum value when it is equal to the transverse force. At this point a moment acts on the ship causing it to heel to an angle θ when the heeling moment is equal to the righting moment.

$$\begin{aligned} \text{Heeling moment} &= F_t \times NL \cos \theta \\ \text{Righting moment} &= \triangle g \times GZ \\ &= \triangle g \times GM \sin \theta \quad \text{if } \theta \text{ is small} \end{aligned}$$

For equilibrium:

$$\text{Righting moment} = \text{heeling moment}$$

$$\triangle g \times GM \sin \theta = F_t \times NL \cos \theta$$

$$\frac{\sin \theta}{\cos \theta} = \frac{F_t \times NL}{\triangle g \times GM}$$

$$\tan \theta = \frac{F_t \times NL}{\triangle g \times GM}$$

From this the angle of heel may be obtained.

The angle of heel due to the force on the rudder is small unless the speed is excessive or the metacentric height small. In most merchant ships this angle is hardly noticeable.

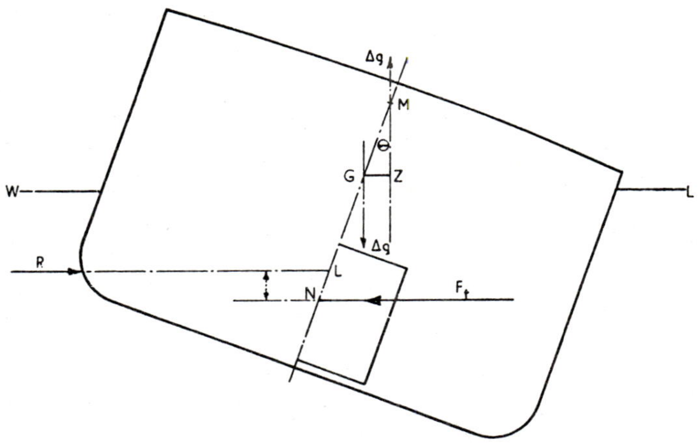

Fig. 81

Example. A ship of 8000 tonne displacement has a rudder of area 18m². The centre of lateral resistance is 4 m above the keel while the centroid of the rudder is 2·35 m above the keel. The maximum rudder angle is 35°. Calculate the angle of heel due to the force on the rudder if the latter is put hard over to port when travelling at 21 knots with a metacentric height of 0·4 m.

$$\text{Given } F = 580\ Av^2 \text{ N}$$

$$\text{Ship speed } v = 21 \times \frac{1852}{3600}$$

$$= 10{\cdot}80 \text{ m/s}$$

$$\text{Transverse force } F_t = 580\ A\ v^2 \sin \alpha \cos \alpha$$
$$= 580 \times 18 \times 10{\cdot}80^2$$
$$\times 0{\cdot}5736 \times 0{\cdot}8192$$

$$= 572\ 200 \text{ N}$$

$$\tan\theta = \frac{F_t \times NL}{\triangle g \times GM}$$

$$= \frac{572\ 200 \times (4{\cdot}0 - 2{\cdot}35)}{8000 \times 10^3 \times 9{\cdot}81 \times 0{\cdot}4}$$

$$= 0{\cdot}03007$$

$$\text{Angle of heel } \theta = 1°43' \text{ to port}$$

ANGLE OF HEEL WHEN TURNING

As the ship commences to turn a centrifugal force acts in addition to the rudder force. The effect of this force is to create a moment opposing the rudder force, i.e. tending to heel the ship in the opposite direction.

It is convenient to ignore the rudder force and consider only the centrifugal force. This force acts at the centre of gravity of the ship and may be calculated from the formula:

$$\text{Centrifugal force } CF = \frac{\triangle v^2}{\rho} \text{ kN}$$

where v is the ship speed in m/s

ρ is the radius of the turning circle in m

A resistance R is again exerted by the water on the ship due to the transverse movement, and has its maximum value when it is equal to the centrifugal force. This resistance is known as the centripetal force. A moment then acts on the ship causing it to heel.

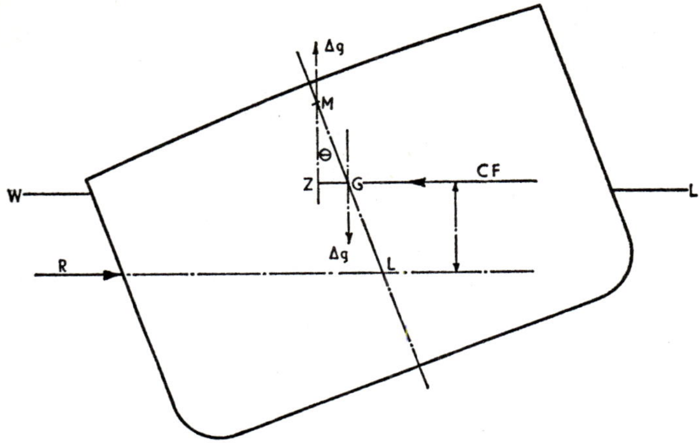

Fig. 82

The ship will be in equilibrium when the heeling moment is equal to the righting moment.

$$\text{Heeling moment} = CF \times LG \cos\theta$$

$$= \frac{\Delta v^2}{\rho} \times LG \cos\theta$$

$$\text{Righting moment} = \Delta g \times GZ$$

$$= \Delta g \times GM \sin\theta \quad \text{if } \theta \text{ is small}$$

$$\therefore \Delta g \times GM \sin\theta = \frac{\Delta v^2}{\rho} \times LG \cos\theta$$

$$\frac{\sin\theta}{\cos\theta} = \frac{\Delta v^2}{\rho} \times \frac{LG}{\Delta g \times GM}$$

$$\tan\theta = \frac{v^2 \times LG}{g \times \rho \times GM}$$

From this expression the angle of heel may be calculated.

It is usual in calculations of heel when turning, to ignore the heel due to the rudder force and consider it to be a small factor of safety, i.e. the actual angle of heel will be less than that calculated. If, when the ship is turning in a circle to port, the rudder is put hard over to starboard, the heel due to the rudder force is added to the previous heel due to centrifugal force, causing an increase in angle of heel. This may prove dangerous, especially in a small, high speed vessel.

Example. A ship with a metacentric height of 0·4 m has a speed of 21 knots. The centre of gravity is 6·2 m above the keel while the centre of lateral resistance is 4 m above the keel. The rudder is put hard over to port and the vessel turns in a circle 1100 m diameter. Calculate the angle to which the ship will heel.

$$\text{Ship speed } v = 21 \times \frac{1852}{3600}$$

$$= 10{\cdot}80 \text{ m/s}$$

$$\tan\theta = \frac{v^2 \times LG}{g \times \rho \times GM}$$

$$= \frac{10{\cdot}80^2 \times (6{\cdot}2 - 4{\cdot}0)}{9{\cdot}81 \times 1100 \times 0{\cdot}4}$$

$$= 0{\cdot}1189$$

$$\text{Angle of heel } \theta = 6°47' \text{ to starboard}$$

Using the details from the previous example, it may be seen that if θ_1 is the final angle of heel:

$$\tan\theta_1 = 0{\cdot}1189 - 0{\cdot}03007$$

$$= 0{\cdot}08883$$

$$\text{Final angle of heel } \theta_1 = 5°\ 5' \text{ to starboard.}$$

TEST EXAMPLES 9

NOTE: In the following questions the rudder force parallel to the streamline should be taken as $580\,Av^2$ N.

1. A ship, whose maximum speed is 18 knots, has a rudder of area 25 m^2. The distance from the centre of stock to the centre of effort of the rudder is 1·2 m and the maximum rudder angle 35°. If the maximum allowable stress in the stock is 85 MN/m^2, calculate the diameter of the stock.

2. The service speed of a ship is 14 knots and the rudder, 13 m^2 in area, has its centre of effort 1·1 m from the centre of stock. Calculate the torque on the stock at 10° intervals of rudder angle up to 40° and estimate the work done in turning the rudder from the centreline to 40°.

3. A ship 150 m long and 8·5 m draught has a rudder whose area is one sixtieth of the middle-line plane and diameter of stock 320 mm. Calculate the maximum speed at which the vessel may travel if the maximum allowable stress is 70 MN/m^2, the centre of stock 0·9 m from the centre of effort and the maximum rudder angle is 35°.

4. A ship displaces 5000 tonne and has a rudder of area 12 m^2. The distance between the centre of lateral resistance and the centre of the rudder is 1·6 m and the metacentric height 0·24 m. Calculate the initial angle of heel if the rudder is put over to 35° when travelling at 16 knots.

5. A vessel travelling at 17 knots turns with a radius of 450 m when the rudder is put hard over. The centre of gravity is 7 m above the keel, the transverse metacentre 7·45 m above the keel and the centre of buoyancy 4 m above the keel. If the centripetal force is assumed to act at the centre of buoyancy, calculate the angle of heel when turning. The rudder force may be ignored.

SOLUTIONS TO TEST EXAMPLES 1

1. (a)

$$\text{Mass of aluminium} = 300 \text{ g}$$
$$= 0{\cdot}300 \text{ kg}$$
$$\text{Volume of aluminium} = 42 \text{ cm}^3$$
$$= 42 \times 10^{-6} \text{ m}^3$$
$$\text{Density of aluminium} = \frac{0{\cdot}300}{42 \times 10^{-6}}$$
$$= \frac{0{\cdot}300 \times 10^6}{42}$$
$$= 7143 \text{ kg/m}^3$$

(b)

$$\text{Mass of equal volume of water} = 42 \times 10^{-3} \text{ kg}$$
$$\text{Relative density of aluminium} = \frac{0{\cdot}300}{42 \times 10^{-3}}$$
$$= 7{\cdot}143$$

Alternatively,

$$\text{Density of water} = 1000 \text{ kg/m}^3$$
$$\text{Relative density of aluminium} = \frac{\text{density of aluminium}}{\text{density of water}}$$
$$= \frac{7143}{1000}$$
$$= 7{\cdot}143$$

(c)

$$\text{Volume of aluminium} = 100 \text{ cm}^3$$
$$= 100 \times 10^{-6} \text{ m}^3$$
$$\text{Mass of aluminium} = 7143 \times 100 \times 10^{-6} \text{ kg/m}^3 \times \text{m}^3$$
$$= 0{\cdot}7143 \text{ kg}$$

2. Load on tank top $= \rho g A h$

$$9{\cdot}6 \times 10^6 = 1025 \times 9{\cdot}81 \times 12 \times 10 \times h$$

$$h = \frac{9{\cdot}6 \times 10^6}{1025 \times 9{\cdot}81 \times 12 \times 10}$$

$$= 7{\cdot}96 \text{ m}$$

3.

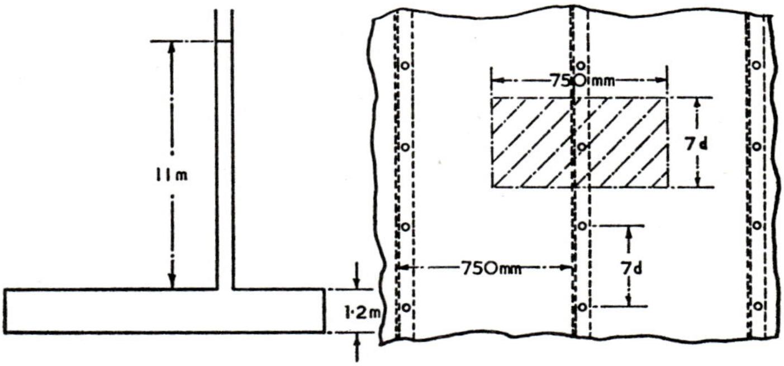

Fig. 83

Pressure on outer bottom $= \rho g h$
$= 0{\cdot}89 \times 10^3 \times 9{\cdot}81 \times (11 + 1{\cdot}2)$
$= 106{\cdot}5 \times 10^3 \text{ N/m}^2$
$= 106{\cdot}5 \text{ kN/m}^2$

Let d = diameter of rivets in mm.

Maximum stress in rivets

$$= \frac{\text{load on one rivet}}{\text{area of one rivet}}$$

$\therefore$ load on one rivet

$$= 30 \times 10^6 \times \frac{\pi}{4} \times d^2 \times 10^{-6}$$

$$= 30 \times \frac{\pi}{4} d^2 \text{ N}$$

But load on one rivet

$$= \text{load on area of tank top supported by one rivet}$$
$$= \rho g A h$$
$$= 0{\cdot}89 \times 10^3 \times 9{\cdot}81 \times 0{\cdot}75 \times 7d \times 10^{-3} \times 11$$
$$= 504{\cdot}3d \text{ N}$$

$$30 \times \frac{\pi}{4} d^2 = 504{\cdot}3d$$

$$d = \frac{504{\cdot}3}{30} \times \frac{4}{\pi}$$

$$= 21{\cdot}4 \text{ mm}$$

4. (a) Load on top $= \rho g A H$

Since H is zero

Load on top $= 0$

$$\text{Load on short side} = 1000 \times 9{\cdot}81 \times 12 \times 1{\cdot}4 \times 0{\cdot}7$$
$$= 115{\cdot}3 \times 10^3 \text{ N}$$
$$= 115{\cdot}3 \text{ kN}$$

(b) $$\text{Load on top} = 1000 \times 9{\cdot}81 \times 15 \times 12 \times 7$$
$$= 12{\cdot}36 \times 10^6 \text{ N}$$
$$= 12{\cdot}36 \text{ MN}$$

$$\text{Load on short side} = 1000 \times 9{\cdot}81 \times 12 \times 1{\cdot}4 \times (7 + 0{\cdot}7)$$
$$= 1{\cdot}268 \times 10^6 \text{ N}$$
$$= 1{\cdot}268 \text{ MN}$$

5. $$\text{Pressure at bottom} = \rho g h$$
$$= 1025 \times 9{\cdot}81 \times 6$$
$$= 60{\cdot}34 \times 10^3 \text{ N/m}^2$$
$$= 60{\cdot}34 \text{ kN/m}^2$$

$$\text{Load on bulkhead} = \rho g A H$$
$$= 1025 \times 9{\cdot}81 \times 9 \times 6 \times 3$$
$$= 1{\cdot}629 \times 10^6 \text{ N}$$
$$= 1{\cdot}629 \text{ MN}$$

6.

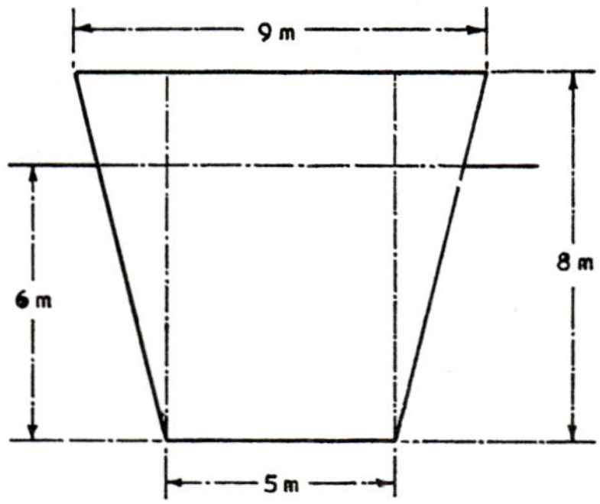

Fig. 84

(a) Width of bulkhead at a depth of 6m

$$= 5 + 4 \times \frac{6}{8}$$
$$= 8\text{m}$$

Divide the bulkhead into a rectangle and two triangles as shown in Fig. 84.

$$\text{Load on rectangle} = 0{\cdot}85 \times 1000 \times 9{\cdot}81 \times 5 \times 6 \times 3$$
$$= 750{\cdot}4 \times 10^3 \text{ N}$$
$$= 750{\cdot}4 \text{ kN}$$

Load on two triangles

$$= 2 \times 0{\cdot}85 \times 1000 \times 9{\cdot}81 \times \frac{1{\cdot}5 \times 6}{2} \times \frac{6}{3}$$
$$= 150{\cdot}1 \times 10^3 \text{ N}$$
$$= 150{\cdot}1 \text{ kN}$$

Total load on bulkhead

$$= 750{\cdot}4 + 150{\cdot}1$$
$$= 900{\cdot}5 \text{ kN}$$

(b) Load on rectangle

$$= 0{\cdot}85 \times 1000 \times 9{\cdot}81 \times 5 \times 8 \times \left(\frac{8}{2} + 4\right)$$
$$= 2{\cdot}669 \times 10^6 \text{ N}$$
$$= 2{\cdot}669 \text{ MN}$$

Load on two triangles

$$= 2 \times 0{\cdot}85 \times 1000 \times 9{\cdot}81 \times \frac{2 \times 8}{2} \times \left(\frac{8}{3} + 4\right)$$

$$= 889{\cdot}6 \times 10^3 \text{ N}$$
$$= 0{\cdot}890 \text{ MN}$$

Total load on bulkhead

$$= 2{\cdot}669 + 0{\cdot}890$$
$$= 3{\cdot}559 \text{ MN}$$

7. (a)Load on bulkhead

$$= \rho g A H$$
$$= 0{\cdot}9 \times 1000 \times 9{\cdot}81 \times 10 \times 12 \times 6$$
$$= 6{\cdot}356 \times 10^6 \text{ N}$$
$$= 6{\cdot}356 \text{ MN}$$

Distance of centre of pressure from top of bulkhead

$$= \tfrac{2}{3}D$$
$$= \tfrac{2}{3} \times 12$$
$$= 8 \text{ m}$$

(b) Load on bulkhead

$$= 0{\cdot}9 \times 1000 \times 9{\cdot}81 \times 10 \times 12 \times \left(\frac{12}{2} + 3\right)$$

$$= 9{\cdot}534 \times 10^6 \text{ N}$$
$$= 9{\cdot}534 \text{ MN}$$

Distance of centre of pressure from surface of oil

$$= \frac{I_{NA}}{AH} + H$$

$$= \frac{\frac{1}{12} 10 \times 12^3}{10 \times 12 \times 9} + 9$$

$$= 1{\cdot}333 + 9$$
$$= 10{\cdot}333 \text{ m}$$

Distance of centre of pressure from top of bulkhead

$$= 10{\cdot}333 - 3$$
$$= 7{\cdot}333 \text{ m}$$

8.

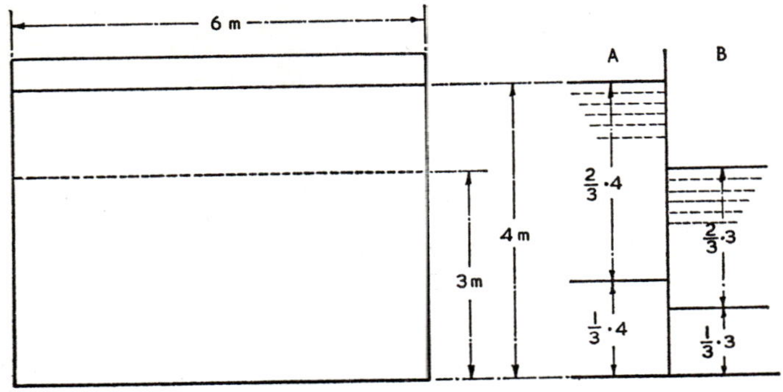

Fig 85

$$\begin{aligned}\text{Load on side A} &= 1025 \times 9{\cdot}81 \times 6 \times 4 \times 2\\ &= 482{\cdot}6 \times 10^3\ \text{N}\\ &= 482{\cdot}6\ \text{kN}\end{aligned}$$

$$\begin{aligned}\text{Centre of pressure on side A} &= \tfrac{2}{3} \times 4\\ &= 2{\cdot}667\ \text{m from top}\\ &= 1{\cdot}333\ \text{m from bottom}\end{aligned}$$

$$\begin{aligned}\text{Load on side B} &= 1000 \times 9{\cdot}81 \times 6 \times 3 \times 1{\cdot}5\\ &= 265{\cdot}0 \times 10^3\ \text{N}\\ &= 265{\cdot}0\ \text{kN}\end{aligned}$$

$$\begin{aligned}\text{Centre of pressure on side B} &= \tfrac{2}{3} \times 3\\ &= 2\ \text{m from top}\\ &= 1\ \text{m from bottom}\end{aligned}$$

Taking moments of load about bottom of gate:

Resultant centre of pressure from bottom

$$= \frac{482{\cdot}6 \times 1{\cdot}333 - 265{\cdot}0 \times 1}{482{\cdot}6 - 265{\cdot}0}$$

$$= \frac{643{\cdot}47 - 265{\cdot}0}{217{\cdot}6}$$

$$= 1{\cdot}740\ \text{m}$$

$$\text{Resultant load} = 217{\cdot}6\ \text{kN}$$

9.

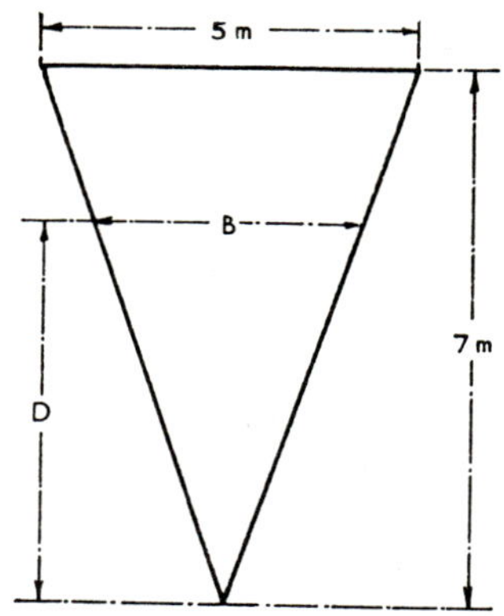

Fig. 86

There are two types of solution possible, one if the water is above the top of the bulkhead and one if the water is below the top of the bulkhead.

Assume water *at* top edge:

$$\text{Load on bulkhead} = 1025 \times 9{\cdot}81 \times \frac{5 \times 7}{2} \times \frac{7}{3}$$

$$= 410{\cdot}6 \times 10^3 \text{ N}$$

$$= 410{\cdot}6 \text{ kN}$$

Since the load on the bulkhead is only 190 kN, the water must be *below* the top of the bulkhead.

$$\text{Width at water level } B = \frac{5}{7} \times D$$

$$190 \times 10^3 = 1025 \times 9{\cdot}81 \times \frac{D}{2} \times \frac{5D}{7} \times \frac{D}{3}$$

$$D^3 = \frac{190 \times 10^3 \times 2 \times 7 \times 3}{1025 \times 9{\cdot}81 \times 5}$$

from which $D = 5{\cdot}414 \text{ m}$

Centre of pressure below surface of water

$$= \tfrac{1}{2} \times D$$

$$= 2{\cdot}707 \text{ m}$$

Centre of pressure below top of bulkhead

$$= 2{\cdot}707 + (7{\cdot}00 - 5{\cdot}414)$$

$$= 4{\cdot}293 \text{ m}$$

10.

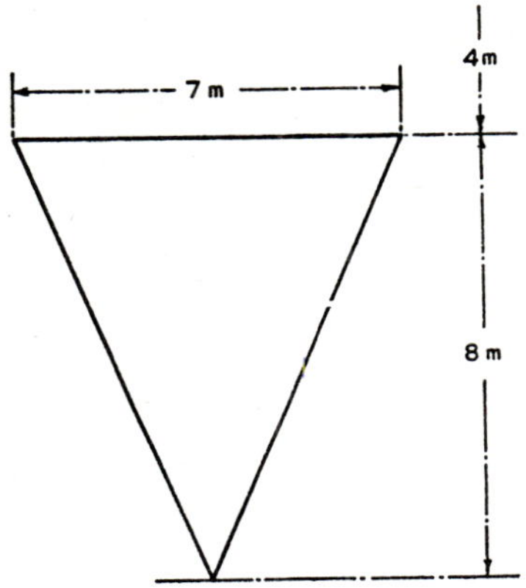

Fig. 87

(a)

$$\text{Load on bulkhead} = 1025 \times 9{\cdot}81 \times \frac{7 \times 8}{2} \times \frac{8}{3}$$

$$= 750{\cdot}8 \times 10^3 \text{ N}$$
$$= 750{\cdot}8 \text{ kN}$$

Centre of pressure from top

$$= \tfrac{1}{2}D$$
$$= 4\text{m}$$

(b)

$$\text{Load on bulkhead} = 1025 \times 9{\cdot}81 \times \frac{7 \times 8}{2} \times \left(\frac{8}{3} + 4\right)$$

$$= 1{\cdot}877 \times 10^6 \text{ N}$$
$$= 1{\cdot}877 \text{ MN}$$

$$\text{For triangle } I_{NA} = \frac{1}{36} BD^3$$

Centre of pressure from surface of water

$$= \frac{I_{NA}}{AH} + H$$

$$= \frac{\frac{1}{36} \times 7 \times 8^3}{\frac{1}{2} \times 7 \times 8 \times (\frac{1}{3} \times 8 + 4)} + (\tfrac{1}{3} \times 8 + 4)$$

$$= \frac{2 \times 7 \times 8^3}{36 \times 7 \times 8 \times 6{\cdot}667} + 6{\cdot}667$$

$$= 7{\cdot}200 \text{ m}$$

Centre of pressure from top of bulkhead

$$= 7{\cdot}200 - 4{\cdot}000$$
$$= 3{\cdot}2 \text{ m}$$

11.

Load on one stiffener $= \rho g A H$

$$= 1025 \times 9{\cdot}81 \times 8 \times 0{\cdot}700 \times \frac{8}{2}$$

$$= 225{\cdot}3 \times 10^3 \text{ N}$$
$$= 225{\cdot}3 \text{ kN}$$

(a) Shearing force at top of stiffener

$$= \tfrac{1}{3} \times \text{load}$$
$$= \tfrac{1}{3} \times 225{\cdot}3$$
$$= 75{\cdot}1 \text{ kN}$$

(b) Shearing force at bottom of stiffener

$$= \tfrac{2}{3} \times \text{load}$$
$$= \tfrac{2}{3} \times 225{\cdot}3$$
$$= 150{\cdot}2 \text{ kN}$$

$\therefore$ Shear force in rivets $= 150{\cdot}2$ kN

Cross-sectional area of rivets $= 10 \times \frac{\pi}{4} \times 20^2 \times 10^{-6}$

$$= 3{\cdot}142 \times 10^{-3} \text{ m}^2$$

Shear stress in rivets $= \dfrac{\text{shear force}}{\text{area}}$

$$= \frac{150{\cdot}2}{3{\cdot}142 \times 10^{-3}}$$

$$= 47{\cdot}75 \times 10^3 \text{ kN/m}^2$$
$$= 47{\cdot}75 \text{ MN/m}^2$$

(c) Position of zero shear $= \dfrac{l}{\sqrt{3}}$

$$= \frac{8}{\sqrt{3}}$$

$$= 4{\cdot}619 \text{ m from top}$$

12. (a) Let l = height of bulkhead

Maximum shearing force in stiffeners

$$= 200 \text{ kN}$$
$$= \tfrac{2}{3} \times \text{load}$$

$$\therefore \text{load} = 200 \times \frac{3}{2}$$

$$= 300 \text{ kN}$$

But load on stiffener $= 1025 \times 9{\cdot}81 \times l \times \dfrac{l}{9} \times \dfrac{l}{2}$

$$l^3 = \frac{300 \times 10^3 \times 9 \times 2}{1025 \times 9{\cdot}81}$$

from which $l = 8{\cdot}128$ m

(b) Shearing force at top $= \tfrac{1}{3} \times \text{load}$
$= \tfrac{1}{3} \times 300$
$= 100$ kN

(c) Position of zero shear $= \dfrac{l}{\sqrt{3}}$

$$= \frac{8{\cdot}128}{\sqrt{3}}$$

$$= 4{\cdot}693 \text{ m from top}$$

SOLUTIONS TO TEST EXAMPLES 2

1. Volume of wood and metal immersed

$$= 3{\cdot}5 \times 10^3 + 250 - 100$$
$$= 3650 \text{ cm}^3$$

$$\text{Mass of wood and metal} = 1{\cdot}000 \times 3650 \text{ g/cm}^3 \times \text{cm}^3$$
$$= 3650 \text{ g}$$

$$\text{Mass of wood} = 3{\cdot}5 \times 10^3 \times 1{\cdot}000 \times 0{\cdot}60$$
$$\doteq 2100 \text{ g}$$

$$\therefore \text{Mass of metal} = 3650 - 2100$$
$$= 1550 \text{ g}$$

Mass of equal volume of fresh water

$$= 250 \times 1{\cdot}000$$
$$= 250 \text{ g}$$

$$\text{Relative density} = \frac{1550}{250}$$

$$= 6{\cdot}20$$

2. $$\text{Mass of raft} = 1000 \times 0{\cdot}7 \times 3 \times 2 \times 0{\cdot}25$$
$$= 1{\cdot}05 \times 10^3 \text{ kg}$$

Mass of raft when completely submerged

$$= 1018 \times 3 \times 2 \times 0{\cdot}25$$
$$= 1{\cdot}527 \times 10^3 \text{ kg}$$

$\therefore$ mass required to submerge raft

$$= 1{\cdot}527 \times 10^3 - 1{\cdot}050 \times 10^3$$
$$= 0{\cdot}477 \times 10^3 \text{ kg}$$
$$= 477 \text{ kg}$$

3. (a) $$\text{Displacement of barge} = 1025 \times 65 \times 12 \times 5{\cdot}5$$
$$= 4397 \times 10^3 \text{ kg}$$
$$= 4397 \text{ tonne}$$

(b) $$\text{Draught in fresh water} = 5{\cdot}5 \times \frac{1{\cdot}025}{1{\cdot}000}$$
$$= 5{\cdot}637 \text{ m}$$

4.

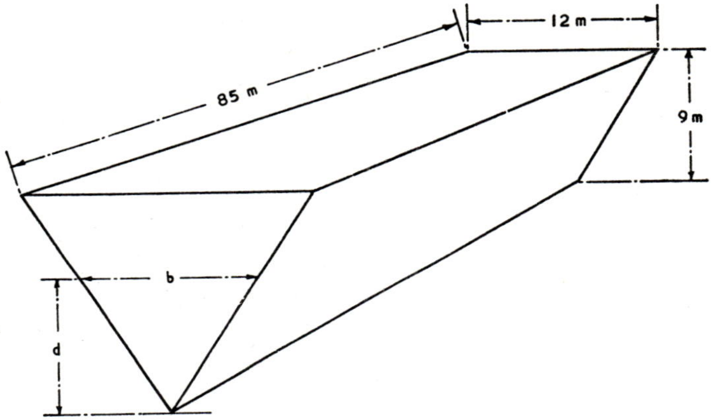

Fig. 88

Let d = draught
b = breadth at waterline

By similar triangles $b = \frac{12}{9} d$

$= \frac{4}{3} d$

At draught d, displacement $= 1{\cdot}025 \times 85 \times \frac{b \times d}{2}$

$= 1{\cdot}025 \times 85 \times \frac{4}{3}d \times \frac{1}{2}d$
$= 58{\cdot}08d^2$ tonne

Tabulating:

draught d	d^2	displacement tone
0	0	0
1·25	1·563	91
2·50	6·250	363
3·75	14·062	817
5·00	25·000	1452
6·25	39·062	2269
7·50	56·250	3267

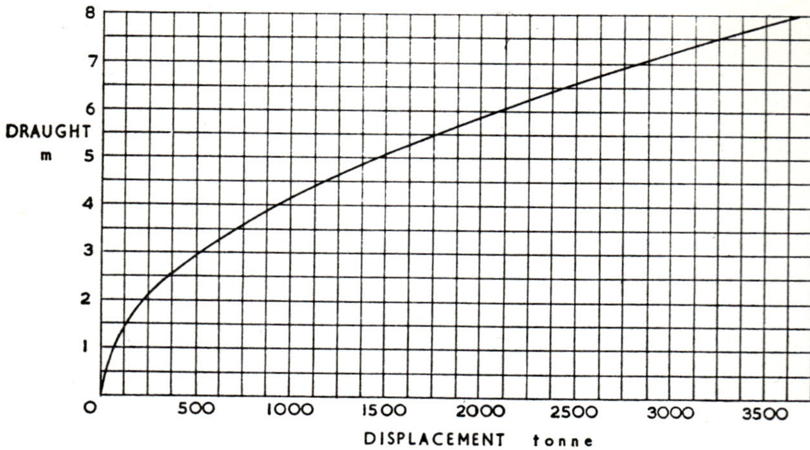

Fig. 89

At 6·50m draught, displacement in sea water is 2450 tonne.

$$\therefore \text{Displacement in fresh water} = 2450 \times \frac{1 \cdot 000}{1 \cdot 025}$$

$$= 2390 \text{ tonne}$$

5. Immersed volume of cylinder $= \frac{1}{2} \times 15 \times \frac{\pi}{4} \times 4^2$

$$= 30\,\pi \text{ m}^3$$

$$\text{Mass of cylinder} = 1 \cdot 025 \times 30\pi$$
$$= 96 \cdot 62 \text{ tonne}$$

6. Mass of water displaced $= 1 \cdot 025 \times 22$

$$= 22 \cdot 55 \text{ tonne}$$

$$\therefore \text{apparent mass of bilge keels} = 36 - 22 \cdot 55$$
$$= 13 \cdot 45 \text{ tonne}$$

$$\text{Increase in draught} = \frac{13 \cdot 45}{20}$$

$$= 0 \cdot 673 \text{ cm}$$

7.

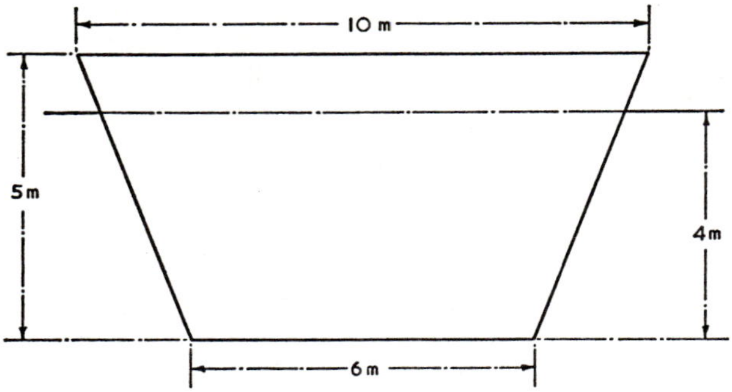

Fig. 90

$$\text{Breadth at waterline} = 6 + \frac{4}{5} \times 4$$

$$= 9{\cdot}2 \text{ m}$$

$$\text{Displacement} = 1{\cdot}025 \times 40 \times \frac{6 + 9{\cdot}2}{2} \times 4$$

$$= 1246 \text{ tonne}$$

8. $\text{TPC} = 0{\cdot}01025 \times Aw$

draught	waterplane area	TPC
7·5	1845	18·91
6·25	1690	17·32
5·00	1535	15·73
3·75	1355	13·89
2·50	1120	11·48

Increase in draught required = 0·20 m
= 20 cm

Mean TPC at 6·20 m = 17·25

Mass required = 17·25 × 20
= 345 tonne

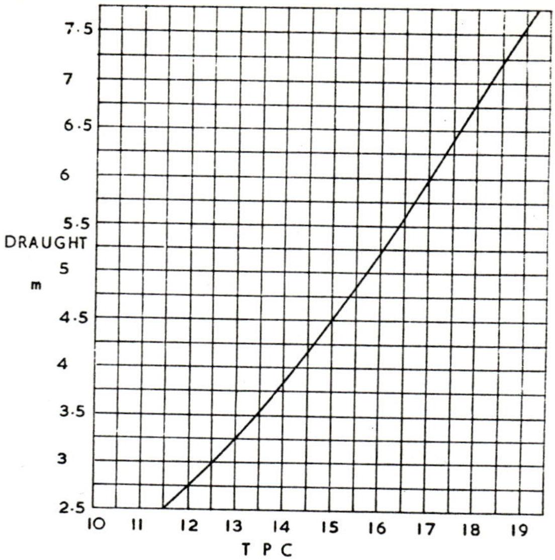

Fig. 91

9. Volume of displacement $= \frac{19\ 500}{1{\cdot}025}$

$= 19\ 024\ \text{m}^3$

Waterplane area $= \frac{\text{TPC}}{0{\cdot}01025}$

$= \frac{26{\cdot}5}{0{\cdot}01025}$

$= 2585\ \text{m}^3$

Block coefficient $= \frac{19\ 024}{150 \times 20{\cdot}5 \times 8}$

$= 0{\cdot}773$

Prismatic coefficient $= \frac{C_b}{C_m}$

$$= \frac{0.773}{0.94}$$

$$= 0.822$$

$$\text{Waterplane area coefficient} = \frac{2585}{150 \times 20.5}$$

$$= 0.841$$

10. Let length of ship $= L$

Then breadth $= 0.13L$

and draught $= \dfrac{0.13L}{2.1}$

$$= 0.0619L$$

$$\text{Volume of displacement} = \frac{9450}{1.025}$$

$$= 9219.5 \text{ m}^3$$

$$C_b = \frac{\nabla}{L \times B \times d}$$

$$0.7 = \frac{9219.5}{L \times 0.13L \times 0.0619L}$$

$$L^3 = \frac{9219.5}{0.7 \times 0.13 \times 0.0619}$$

Length of ship $L = 117.9$ m

$$C_p = \frac{9219.5}{117.9 \times 106}$$

Prismatic coefficient $= 0.738$

11. Let length of ship $= L$

Then draught $= \dfrac{L}{18}$

and breadth $= 2.1 \times$ draught

$$= \frac{2.1}{18}L$$

TPC sea water $= 0.01025\ A_w$

TPC fresh water $= 0.0100\ A_w$

TPC sea water − TPC fresh water

$$= 0{\cdot}7$$

$$\therefore 0{\cdot}01025\,A_w - 0{\cdot}0100\,A_w = 0{\cdot}7$$

$$0{\cdot}00025\,A_w = 0{\cdot}7$$

$$A_w = \frac{0{\cdot}7}{0{\cdot}00025}$$

$$= 2800\ \text{m}^2$$

But $A_w = C_w \times L \times B$

$$2800 = 0{\cdot}83 \times L \times \frac{2{\cdot}1}{18}L$$

$$L^2 = \frac{2800 \times 18}{0{\cdot}83 \times 2{\cdot}1}$$

From which $L = 170$ m

TPC fresh water $= 0{\cdot}0100 \times 2800$

$= 28$

12.

½ girth	SM	product
2·1	1	2·1
6·6	4	26·4
9·3	2	18·6
10·5	4	42·0
11·0	2	22·0
11·0	4	44·0
11·0	2	22·0
9·9	4	39·6
7·5	2	15·0
3·9	4	15·6
0	1	0
		247·3

Common interval = 9 m

Wetted surface area $= \frac{2}{3} \times 9 \times 247{\cdot}3$

$= 1483{\cdot}8\ \text{m}^2$

½% $= 7{\cdot}42\ \text{m}^2$

Appendages $= 30{\cdot}00\ \text{m}^2$

Total wetted surface area $= 1521{\cdot}22\ \text{m}^2$

13. (a)

$$\text{Volume of displacement} = \frac{14\,000}{1{\cdot}025}$$

$$= 13\,658 \text{ m}^3$$

$$S = 1{\cdot}7Ld + \frac{\nabla}{d}$$

$$= 1{\cdot}7 \times 130 \times 8 + \frac{13\,658}{8}$$

$$= 1768{\cdot}0 + 1707{\cdot}3$$
$$= 3475{\cdot}3 \text{ m}^2$$

(b)

$$S = c\sqrt{\triangle L}$$
$$= 2{\cdot}58\sqrt{14\,000 \times 130}$$
$$= 3480 \text{ m}^2$$

14. Volume of existing barge $= 75 \times 9 \times 6$
$$= 4050 \text{ m}^3$$

Volumes of similar ships $\propto$ length^3

i.e.
$$\frac{V_1}{V_2} = \left(\frac{L_1}{L_2}\right)^3$$

$$L_2 = L_1 \sqrt[3]{\frac{V_2}{V_1}}$$

$$= 75 \sqrt[3]{\frac{3200}{4050}}$$

New length $L_2 = 69{\cdot}34$ m

new beam
$$B_2 = 9 \sqrt[3]{\frac{3200}{4050}}$$
$$= 8{\cdot}32 \text{ m}$$

New depth
$$D_2 = 6 \sqrt[3]{\frac{3200}{4050}}$$
$$= 5{\cdot}55 \text{ m}$$

15. Let S = wetted surface area of small ship
$2S$ = wetted surface area of large ship
$\triangle$ = displacement of small ship
$\triangle + 2000$ = displacement of large ship

It was shown that for similar ships:

$$\triangle \propto S^{\frac{3}{2}}$$

Hence
$$\frac{\triangle}{\triangle + 2000} = \left(\frac{S}{2S}\right)^{\frac{3}{2}}$$

$$= (\tfrac{1}{2})^{\frac{3}{2}}$$

$$2^{\frac{3}{2}} \times \triangle = \triangle + 2000$$
$$2{\cdot}828\triangle = \triangle + 2000$$
$$1{\cdot}828\triangle = 2000$$

$$\triangle = \frac{2000}{1{\cdot}828}$$

Displacement of smaller ship $\triangle$

$$= 1094 \text{ tonne}$$

16. For similar ships: $\triangle \propto L^3$

$$\therefore \frac{\triangle_1}{\triangle_2} = \left(\frac{L_1}{L_2}\right)^3$$

Displacement of model $\triangle_2 = 11\,000 \times \left(\frac{6}{120}\right)^3$

$$= 1{\cdot}375 \text{ tonne}$$

$$S \propto L^2$$

$$\therefore \frac{S_1}{S_2} = \left(\frac{L_1}{L_2}\right)^2$$

Wetted surface area of model

$$S_2 = 2500 \times \left(\frac{6}{120}\right)^2$$

$$= 6{\cdot}25 \text{ m}^2$$

SOLUTIONS TO TEST EXAMPLES 3

1.

½ width	SM	Product for area
1	1	1
7·5	4	30
12	2	24
13·5	4	54
14	2	28
14	4	56
14	2	28
13·5	4	54
12	2	24
7	4	28
0	1	0
		327 = ΣA

$$\text{Common interval } h = \frac{180}{10} = 18 \text{ m}$$

(a)

$$\text{Waterplane area} = \frac{h}{3}\Sigma A \times 2 = \frac{18}{3} \times 327 \times 2 = 3924 \text{ m}^2$$

(b)

$$\text{TPC} = \frac{\text{waterplane area} \times \text{density}}{100} = \frac{3924 \times 1{\cdot}025}{100} = 40{\cdot}22$$

(c)

$$\text{Waterplane area coefficient} = \frac{\text{waterplane area}}{\text{length} \times \text{breadth}} = \frac{3924}{180 \times 28} = 0{\cdot}778$$

2.

Waterplane area	SM	Product for volume
865	1	865
1735	4	6940
1965	2	3930
2040	4	8160
2100	2	4200
2145	4	8580
2215	1	2215
		34 890 $= \Sigma\nabla$

$$\text{Volume of displacement} = \frac{h}{3}\Sigma\nabla$$

$$\text{Displacement} = \frac{h}{3}\Sigma\nabla \times \rho$$

$$= \frac{1{\cdot}5}{3} \times 34\,890 \times 1{\cdot}025$$

$$= 17\,881 \text{ tonne}$$

3.

Cross-sectional area	SM	Product for volume
5	1	5
60	4	240
116	2	232
145	4	580
152	2	304
153	4	612
153	2	306
151	4	604
142	2	284
85	4	340
0	1	0
		3507 $= \Sigma\nabla$

$$\text{Common interval} = \frac{140}{10}$$

$$= 14 \text{ m}$$

(a)

$$\text{Volume of displacement} = \frac{h}{3}\Sigma\nabla$$

$$= \frac{14}{3} \times 3507$$

$$= 16\,366 \text{ m}^3$$

$$\text{Displacement} = \text{volume} \times \text{density}$$
$$= 16\,366 \times 1{\cdot}025$$
$$= 16\,775 \text{ tonne}$$

(b)

$$\text{Block coefficient} = \frac{\text{volume of displacement}}{\text{length} \times \text{breadth} \times \text{draught}}$$

$$= \frac{16\,366}{140 \times 18 \times 9}$$

$$= 0{\cdot}722$$

(c)

$$\text{Midship section area coefficient} = \frac{\text{midship section area}}{\text{breadth} \times \text{draught}}$$

$$= \frac{153}{18 \times 9}$$

$$= 0{\cdot}944$$

(d)

$$\text{Prismatic coefficient} = \frac{\text{volume of displacement}}{\text{length} \times \text{midship section area}}$$

$$= \frac{16\,366}{140 \times 153}$$

$$= 0{\cdot}764$$

Alternatively

$$\text{Prismatic coefficient} = \frac{\text{block coefficient}}{\text{midship section area coefficient}}$$

$$= \frac{0{\cdot}722}{0{\cdot}944}$$

$$= 0{\cdot}764$$

4.

Section	$\frac{1}{2}$ ordinate	SM	Product for area	Lever	Product for 1st moment
AP	1·2	$\frac{1}{2}$	0·6	+5	+ 3·0
$\frac{1}{2}$	3·5	2	7·0	+4$\frac{1}{2}$	+31·5
1	5·3	1	5·3	+4	+21·2
1$\frac{1}{2}$	6·8	2	13·6	+3$\frac{1}{2}$	+47·6
2	8·0	1$\frac{1}{2}$	12·0	+3	+36·0
3	8·3	4	33·2	+2	+66·4
4	8·5	2	17·0	+1	+17·0
					+222·7 = ΣMA
5	8·5	4	34·0	0	
6	8·5	2	17·0	−1	−17·0
7	8·4	4	33·6	−2	−67·2
8	8·2	1$\frac{1}{2}$	12·3	−3	−36·9
8$\frac{1}{2}$	7·9	2	15·8	−3$\frac{1}{2}$	−55·3
9	6·2	1	6·2	−4	−24·8
9$\frac{1}{2}$	3·5	2	7·0	−4$\frac{1}{2}$	−31·5
FP	0	$\frac{1}{2}$	0	−5	− 0
			214·6 = ΣA		−232·7 = ΣMF

$$\text{Common interval} = \frac{120}{10}$$

$$= 12 \text{ m}$$

(a)

$$\text{Waterplane area} = \frac{h}{3} \Sigma A \times 2$$

$$= \frac{12}{3} \times 214{\cdot}6 \times 2$$

$$= 1716{\cdot}8 \text{ m}^2$$

Since ΣMF exceeds ΣMA the centroid will be *forward* of midships.

$$\text{Distance of centroid from midships} = \frac{h(\Sigma MA + \Sigma MF)}{\Sigma A}$$

$$= \frac{12(222{\cdot}7 - 232{\cdot}7)}{214{\cdot}6}$$

$$= 0{\cdot}559 \text{ m forward}$$

5.

TPC	SM	Product for displacement	Lever	Product for 1st moment
4·0	1	4·0	0	0
6·1	4	24·4	1	24·4
7·8	2	15·6	2	31·2
9·1	4	36·4	3	109·2
10·3	2	20·6	4	82·4
11·4	4	45·6	5	228·0
12·0	1	12·0	6	72·0
		$158{\cdot}6 = \Sigma\Delta$		$547{\cdot}2 = \Sigma M$

$$\text{Common interval} = 1{\cdot}5 \text{ m} = 150 \text{ cm}$$

(a)
$$\text{Displacement} = \frac{h}{3}\Sigma\Delta \quad (h \text{ in cm})$$
$$= \frac{150}{3} \times 158{\cdot}6$$
$$= 7930 \text{ tonne}$$

(b)
$$KB = \frac{h\,\Sigma M}{\Sigma\Delta} \quad (h \text{ in m})$$
$$= \frac{1{\cdot}5 \times 547{\cdot}2}{158{\cdot}6}$$
$$= 5{\cdot}175 \text{ m}$$

6.

½ breadths	SM	Product for area	Lever	Product for 1st moment	Lever	Product for 2nd moment
0·3	1	0·3	+5	+ 1·5	+5	+ 7·5
3·8	4	15·2	+4	+ 60·8	+4	+ 243·2
6·0	2	12·0	+3	+ 36·0	+3	+ 108·0
7·7	4	30·8	+2	+ 61·6	+2	+ 123·2
8·3	2	16·6	+1	+ 16·6	+1	+ 16·6
9·0	4	36·0	0	$+176{\cdot}5 = \Sigma M_A$	0	0

8·4	2	16·8	−1	− 16·8	−1	+ 16·8
7·8	4	31·2	−2	− 62·4	−2	+ 124·8
6·9	2	13·8	−3	− 41·4	−3	+ 124·2
4·7	4	18·8	−4	− 75·2	−4	+ 300·8
0	1	0	−5	0	−5	0
		191·5 = ΣA		−195·8 = ΣMF		+1065·1 = ΣI

Common interval = 15 m

(a) Waterplane area $= \frac{h}{3} \Sigma A \times 2$

$$= \frac{15}{3} \times 191{\cdot}5 \times 2$$

$$= 1915 \text{ m}^2$$

(b)
Distance of centroid from midships

$$= \frac{h(\Sigma MA + \Sigma MF)}{\Sigma A}$$

$$= \frac{15(176{\cdot}5 - 195{\cdot}8)}{191{\cdot}5}$$

$$= 1{\cdot}512 \text{ m forward}$$

Second moment of area about midships

$$= \frac{h^3}{3} \Sigma I \times 2$$

$$= \frac{15^3}{3} \times 1065{\cdot}1 \times 2$$

$$= 2\,396\,475 \text{ m}^4$$

Second moment of area about centroid

$$= 2\,396\,475 - 1915 \times 1{\cdot}512^2$$
$$= 2\,396\,475 - 4378$$
$$= 2\,392\,097 \text{ m}^4$$

7.

Draught	Displacement	SM	Product for moment
0	0	1	0
1	189	4	756
2	430	2	860
3	692	4	2768
4	977	1	977
			5361 = Σ_M

Common interval = 1 m

$$\text{Area of curve} = \frac{h}{3}\Sigma_M$$

$$= \tfrac{1}{3} \times 5361$$
$$= 1787 \text{ tonne m}$$

$$\text{VCB below waterline} = \frac{\text{area of curve}}{\text{displacement}}$$

$$= \frac{1787}{977}$$

$$= 1{\cdot}829 \text{ m}$$
$$KB = 4 - 1{\cdot}829$$
$$= 2{\cdot}171 \text{ m}$$

8.

Width	SM	Product for area	Lever	Product for 1st moment	Lever	Product for 2nd moment
8·0	1	8·0	0	0	0	0
7·5	4	30·0	1	30·0	1	30·0
6·5	2	13·0	2	26·0	2	52·0
5·7	4	22·8	3	68·4	3	205·2
4·7	2	9·4	4	37·6	4	150·4
3·8	4	15·2	5	76·0	5	380·0
3·0	1	3·0	6	18·0	6	108·0
				256·0 = Σ_M		925·6 = Σ_I

$$\text{Common interval} = 1{\cdot}2 \text{ m}$$

$$\text{Load on bulkhead} = \text{density} \times g \times \text{1st moment}$$

$$= \rho g \times \frac{h^2}{3} \Sigma \text{M}$$

$$= 1{\cdot}025 \times 9{\cdot}81 \times \frac{1{\cdot}2^2}{3} \times 256{\cdot}0$$

$$= 1235 \text{ kN}$$

$$\text{Centre of pressure from top} = \frac{\text{2nd moment about top}}{\text{1st moment about top}}$$

$$= \frac{h\,\Sigma \text{I}}{\Sigma \text{M}}$$

$$= \frac{1{\cdot}2 \times 925{\cdot}6}{256{\cdot}0}$$

$$= 4{\cdot}339 \text{ m}$$

9.

Width	SM	Product for area	(Width)^2	SM	Product for 1st moment	(Width)^3	SM	Product for 2nd moment
10	1	10	100	1	100	1000	1	1000
9	4	36	81	4	324	729	4	2916
7	2	14	49	2	98	343	2	686
4	4	16	16	4	64	64	4	256
1	1	1	1	1	1	1	1	1
		77 = Σa			587 = Σm			4859 = Σi

$$\text{Common interval} = \frac{12}{4}$$

$$= 3 \text{ m}$$

Area of surface

$$a = \frac{h}{3} \Sigma \text{a}$$

$$= \frac{3}{3} \times 77$$

$$= 77 \text{ m}^2$$

Distance of centroid from longitudinal bulkhead

$$= \bar{y} \frac{\Sigma m}{2\Sigma a}$$

$$= \frac{587}{2 \times 77}$$

$$= 3{\cdot}812 \text{ m}$$

Second moment of area about longitudinal bulkhead

$$i_b = \frac{h}{9}\Sigma i$$

$$= \frac{3}{9} \times 4859$$

$$= 1619{\cdot}7 \text{ m}^4$$

Second moment of area about centroid

$$\begin{aligned} i_g &= i_b - ay^2 \\ &= 1619{\cdot}7 - 77 \times 3{\cdot}812^2 \\ &= 1619{\cdot}7 - 1118{\cdot}7 \\ &= 501{\cdot}0 \text{ m}^4 \end{aligned}$$

10.

½ ordinate	(½ ordinate)³	SM	Product for 2nd moment
1·6	4·1	1	4·1
5·7	185·2	4	740·8
8·8	681·5	2	1363·0
10·2	1061·2	4	4244·8
10·5	1157·6	2	2315·2
10·5	1157·6	4	4630·4
10·5	1157·6	2	2315·2
10·0	1000·0	4	4000·0
8·0	512·0	2	1024·0
5·0	125·0	4	500·0
0	0	1	0
			21 137·5 = ΣI

Common interval = 16 m

Second moment of area about centreline

$$= \frac{h}{9} \times \Sigma \text{I} \times 2$$

$$= \frac{16}{9} \times 21\,137{\cdot}5 \times 2$$

$$= 75\,155 \text{ m}^4$$

11.

Cross-sectional area	SM	Product for volume	Lever	Product for 1st moment
2	1	2	+5	+ 10
40	4	160	+4	+ 640
79	2	158	+3	+ 474
100	4	400	+2	+ 800
103	2	206	+1	+ 206
104	4	416	0	+2130 = ΣMA
104	2	208	−1	− 208
103	4	412	−2	− 824
97	2	194	−3	− 582
58	4	232	−4	− 928
0	1	0	−5	0
		2388 = Σ∇		−2542 = ΣMF

Common interval = 12 m

(a) Displacement $= \rho \times \frac{h}{3} \Sigma \nabla$

$$= 1{\cdot}025 \times \frac{12}{3} \times 2388$$

$$= 9790{\cdot}8 \text{ tonne}$$

(b) Centre of buoyancy from midships

$$= \frac{h(\Sigma \text{MA} + \Sigma \text{MF})}{\Sigma \nabla}$$

$$= \frac{12(2130 - 2542)}{2388}$$

$$= 2{\cdot}070 \text{ m forward}$$

12.

½ ordinate	SM	Product for area
1·6	1	1·6
6·0	3	18·0
9·2	3	27·6
10·5	2	21·0
11·0	3	33·0
11·0	3	33·0
10·2	2	20·4
8·3	3	24·9
5·1	3	15·3
0	1	0
		194·8 = ΣA

$$\text{Common interval} = \frac{180}{9}$$

$$= 20 \text{ m}$$

$$\text{Waterplane area} = \frac{3}{8} h\, \Sigma A \times 2$$

$$= \frac{3}{8} \times 20 \times 194{\cdot}8 \times 2$$

$$= 2922 \text{ m}^2$$

SOLUTIONS TO TEST EXAMPLES 4

1.

Mass	Kg	Vertical moment
4000	6·0	24 000
1000	0·8	800
200	1·0	200
5000	5·0	25 000
3000	9·5	28 500
13 200		78 500

Thus displacement = 13 200 tonne

$$\text{Centre of gravity above keel} = \frac{78\ 500}{13\ 200}$$

$$= 5{\cdot}947 \text{ m}$$

2.

Mass	Kg	Vertical moment	LCG from midships	Longitudinal moment
5000	6·0	30 000	1·5 ford	7 500 ford
500	10·0	5 000	36·0 aft	18 000 aft
5500		35 000		10 500 aft

$$\text{Centre of gravity above keel} = \frac{35\ 000}{5500}$$

$$= 6{\cdot}364 \text{ m}$$

$$\text{Centre of gravity from midships} = \frac{10\ 500}{5500}$$

$$= 1{\cdot}909 \text{ m aft}$$

3. $$\text{Shift in centre of gravity} = \frac{300 \times (24 + 40)}{6000}$$

$$= 3{\cdot}2 \text{ m aft}$$

$$\text{New position of centre of gravity} = 1{\cdot}2 - 3{\cdot}2$$

$$= -\ 2{\cdot}0 \text{ m}$$

$$\text{or} = 2{\cdot}0 \text{ m aft of midships}$$

4. (a) Shift in centre of gravity due to transfer of oil

$$= \frac{250 \times (75 + 50)}{17\,000}$$

$$= 1{\cdot}839 \text{ m aft}$$

$\therefore$ new position of centre of gravity $= 1{\cdot}0 + 1{\cdot}839$
$= 2{\cdot}839$ m aft of midships

(b) Taking moments about midships:

$$\text{new position of centre of gravity} = \frac{17\,000 \times 2{\cdot}839 - 200 \times 50}{17\,000 - 200}$$

$$= \frac{48\,260 - 10\,000}{16\,800}$$

$$= 2{\cdot}278 \text{ m aft of midships}$$

5. Shift in centre of gravity due to lowered cargo

$$= \frac{500 \times 3}{3000}$$

$$= 0{\cdot}5 \text{ m down}$$

Taking moments about the new position of the centre of gravity:
Shift in centre of gravity due to added cargo

$$= \frac{3000 \times 0 + 500 \times 3{\cdot}5}{3000 + 500}$$

$$= \frac{1750}{3500}$$

$$= 0{\cdot}5 \text{ m up}$$

i.e. the position of the centre of gravity does not change.

6.

	Mass	*Kg*	Vertical moment
	2000	1·5	3000
	300	4·5	1350
	50	6·0	300
Total removed	2350		4650
Original	10 000	3·0	30 000
Final	7650		25 350

$$\text{New displacement} = 7650 \text{ tonne}$$

$$\text{New centre of gravity} = \frac{25\,350}{7650}$$

$$= 3{\cdot}313 \text{ m above keel}$$

7.

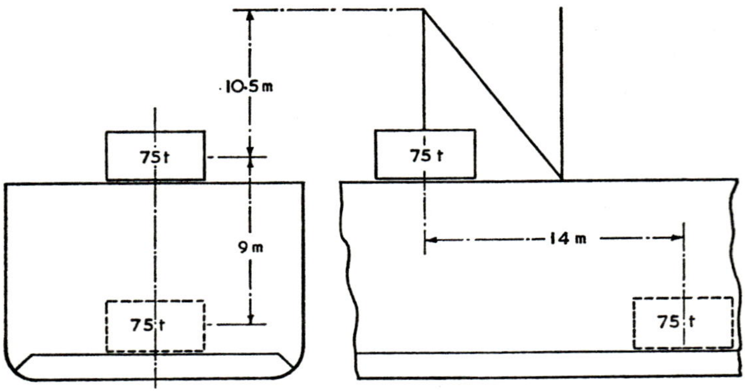

Fig. 92

(a) Since the centre of gravity of a suspended mass is at the point of suspension, the mass is virtually raised to the derrick head.

$$\text{Rise in centre of gravity} = \frac{75 \times 10{\cdot}5}{8000}$$

$$= 0{\cdot}0984 \text{ m}$$

(b) When the mass is at the derrick head there is no further movement of the centre of gravity.
i.e. rise in centre of gravity $= 0{\cdot}0984$ m

(c) When the mass is in its final position the centre of gravity moves down and forward.

$$\text{Vertical shift in centre of gravity} = \frac{75 \times 9}{8000}$$

$$= 0{\cdot}0844 \text{ m down}$$

$$\text{Longitudinal shift in centre of gravity} = \frac{75 \times 14}{8000}$$

$$= 0{\cdot}1313 \text{ m forward}$$

SOLUTIONS TO TEST EXAMPLES 5

1.
$$GM = KB + BM - KG$$
$$BM = \frac{I}{\nabla}$$
$$= \frac{42{\cdot}5 \times 10^3 \times 1{\cdot}025}{12\,000}$$
$$= 3{\cdot}630 \text{ m}$$
$$GM = 3{\cdot}60 + 3{\cdot}63 - 6{\cdot}50$$
$$= 0{\cdot}73 \text{ m}$$

2.
$$BM = \frac{I}{\nabla}$$
$$= \frac{60 \times 10^3 \times 1{\cdot}025}{10\,000}$$
$$= 6{\cdot}150 \text{ m}$$
$$KG = \frac{4000 \times 6{\cdot}30 + 2000 \times 7{\cdot}50 + 4000 \times 9{\cdot}15}{4000 + 2000 + 4000}$$
$$= \frac{25\,200 + 15\,000 + 36\,600}{10\,000}$$
$$= \frac{76\,800}{10\,000}$$
$$= 7{\cdot}680 \text{ m}$$
$$GM = 2{\cdot}750 + 6{\cdot}150 - 7{\cdot}680$$
$$= 1{\cdot}22 \text{ m}$$

3.

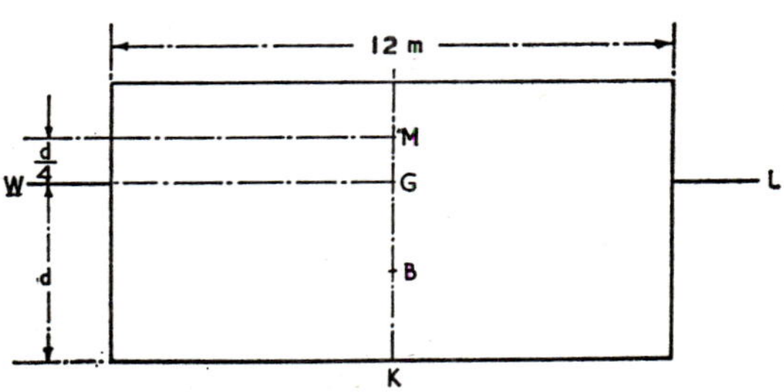

Fig. 93

Let d = draught

Then
$$KB = \frac{d}{2}$$
$$BM = \frac{B^2}{12d}$$
$$= \frac{12}{d}$$
$$KG = d$$
$$GM = \frac{d}{4}$$
$$KG + GM = KB + BM$$
$$d + \frac{d}{4} = \frac{d}{2} + \frac{12}{d}$$
$$\tfrac{3}{4}d = \frac{12}{d}$$
$$d^2 = \frac{4}{3} \times 12$$
$$= 16$$

$\therefore$ Draught $d = 4$ m

4.

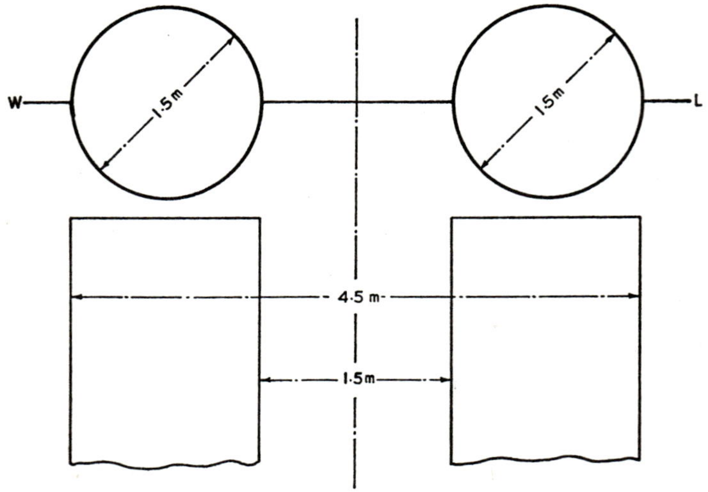

Fig. 94

Although the raft is formed by cylinders, the waterplane consists of two rectangles, a distance of 1·5 m apart.

Second moment of area of waterplane about centreline

$$I = \frac{1}{12} \times 6 \times 4{\cdot}5^3 - \frac{1}{12} \times 6 \times 1{\cdot}5^3$$

$$= \frac{1}{12} \times 6\,(4{\cdot}5^3 - 1{\cdot}5^3)$$

$$= \tfrac{1}{2} \times 87{\cdot}75$$

$$= 43{\cdot}875 \text{ m}^4$$

$$\text{Volume of displacement } \nabla = 2 \times 6 \times \frac{\pi}{4} \times 1{\cdot}5^2 \times \tfrac{1}{2}$$

$$= 10{\cdot}603 \text{ m}^3$$

$$BM = \frac{I}{\nabla}$$

$$= \frac{43{\cdot}875}{10{\cdot}603}$$

$$= 4{\cdot}138 \text{ m}$$

5.

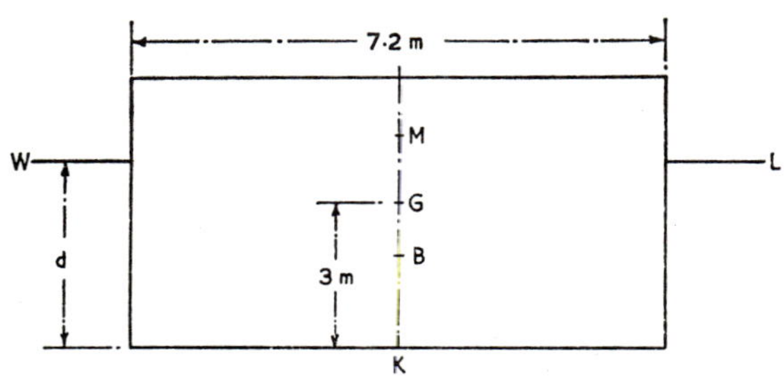

Fig. 95

(a)

$$\text{Let } d = \text{draught}$$

$$KB = \frac{d}{2}$$

$$BM = \frac{B^2}{12d}$$

$$= \frac{4{\cdot}32}{d}$$

d	*KB*	*BM*	*KM*
0	0	∞	∞
0·5	0·25	8·640	8·890
1·0	0·50	4·320	4·820
1·5	0·75	2·880	3·630
2·0	1·00	2·160	3·160
2·5	1·25	1·728	2·998
3·0	1·50	1·440	2·940
3·5	1·75	1·234	2·984
4·0	2·00	1·080	3·080

(b) Between draughts of 2·4 m and 3·6 m the vessel will be unstable.

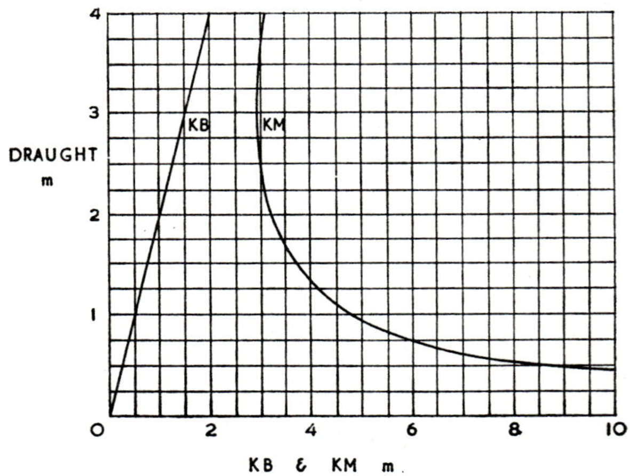

Fig. 96

6.

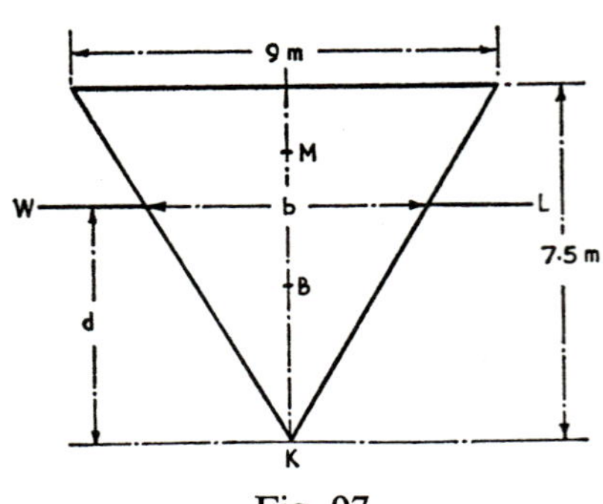

Fig. 97

Let d = draught

b = breadth at waterline

By similar triangles

$$b = \frac{9}{7 \cdot 5} \times d$$

$$= 1 \cdot 2d$$

$$KB = \tfrac{2}{3}d$$

$$BM = \frac{b^2}{6d}$$

$$= \frac{(1 \cdot 2d)^2}{6d}$$

$$= 0 \cdot 24d$$

d	*KB*	*BM*	*KM*
0	0	0	0
0·5	0·333	0·120	0·453
1·0	0·667	0·240	0·907
1·5	1·000	0·360	1·360
2·0	1·333	0·480	1·813
2·5	1·667	0·600	2·267
3·0	2·000	0·720	2·720

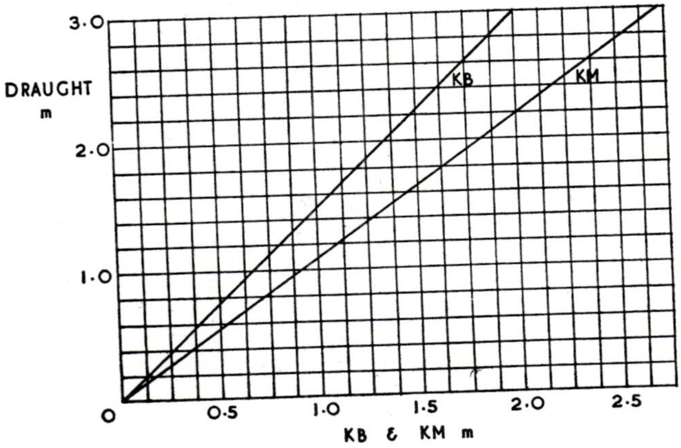

Fig. 98

7.
$$GM = \frac{m \times d}{\triangle \tan \theta}$$

$$\tan \theta = \frac{\text{deflection of pendulum}}{\text{length of pendulum}}$$

$$= \frac{0{\cdot}110}{8{\cdot}50}$$

$$GM = \frac{10 \times 14 \times 8{\cdot}50}{8000 \times 0{\cdot}110}$$

$$= 1{\cdot}352 \text{ m}$$
$$KG = KM - GM$$
$$= 7{\cdot}150 - 1{\cdot}352$$
$$= 5{\cdot}798 \text{ m}$$

8. Mean deflection

$$= \tfrac{1}{8}(81 + 78 + 85 + 83 + 79 + 82 + 84 + 80)$$
$$= 81{\cdot}5 \text{ mm}$$

$$GM = \frac{6 \times 13{\cdot}5 \times 7{\cdot}5}{4000 \times 0{\cdot}0815}$$

$$= 1{\cdot}863 \text{ m}$$

9. Virtual reduction in GM due to free surface

$$= \frac{\rho_1 i}{\rho \nabla}$$

$$= \frac{i}{\nabla} \qquad \text{since } \rho = \rho_1$$

$$i = \frac{1}{12} \times 12 \times 9^3$$

$$= 729$$

$$\nabla = \frac{5000}{1{\cdot}025}$$

$$\text{Free surface effect} = \frac{729 \times 1{\cdot}025}{5000}$$

$$= 0{\cdot}149 \text{ m}$$

10. (a)
$$GM = KM - KG$$
$$= 5{\cdot}00 - 4{\cdot}50$$
$$= 0{\cdot}50 \text{ m}$$

But
$$GM = \frac{m \times d}{\triangle \tan \theta}$$

$$\therefore \quad \tan \theta = \frac{m \times d}{\triangle \times GM}$$

$$= \frac{10 \times 12}{8000 \times 0{\cdot}50}$$

$$= 0{\cdot}030$$

Angle of heel
$$\theta = 1° \, 43'$$

(b)
$$\text{Free surface effect} = \frac{i}{\nabla}$$

$$= \frac{7{\cdot}5 \times 15^3 \times 1{\cdot}025}{12 \times 8000}$$

$$= 0{\cdot}270 \text{ m}$$

Virtual
$$GM = 0{\cdot}50 - 0{\cdot}27$$
$$= 0{\cdot}23 \text{ m}$$

$$\tan \theta = \frac{10 \times 12}{8000 \times 0{\cdot}23}$$

$$= 0{\cdot}0652$$

Angle of heel
$$\theta = 3° \, 44'$$

11.

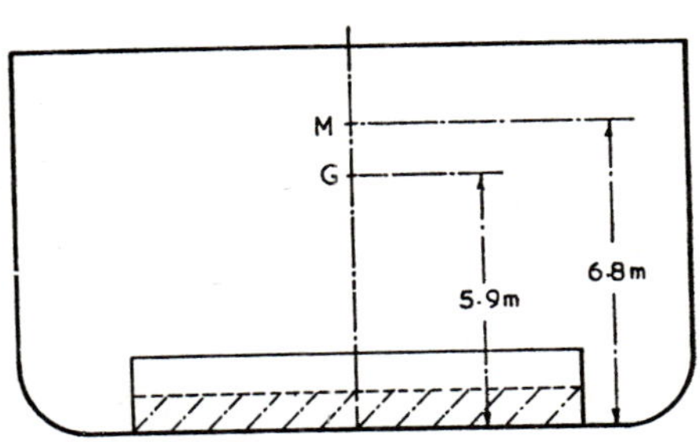

Fig. 99

$$\text{Mass of water added} = 10.5 \times 12 \times 0.6 \times 1.025$$
$$= 77.49 \text{ tonne}$$

$$\text{New } KG = \frac{6000 \times 5.9 + 77.49 \times 0.3}{6000 + 77.49}$$

$$= \frac{35\,400 + 23.25}{6077.49}$$

$$= 5.829 \text{ m}$$

$$\text{Free surface effect} = \frac{\rho i}{\Delta}$$

$$= \frac{1.025 \times 10.5 \times 12^3}{12 \times 6077.49}$$

$$= 0.255 \text{ m}$$

$$\text{Virtual } KG = 5.829 + 0.255$$

$$= 6.084 \text{ m}$$

$$GM = 6.80 - 6.084$$

$$= 0.716 \text{ m}$$

12(a)

$$\text{Free surface effect} = \frac{\rho i}{\Delta}$$

$$= \frac{0.8 \times 9 \times 10 \times 24^3}{12 \times 25\,000}$$

$$= 3.226 \text{ m}$$

(b) With centreline bulkhead, free surface effect is reduced to one quarter of (a).

$$\text{Free surface effect} = \tfrac{1}{4} \times 3{\cdot}226$$
$$= 0{\cdot}807 \text{ m}$$

(c) With twin longitudinal bulkheads and equal tanks, free surface effect is reduced to one ninth of (a).

$$\text{Free surface effect} = \tfrac{1}{9} \times 3{\cdot}226$$
$$= 0{\cdot}358 \text{ m}$$

(d) The only change is in the second moment of area.

For centre tank $$i = \frac{9 \times 10 \times 12^3}{12}$$

$$= \frac{9 \times 10}{12} \times 1728$$

For wing tanks $$i = \frac{9 \times 10 \times 6^3}{12} \times 2$$

$$= \frac{9 \times 10}{12} \times 432$$

Thus for both tanks $$i = \frac{9 \times 10}{12}(1728 + 432)$$

$$= \frac{30}{4} \times 2160$$

Total free surface effect $$= \frac{0.8 \times 30 \times 2160}{4 \times 25\,000}$$

$$= 0{\cdot}518 \text{ m}$$

13.

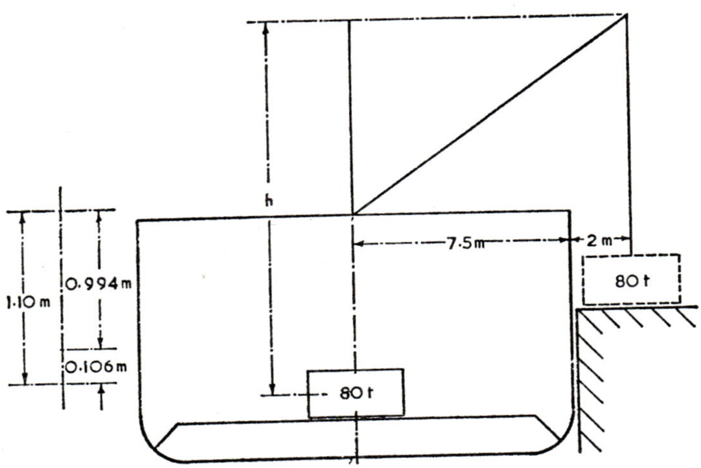

Fig. 100

$$GM = \frac{m \times d}{\triangle \tan \theta}$$

$$= \frac{80 \times 9{\cdot}5}{12\,500 \times \tan 3{\cdot}5°}$$

$$= 0{\cdot}994 \text{ m}$$

Thus the metacentric height is reduced from 1·10 m to 0·994 m when the mass is suspended over the quay. Since the draught does not alter, and hence the transverse metacentre remains in the same position, this reduction in metacentric height must be due to a rise in the centre of gravity. This rise is due to the effect of the suspended mass.

$$\text{Rise in centre of gravity} = 1{\cdot}10 - 0{\cdot}994$$
$$= 0{\cdot}106 \text{ m}$$

Let h be the distance from the centre of gravity of the mass to the derrick head.

Then $\text{rise in centre of gravity} = \dfrac{\text{mass} \times h}{\text{displacement}}$

$$0{\cdot}106 = \frac{80 \times h}{12\,500}$$

$$h = \frac{0{\cdot}106 \times 12\,500}{80}$$

$$= 16{\cdot}56 \text{ m}$$

14.

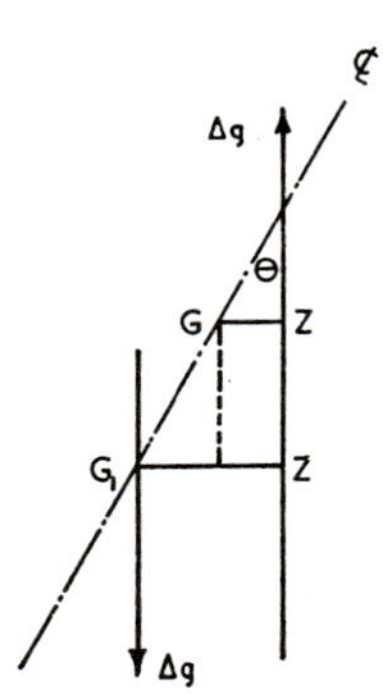

Fig. 101

Since the actual centre of gravity G_1 is *below* the assumed centre of gravity G, the ship is *more* stable, and

$$G_1Z = GZ + GG_1 \sin \theta$$
$$GG_1 = 3{\cdot}50 - 3{\cdot}00$$
$$= 0{\cdot}50 \text{ m}$$

θ	sin θ	GG_1 sin θ	GZ	G_1Z
0	0	0	0	0
15°	0·2588	0·129	0·25	0·379
30°	0·500	0·250	0·46	0·710
45°	0·7071	0·353	0·51	0·863
60°	0·8660	0·433	0·39	0·823
75°	0·9659	0·483	0·10	0·583
90°	1·000	0·500	−0·38	0·120

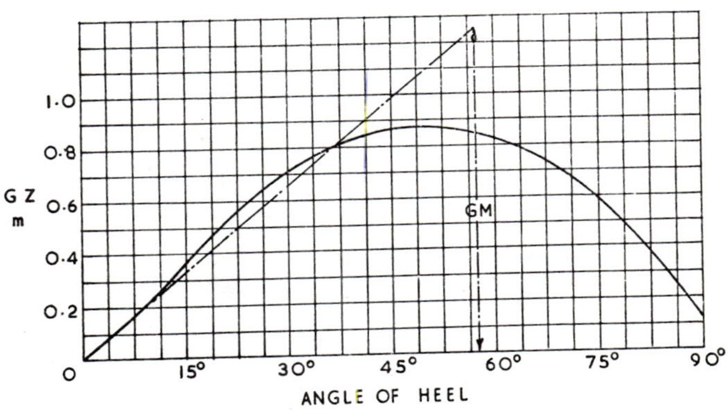

Fig. 102

15. The dynamical stability of a ship to any given angle is represented by the area under the righting moment curve to that angle.

θ	Righting moment	SM	Product
0	0	1	0
15°	1690	4	6760
30°	5430	2	10 860
45°	9360	4	37 440
60°	9140	1	9140
			64 200

$$\text{Common interval} = 15°$$

$$= \frac{15}{57 \cdot 3} \text{ radians}$$

$$\text{Dynamical stability} = \frac{1}{3} \times \frac{15}{57 \cdot 3} \times 64\,200$$

$$= 5602 \text{ kJ}$$

16. Wall-sided formula $GZ = \sin\theta\,(GM + \tfrac{1}{2}BM\tan^2\theta)$

$$BM = \frac{I}{\nabla}$$

$$= \frac{82 \times 10^3 \times 1 \cdot 025}{18\,000}$$

$$= 4 \cdot 67 \text{ m}$$

$$GM = KB + BM - KG$$

$$= 5 \cdot 25 + 4 \cdot 67 - 9 \cdot 24$$

$$= 0 \cdot 68 \text{ m}$$

θ	$\tan\theta$	$\tan^2\theta$	$\frac{1}{2}BM\tan^2\theta$	$GM+\frac{1}{2}BM\tan^2\theta$	$\sin\theta$	GZ
0	0·0	0·0	—	0·68	0·0	0
5°	0·0875	0·00766	0·018	0·698	0·0872	0·061
10°	0·1763	0·03109	0·073	0·753	0·1736	0·131
15°	0·2680	0·07180	0·168	0·848	0·2588	0·219
20°	0·3640	0·13247	0·309	0·989	0·3420	0·338

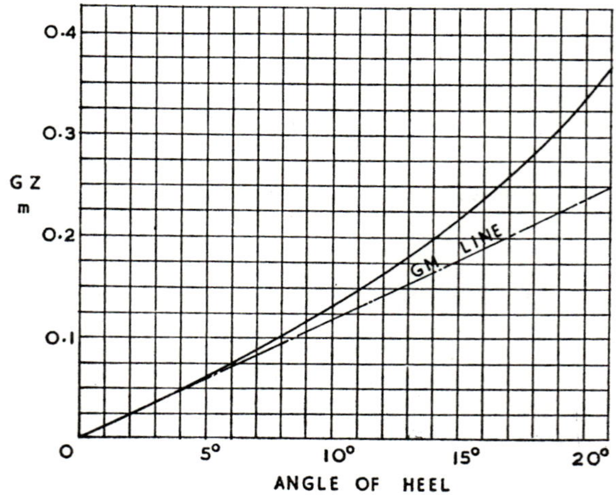

Fig. 103

17. $$\text{New } KG = \frac{7200 \times 5{\cdot}20 - 300 \times 0{\cdot}60}{7200 - 300}$$

$$= \frac{37\,260}{6900}$$

$$\begin{aligned} &= 5{\cdot}40 \text{ m} \\ GM &= KM - KG \\ &= 5{\cdot}35 - 5{\cdot}40 \\ &= -\ 0{\cdot}05 \text{ m} \\ BM &= KM - KB \\ &= 5{\cdot}35 - 3{\cdot}12 \\ &= 2{\cdot}23 \text{ m} \end{aligned}$$

$$\tan\theta = \pm\sqrt{\frac{-2GM}{BM}}$$

$$= \pm\sqrt{\frac{0{\cdot}10}{2{\cdot}23}}$$

$$= \pm\, 0{\cdot}2118$$

Angle of loll $\theta = \pm\, 11^\circ\, 57'$

SOLUTIONS TO TEST EXAMPLES 6

1. (a)
$$\text{Change in trim} = \frac{\text{trimming moment}}{\text{MCT1 cm}}$$

$$\therefore \quad \text{MCT1 cm} = \frac{100 \times 75}{65}$$

$$= 115{\cdot}4 \text{ tonne m}$$

(b)
$$\text{MCT1 cm} = \frac{\triangle \times GM_L}{100L}$$

$$\therefore \quad GM_L = \frac{115{\cdot}4 \times 100 \times 125}{12\,000}$$

$$= 120{\cdot}2 \text{ m}$$

(c)
$$GG_1 = \frac{m \times d}{\triangle}$$

$$= \frac{100 \times 75}{12\,000}$$

$$= 0{\cdot}625 \text{ m}$$

2.
$$\text{Bodily sinkage} = \frac{\text{mass added}}{\text{TPC}}$$

$$= \frac{110}{3}$$

$$\text{Change in trim} = \frac{110 \times (24 + 2{\cdot}5)}{80}$$

$$= 36{\cdot}44 \text{ cm by the stern}$$

$$\text{Change forward} = -\frac{36{\cdot}44}{120}\left(\frac{120}{2} - 2{\cdot}5\right)$$

$$= -\ 17{\cdot}46 \text{ cm}$$

$$\text{Change aft} = +\frac{36\cdot44}{120}\left(\frac{120}{2} + 2\cdot5\right)$$

$$\begin{aligned}
&= +\ 18\cdot98 \text{ cm} \\
\text{New draught forward} &= 5\cdot50 + 0\cdot085 - 0\cdot175 \\
&= 5\cdot410 \text{ m} \\
\text{New draught aft} &= 5\cdot80 + 0\cdot085 + 0\cdot190 \\
&= 6\cdot075 \text{ m}
\end{aligned}$$

3.

$$\text{Bodily rise} = \frac{180}{18}$$

$$= 10 \text{ cm}$$

$$\text{MCT1 cm} = \frac{14\,000 \times 125}{100 \times 130}$$

$$= 134\cdot6 \text{ tonne m}$$

$$\text{Change in trim} = -\ \frac{180 \times (40 - 3)}{134\cdot6}$$

$$\begin{aligned}
&= -\ 49\cdot48 \text{ cm by the stern} \\
&= +\ 49\cdot48 \text{ cm by the head}
\end{aligned}$$

$$\text{Change forward} = +\ \frac{49\cdot48}{130}\left(\frac{130}{2} + 3\right)$$

$$= +\ 25\cdot88 \text{ cm}$$

$$\text{Change aft} = -\ \frac{49\cdot48}{130}\left(\frac{130}{2} - 3\right)$$

$$\begin{aligned}
&= -\ 23\cdot60 \text{ cm} \\
\text{New draught forward} &= 7\cdot50 - 0\cdot10 + 0\cdot259 \\
&= 7\cdot659 \text{ cm} \\
\text{New draught aft} &= 8\cdot10 - 0\cdot10 - 0\cdot236 \\
&= 7\cdot764 \text{ cm}
\end{aligned}$$

4. If the draught aft remains constant, the reduction in draught aft due to change in trim must equal the bodily sinkage.

$$\text{Bodily sinkage} = \frac{180}{11}$$

$$= 16{\cdot}36 \text{ cm}$$

$$\text{Change in trim aft} = -\frac{t}{L}\text{WF}$$

$$16{\cdot}36 = \frac{t}{90}\left(\frac{90}{2} - 2\right)$$

$$t = \frac{16{\cdot}36 \times 90}{43}$$

$$= 34{\cdot}24 \text{ cm by the head}$$

But $$\text{Change in trim } t = \frac{m \times d}{\text{MCT1 cm}}$$

$$34{\cdot}24 = \frac{180 \times d}{50}$$

$$d = \frac{34{\cdot}24 \times 50}{180}$$

$$= 9{\cdot}511 \text{ m}$$

Thus the mass must be placed 9·511 m forward of the centre of flotation or 7·511 m forward of midships.

$$\text{Change in trim forward} = +\frac{34{\cdot}24}{90}\left(\frac{90}{2} + 2\right)$$

$$= 17{\cdot}88 \text{ cm}$$

$$\text{New draught forward} = 5{\cdot}80 + 0{\cdot}164 + 0{\cdot}179$$

$$= 6{\cdot}143 \text{ m}$$

5. Change in trim required $= 8{\cdot}90 - 8{\cdot}20$
$= 0{\cdot}70$ m
$= 70$ cm by the head

$$\therefore \quad 70 = \frac{m \times d}{\text{MCT1 cm}}$$

$$= \frac{m \times (60 + 1{\cdot}5)}{260}$$

$$m = \frac{70 \times 260}{61{\cdot}5}$$

$$= 296 \text{ tonne}$$

$$\text{Bodily sinkage} = \frac{296}{28}$$

$$= 10{\cdot}57 \text{ cm}$$

$$\text{Change forward} = + \frac{70}{150}\left(\frac{150}{2} + 1{\cdot}5\right)$$

$$= + 35{\cdot}70 \text{ cm}$$

$$\text{Change aft} = - \frac{70}{150}\left(\frac{150}{2} - 1{\cdot}5\right)$$

$$= - 34{\cdot}30 \text{ cm}$$

New draught forward $= 8{\cdot}20 + 0{\cdot}106 + 0{\cdot}357$
$= 8{\cdot}663$ m

New draught aft $= 8{\cdot}90 + 0{\cdot}106 - 0{\cdot}343$
$= 8{\cdot}663$ m

(It is not necessary to calculate both draughts but this method checks the calculation.)

6.

Mass	distance from F	moment forward	moment aft
20	41·5A	—	830
50	24·5A	—	1225
30	3·5A	—	105
70	4·5F	315	—
15	28·5F	427·5	—
185		742·5	2160

$$\text{Bodily sinkage} = \frac{185}{9}$$

$$= 20{\cdot}55 \text{ cm}$$

$$\text{Excess moment aft} = 2160 - 742{\cdot}5$$

$$= 1417{\cdot}5 \text{ tonne m}$$

$$\text{Change in trim} = \frac{1417{\cdot}5}{55}$$

$$= 25{\cdot}77 \text{ cm by the stern}$$

$$\text{Change forward} = -\frac{25{\cdot}77}{110}\left(\frac{110}{2} - 1{\cdot}5\right)$$

$$= -\ 12{\cdot}53 \text{ cm}$$

$$\text{Change aft} = +\frac{25{\cdot}77}{110}\left(\frac{110}{2} + 1{\cdot}5\right)$$

$$= +\ 13{\cdot}24 \text{ cm}$$

$$\text{New draught forward} = 4{\cdot}20 + 0{\cdot}205 - 0{\cdot}125$$

$$= 4{\cdot}280 \text{ m}$$

$$\text{New draught aft} = 4{\cdot}45 + 0{\cdot}205 + 0{\cdot}132$$

$$= 4{\cdot}787 \text{ m}$$

7.

Mass	distance from F	moment forward	moment aft
+160	66·5A	—	+10 640
+200	23·5F	+4700	—
−120	78·5A	—	− 9420
− 70	19·5A	—	− 1365
+170		+4700	− 145

$$\text{Bodily sinkage} = \frac{170}{28}$$

$$= 6{\cdot}07 \text{ cm}$$

$$\text{Excess moment forward} = 4700 - (-145)$$
$$= 4845 \text{ tonne m}$$

$$\text{Change in trim} = \frac{4845}{300}$$

$$= 16{\cdot}15 \text{ cm by the head}$$

$$\text{Change forward} = \frac{16{\cdot}15}{170}\left(\frac{170}{2} - 3{\cdot}5\right)$$

$$= +7{\cdot}74 \text{ cm}$$

$$\text{Change aft} = -\frac{16{\cdot}15}{170}\left(\frac{170}{2} + 3{\cdot}5\right)$$

$$= -8{\cdot}41 \text{ cm}$$

$$\text{New draught forward} = 6{\cdot}85 + 0{\cdot}061 + 0{\cdot}077$$
$$= 6{\cdot}988 \text{ m}$$

$$\text{New draught aft} = 7{\cdot}50 + 0{\cdot}061 - 0{\cdot}084$$
$$= 7{\cdot}477 \text{ m}$$

8.

Item	mass	Lcg	moment forward	moment aft
Lightweight	1050	4·64A	—	4872·0
Cargo	2150	4·71F	10 126·5	—
Fuel	80	32·55A	—	2604·0
Water	15	32·90A	—	493·5
Stores	5	33·60F	168·0	—
	3300		10 294·5	7969·5

$$\text{Excess moment forward} = 10\,294{\cdot}5 - 7969{\cdot}5$$
$$= 2325 \text{ tonne m}$$

$$\text{LCG of loaded ship} = \frac{2325}{3300}$$

$$= 0{\cdot}704 \text{ m forward of midships}$$

The mean draught, MCT1 cm, LCB and LCF may be found by interpolation from the tabulated values.

Displacement difference 4·50 m to 5·00 m

$$= 3533 - 3172$$
$$= 361 \text{ tonne}$$

$$\text{Actual difference required} = 3300 - 3172$$
$$= 128 \text{ tonne}$$

$$\text{Proportion of draught difference} = \frac{128}{361}$$

$$\text{Actual draught difference} = 0{\cdot}50 \times \frac{128}{361}$$

$$= 0{\cdot}50 \times 0{\cdot}355$$
$$= 0{\cdot}177 \text{ m}$$

$$\therefore \quad \text{mean draught} = 4{\cdot}50 + 0{\cdot}177$$
$$= 4{\cdot}677 \text{ m}$$

MCT1 cm difference 4·50 m to 5·00 m

$$= 43{\cdot}10 - 41{\cdot}26$$
$$= 1{\cdot}84 \text{ tonne m}$$

$$\text{Actual difference} = 1{\cdot}84 \times 0{\cdot}355$$
$$= 0{\cdot}65 \text{ tonne m}$$

$$\therefore \quad \text{MCT1 cm} = 41{\cdot}26 + 0{\cdot}65$$
$$= 41{\cdot}91 \text{ tonne m}$$

$$\text{LCB difference } 4{\cdot}50 \text{ m to } 5{\cdot}00 \text{ m} = -\,0{\cdot}24 \text{ m}$$

$$\text{Actual difference} = -\,0{\cdot}24 \times 0{\cdot}355$$
$$= -\,0{\cdot}085$$

$$\therefore \quad \text{LCB} = 1{\cdot}24 - 0{\cdot}085$$
$$= 1{\cdot}155 \text{ m forward of midships}$$

$$\text{LCF difference } 4{\cdot}50 \text{ m to } 5{\cdot}00 \text{ m} = 0{\cdot}43 \text{ m}$$

$$\text{Actual difference} = 0{\cdot}43 \times 0{\cdot}355$$
$$= 0{\cdot}153$$

$$\therefore \quad \text{LCF} = 0{\cdot}84 + 0{\cdot}153$$
$$= 0{\cdot}993 \text{ m aft of midships}$$

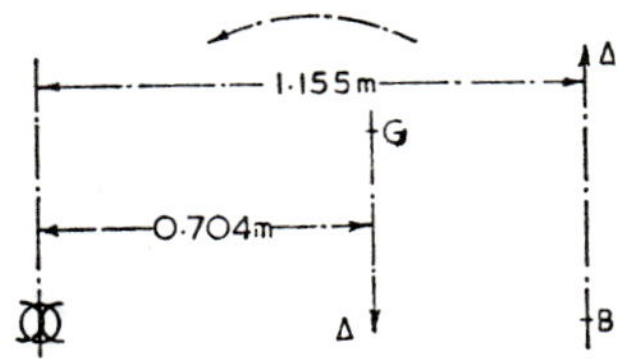

Fig. 104

$$\text{Trimming lever} = 1{\cdot}155 - 0{\cdot}704$$
$$= 0{\cdot}451 \text{ m by the stern}$$

$$\text{Trimming moment} = 3300 \times 0{\cdot}451$$

$$\text{Trim} = \frac{3300 \times 0{\cdot}451}{41{\cdot}91}$$

$$= 35{\cdot}51 \text{ cm by the stern}$$

$$\text{Change forward} = -\frac{35{\cdot}51}{80}\left(\frac{80}{2} + 0{\cdot}993\right)$$

$$= -\ 18{\cdot}20 \text{ cm}$$

$$\text{Change aft} = +\frac{35{\cdot}51}{80}\left(\frac{80}{2} - 0{\cdot}993\right)$$

$$= +\ 17{\cdot}31 \text{ cm}$$

$$\text{Draught forward} = 4{\cdot}677 - 0{\cdot}182$$
$$= 4{\cdot}495 \text{ m}$$

$$\text{Draught aft} = 4{\cdot}677 + 0{\cdot}173$$
$$= 4{\cdot}850 \text{ m}$$

9. Change in mean draught $= \frac{100\Delta}{A_w}\left(\frac{\rho_S - \rho_R}{\rho_S \times \rho_R}\right)$

$$= \frac{100 \times 15\,000}{1950}\left(\frac{1{\cdot}022 - 1{\cdot}005}{1{\cdot}005 \times 1{\cdot}022}\right)$$

$$= 12{\cdot}73 \text{ cm}$$

Since the vessel moves from river water into sea water, the draught will be *reduced.*

10. Let ρ_S = density of sea water in kg/m³

$$10 = \frac{7000 \times 100}{1500}\left(\frac{\rho_S - 1005}{1005 \times \rho_S}\right) \times 1000$$

$$\frac{\rho_S - 1005}{1005\rho_S} = \frac{10 \times 1500}{7000 \times 100 \times 1000}$$

$$\rho_S - 1005 = \frac{10 \times 1500 \times 1005\rho_S}{7000 \times 100 \times 1000}$$

$$= 0{\cdot}02153\ \rho_S$$

$$\rho_S(1 - 0{\cdot}02153) = 1005$$

$$\rho_S = \frac{1005}{0{\cdot}97847}$$

$$= 1027 \text{ kg/m}^3$$

11.

½ ordinate	SM	product
0	1	0
2·61	4	10·44
3·68	2	7·36
4·74	4	18·96
5·84	2	11·68
7·00	4	28·00
7·30	2	14·60
6·47	4	25·88
5·35	2	10·70
4·26	4	17·04
3·16	2	6·32
1·88	4	7·52
0	1	0
		158·50

$$\text{Common interval} = \frac{90}{12}\text{ m}$$

$$\text{Waterplane area} = \frac{2}{3} \times \frac{90}{12} \times 158{\cdot}50$$

$$= 792{\cdot}5\text{ m}^2$$

(a) $$\text{TPC} = \frac{792{\cdot}5 \times 1{\cdot}024}{100}$$

$$= 8{\cdot}115$$

(b) $$\text{Mass required} = 12 \times 8{\cdot}115$$
$$= 97{\cdot}38\text{ tonne}$$

(c) $$\text{Change in mean draught} = \frac{8200 \times 100}{792{\cdot}5}\left(\frac{1{\cdot}024 - 1{\cdot}005}{1{\cdot}005 \times 1{\cdot}024}\right)$$

$$= 19{\cdot}10\text{ cm increase}$$

12. Let $\triangle$ = displacement in sea water

Then $(\triangle - 360)$ = displacement in fresh water

Volume of displacement in sea water

$$= \frac{\triangle}{1{\cdot}025}\ \text{m}^3$$

Volume of displacement in fresh water

$$= \frac{\triangle - 360}{1{\cdot}000}\ \text{m}^3$$

Since the draught remains constant, these two volumes must be equal

$$\frac{\triangle - 360}{1{\cdot}000} = \frac{\triangle}{1{\cdot}025}$$

$$1{\cdot}025\,\triangle - 1{\cdot}025 \times 360 = \triangle$$

$$0{\cdot}025\triangle = 1{\cdot}025 \times 360$$

$$\triangle = \frac{1{\cdot}025 \times 360}{0{\cdot}025}$$

Displacement in sea water = 14 760 tonne

13. Change in mean draught $= \dfrac{5200 \times 100}{1100}\left(\dfrac{1{\cdot}023 - 1{\cdot}002}{1{\cdot}002 \times 1{\cdot}023}\right)$

$= 9{\cdot}68$ cm increase

$$\text{MCT1 cm} = \frac{5200 \times 95}{100 \times 90}$$

$= 54{\cdot}88$ tonne nm

$$FB = 0{\cdot}6 + 2{\cdot}2$$
$$= 2{\cdot}8\ \text{m}$$

$$\text{Change in trim} = \frac{\triangle \times FB\,(\rho_S - \rho_R)}{\text{MCT1 cm} \times \rho_S}$$

$$= \frac{5200 \times 2{\cdot}8\,(1{\cdot}023 - 1{\cdot}002)}{54{\cdot}88 \times 1{\cdot}023}$$

$= 5{\cdot}45$ cm by the head

$$\text{Change forward} = +\frac{5 \cdot 45}{90}\left(\frac{90}{2} + 2 \cdot 2\right)$$

$$= + 2 \cdot 86 \text{ cm}$$

$$\text{Change aft} = -\frac{5 \cdot 45}{90}\left(\frac{90}{2} - 2 \cdot 2\right)$$

$$= - 2 \cdot 59 \text{ cm}$$

$$\text{New draught forward} = 4 \cdot 95 + 0 \cdot 097 + 0 \cdot 029$$
$$= 5 \cdot 076 \text{ cm}$$

$$\text{New draught aft} = 5 \cdot 35 + 0 \cdot 097 - 0 \cdot 026$$
$$= 5 \cdot 421 \text{ cm}$$

14. $$\text{Change in mean draught} = \frac{22\,000 \times 100}{3060}\left(\frac{1 \cdot 026 - 1 \cdot 007}{1 \cdot 007 \times 1 \cdot 026}\right)$$

$$= 13 \cdot 22 \text{ cm reduction}$$

$$\text{Change in trim} = \frac{22\,000 \times 3\,(1 \cdot 026 - 1 \cdot 007)}{280 \times 1 \cdot 007}$$

$$= 4 \cdot 45 \text{ cm by the stern}$$

$$\text{Change forward} = -\frac{4 \cdot 45}{160}\left(\frac{160}{2} + 4\right)$$

$$= - 2 \cdot 34 \text{ cm}$$

$$\text{Change aft} = +\frac{4 \cdot 45}{160}\left(\frac{160}{2} - 4\right)$$

$$= + 2 \cdot 11 \text{ cm}$$

$$\text{New draught forward} = 8 \cdot 15 - 0 \cdot 132 - 0 \cdot 023$$
$$= 7 \cdot 995 \text{ m}$$

$$\text{New draught aft} = 8 \cdot 75 - 0 \cdot 132 + 0 \cdot 021$$
$$= 8 \cdot 639 \text{ m}$$

15. (a)

$$\text{Volume of lost buoyancy} = 12 \times 10 \times 4 \text{ m}^3$$

$$\text{Area of intact waterplane} = (60 - 12) \times 10 = 48 \times 10 \text{ m}^2$$

$$\text{Increase in draught} = \frac{12 \times 10 \times 4}{48 \times 10} = 1 \text{ m}$$

$$\therefore \quad \text{New draught} = 4 + 1 = 5 \text{ m}$$

(b)

$$\text{Volume of lost buoyancy} = 0{\cdot}85 \times 12 \times 10 \times 4 \text{ m}^3$$

$$\text{Area of intact waterplane} = (60 - 0{\cdot}85 \times 12) \times 10 = 49{\cdot}8 \times 10 \text{ m}^2$$

$$\text{Increase in draught} = \frac{0{\cdot}85 \times 12 \times 10 \times 4}{49{\cdot}8 \times 10} = 0{\cdot}819 \text{ m}$$

$$\therefore \quad \text{New draught} = 4 + 0{\cdot}819 = 4{\cdot}819 \text{ m}$$

(c)

$$\text{Volume of lost buoyancy} = 0{\cdot}60 \times 12 \times 10 \times 4 \text{ m}^3$$

$$\text{Area of intact waterplane} = (60 - 0{\cdot}60 \times 12) \times 10 = 52{\cdot}8 \times 10 \text{ m}^2$$

$$\text{Increase in draught} = \frac{0{\cdot}60 \times 12 \times 10 \times 4}{52{\cdot}8 \times 10} = 0{\cdot}545 \text{ m}$$

$$\therefore \quad \text{New draught} = 4 + 0{\cdot}545 = 4{\cdot}545 \text{ m}$$

16. Density of coal $= 1{\cdot}000 \times 1{\cdot}28$
$= 1{\cdot}28$ t/m³

Volume of 1 tonne of solid coal

$$= \frac{1}{1{\cdot}28}$$

$$= 0{\cdot}781 \text{ m}^3$$

Volume of 1 tonne of stowed coal
$= 1{\cdot}22$ m³

$\therefore$ in every $1{\cdot}22$ m³ of volume, $0{\cdot}439$ m³ is available for water

Hence Permeability $\mu = \dfrac{0{\cdot}439}{1{\cdot}22}$

$$= 0{\cdot}3598$$

$$\text{Increase in draught} = \frac{0{\cdot}3598 \times 9 \times 8 \times 3}{(50 - 0{\cdot}3598 \times 9) \times 8}$$

$$= 0{\cdot}208 \text{ m}$$

$\therefore$ New draught $= 3 + 0{\cdot}208$
$= 3{\cdot}208$ m

17.

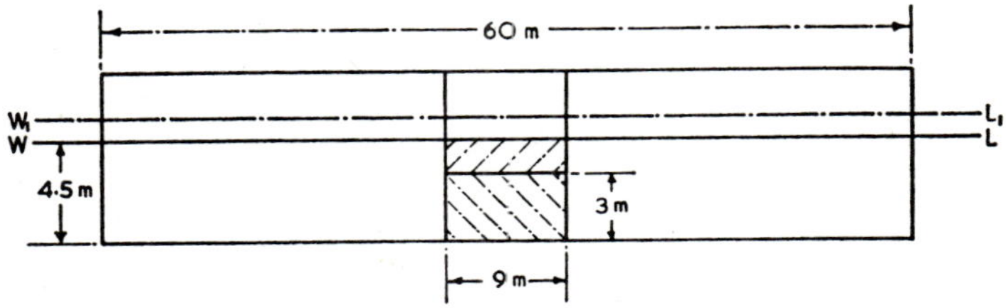

Fig. 105

(a) Volume of lost buoyancy $= 9 \times 12 \times 3 \text{ m}^3$

Area of intact waterplane $= 60 \times 12 \text{ m}^2$
(Since the water is restricted at the flat, the whole of the waterplane is intact.)

$$\text{Increase in draught} = \frac{9 \times 12 \times 3}{60 \times 12}$$

$$= 0{\cdot}45 \text{ m}$$

$$\text{New draught} = 4{\cdot}5 + 0{\cdot}45$$
$$= 4{\cdot}95 \text{ m}$$

(b) Volume of lost buoyancy $= 9 \times 12 \times (4{\cdot}5 - 3) \text{ m}^3$

Area of intact waterplane $= (60 - 9) \times 12 \text{ m}^2$

$$\text{Increase in draught} = \frac{9 \times 12 \times 1{\cdot}5}{51 \times 12}$$

$$= 0{\cdot}265 \text{ m}$$

$$\text{New draught} = 4{\cdot}5 + 0{\cdot}265$$
$$= 4{\cdot}765 \text{ m}$$

18. Mass of barge and teak $= 25 \times 4 \times 1{\cdot}2 \times 1{\cdot}000$
$= 120$ tonne

Mass of teak $= 25 \times 4 \times 0{\cdot}120 \times 0{\cdot}805$
$= 9{\cdot}66$ tonne

$\therefore$ Mass of barge $= 120 - 9{\cdot}66$
$= 110{\cdot}34$ tonne

If it is assumed that the teak is first removed and then the compartment bilged:

$$\text{Draught when teak removed} = \frac{110{\cdot}34}{1{\cdot}000 \times 25 \times 4}$$

$$= 1{\cdot}1034$$

Increase in draught due to bilging

$$= \frac{5 \times 4 \times 1{\cdot}1034}{(25 - 5)\,4}$$

$$= 0{\cdot}2758 \text{ m}$$

$$\text{Final draught} = 1{\cdot}1034 + 0{\cdot}2758$$
$$= 1{\cdot}379 \text{ m}$$

19. It has been convenient to consider this as a three-part question, but any *one* part could constitute an examination question.

(a)

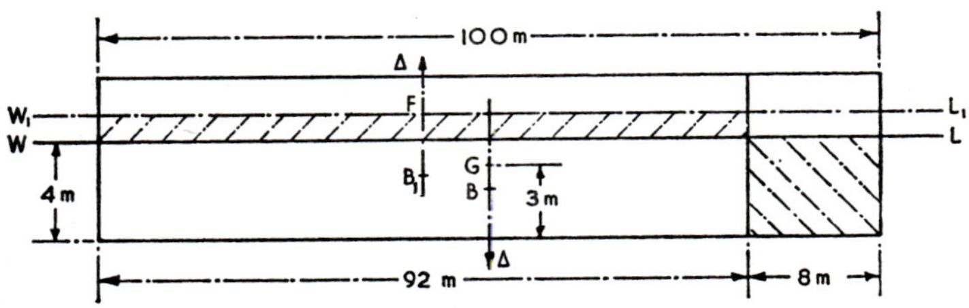

Fig. 106

$$\text{Increase in mean draught} = \frac{8 \times 12 \times 4}{(100 - 8) \times 12}$$

$$= 0{\cdot}348 \text{ m}$$

$$\text{New mean draught} = 4 + 0{\cdot}348$$
$$= 4{\cdot}348$$

$$KB_1 = \frac{4{\cdot}348}{2}$$

$$= 2{\cdot}174$$

$$B_1M_L = \frac{92^2}{12 \times 4{\cdot}348}$$

$$= 162{\cdot}22$$

$$GM_L = 2{\cdot}17 + 162{\cdot}22 - 3{\cdot}00$$
$$= 161{\cdot}39$$

$$\text{Change in trim} = \frac{50\,L\,l}{GM_L}$$

$$= \frac{50 \times 100 \times 8}{161{\cdot}39}$$

$$= 247{\cdot}8 \text{ cm by the head}$$

$$\text{Change forward} = +\frac{247{\cdot}8}{100}\left(\frac{100}{2} + 4\right)$$

$$= +\ 133{\cdot}8 \text{ cm}$$

$$\text{Change aft} = -\ \frac{247{\cdot}8}{100}\left(\frac{100}{2} - 4\right)$$

$$= 114{\cdot}0 \text{ cm}$$

$$\text{New draught forward} = 4{\cdot}348 + 1{\cdot}338$$
$$= 5{\cdot}686 \text{ m}$$

$$\text{New draught aft} = 4{\cdot}348 - 1{\cdot}140$$
$$= 3{\cdot}208 \text{ m}$$

(b)

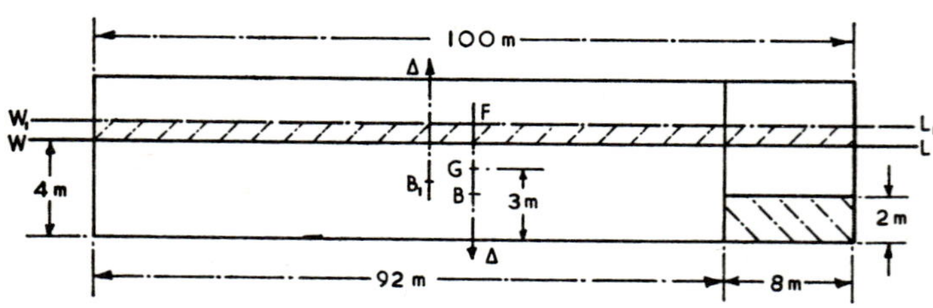

Fig. 107

$$\text{Volume of lost buoyancy} = 8 \times 12 \times 2 \text{ m}^3$$

$$\text{Area of intact waterplane} = 100 \times 12 \text{ m}^2$$

$$\text{Increase in draught} = \frac{8 \times 12 \times 2}{100 \times 12}$$

$$= 0{\cdot}16 \text{ m}$$

$$\text{New draught} = 4{\cdot}16 \text{ m}$$

$$KB_1 = 2{\cdot}08 \text{ m (approx.)}$$

$$I_F = \frac{1}{12} \mathrm{L}^3 \mathrm{B}$$

$$= \frac{1}{12} \times 100^3 \times 12$$

$$= 1{\cdot}0 \times 10^6 \text{ m}^4$$

$$\bigtriangledown = L\,B\,d$$

$$= 100 \times 12 \times 4$$

$$B_1M_L = \frac{1{\cdot}0 \times 10^6}{100 \times 12 \times 4}$$

$$= 208{\cdot}33 \text{ m}$$

$$GM_L = 2{\cdot}08 + 208{\cdot}33 - 3{\cdot}00$$

$$= 207{\cdot}41 \text{ m}$$

To find the longitudinal shift in the centre of buoyancy, consider the volume of lost buoyancy moved to the centre of the intact waterplane, which in this case is midships.

$$BB_1 = \frac{8 \times 12 \times 2 \times (50 - 4)}{100 \times 12 \times 4}$$

$$= 1{\cdot}84 \text{ m}$$

$$\text{Trimming moment} = \triangle . BB_1$$

$$\text{MCT1 cm} = \frac{\triangle \times GM_L}{100\,L}$$

$$= \frac{\triangle \times 209{\cdot}49}{100 \times 100}$$

$$\text{Change in trim} = \frac{1{\cdot}84 \times 100 \times 100}{207{\cdot}41}$$

$$= 88{\cdot}71 \text{ cm by the head}$$

$$\text{Change forward} = + \frac{88 \cdot 71}{100} \times 50$$

$$= + 44 \cdot 36 \text{ cm}$$

$$\text{Change aft} = - \frac{88 \cdot 71}{100} \times 50$$

$$= - 44 \cdot 36 \text{ cm}$$

$$\text{New draught forward} = 4 \cdot 16 + 0 \cdot 444$$
$$= 4 \cdot 604 \text{ m}$$

$$\text{New draught aft} = 4 \cdot 16 - 0 \cdot 444$$
$$= 3 \cdot 716 \text{ m}$$

(c)

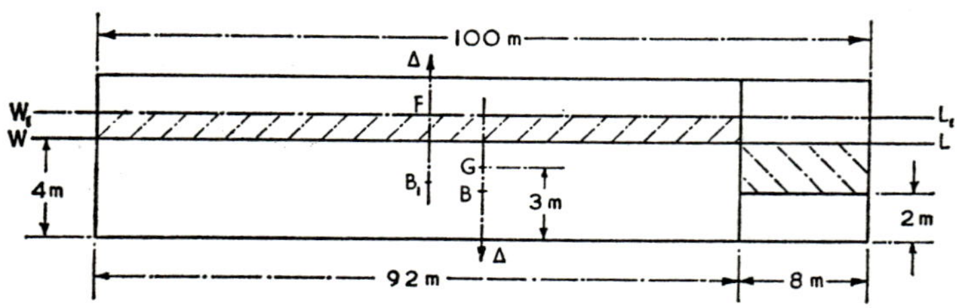

Fig. 108

$$\text{Volume of lost buoyancy} = 8 \times 12 \times 2 \text{ m}^3$$

$$\text{Area of intact waterplane} = (100 - 8) \times 12 \text{ m}^2$$

$$\text{Increase in draught} = \frac{8 \times 12 \times 2}{(100 - 8) \times 12}$$

$$= 0 \cdot 174 \text{ m}$$

$$\text{New draught} = 4 \cdot 174 \text{ m}$$

$$KB_1 = 2 \cdot 087 \text{ m}$$

$$I_F = \frac{1}{12} \times 92^3 \times 12$$

$$= 778\ 688$$
$$\nabla = 100 \times 12 \times 4$$

$$B_1M_L = \frac{778\ 688}{4800}$$

$$= 162{\cdot}22 \text{ m}$$

$$GM_L = 2{\cdot}09 + 162{\cdot}22 - 3{\cdot}00$$
$$= 161{\cdot}31 \text{ m}$$

$$\text{Shift in centre of buoyancy } BB_1 = \frac{8 \times 12 \times 2 \times (46 + 4)}{100 \times 12 \times 4}$$

$$= 2{\cdot}00 \text{ m}$$

$$\text{Change in trim} = \frac{2{\cdot}00 \times 100 \times 100}{161{\cdot}31}$$

$$= 124 \text{ cm by the head}$$

$$\text{Change forward} = +\frac{124}{100}\left(\frac{100}{2} + 4\right)$$

$$= +\ 67 \text{ cm}$$

$$\text{Change aft} = -\frac{124}{100}\left(\frac{100}{2} - 4\right)$$

$$= -\ 57 \text{ cm}$$

$$\text{New draught forward} = 4{\cdot}174 + 0{\cdot}670$$
$$= 4{\cdot}844 \text{ m}$$

$$\text{New draught aft} = 4{\cdot}174 - 0{\cdot}570$$
$$= 3{\cdot}604 \text{ m}$$

SOLUTIONS TO TEST EXAMPLES 7

1.
$$\begin{aligned} R_f &= f\,S\,V^n \\ &= 0{\cdot}424 \times 3200 \times 17^{1\cdot 825} \\ &= 238\ 900\ \text{N} \\ &= 238{\cdot}9\ \text{kN} \\ \text{Power} &= R_f \times v \\ &= 238{\cdot}9 \times 17 \times \frac{1852}{3600} \\ &= 208{\cdot}93\ \text{kW} \end{aligned}$$

2. At 3 m/s
$$R_f = 13\ \text{N/m}^2$$

At 15 knots
$$\begin{aligned} R_f &= 13 \times \left(\frac{15}{3} \times \frac{1852}{3600}\right)^{1\cdot 97} \\ &= 83{\cdot}605\ \ \text{N/m}^2 \end{aligned}$$

$\therefore$
$$\begin{aligned} R_f &= 83{\cdot}605 \times 3800 \\ &= 317\ 700\ \text{N} \\ \text{Power} &= 317\ 700 \times 15 \times \frac{1852}{3600} \\ &= 2\ 451\ 500\ \text{W} \\ &= 2451{\cdot}5\ \text{kW} \end{aligned}$$

3. At 180 m/min
$$R_f = 12\ \text{N/m}^2$$

At 14 knots
$$\begin{aligned} R_f &= 12 \times \left(\frac{14}{180} \times \frac{1852}{60}\right)^{1\cdot 9} \\ &= 63{\cdot}36\ \text{N/m}^2 \end{aligned}$$

$\therefore$
$$\begin{aligned} R_f &= 63{\cdot}36 \times 4000 \\ &= 253\ 400\ \text{N} \\ R_t &= \frac{R_f}{0{\cdot}7} \\ &= \frac{253\ 400}{0{\cdot}7} \\ &= 362\ 000\ \text{N} \end{aligned}$$

$$\text{Effective power} = R_t \times v$$

$$= 362\,000 \times 14 \times \frac{1852}{3600}$$

$$= 2\,547\,800 \text{ W}$$
$$= 2547{\cdot}8 \text{ kW}$$

4. Wetted surface area $S = c\sqrt{\triangle L}$

$$= 2{\cdot}55\sqrt{125 \times 16 \times 7{\cdot}8 \times 0{\cdot}72 \times 1{\cdot}025 \times 125}$$

$$= 3059 \text{ m}^2$$

$$R_f = f\,S\,V^n$$
$$= 0{\cdot}423 \times 3059 \times 17{\cdot}5^{1{\cdot}825}$$
$$= 240\,140 \text{ N}$$

$$\text{Power} = 240\,140 \times 17{\cdot}5 \times \frac{1852}{3600}$$

$$= 2\,161\,900 \text{ W}$$
$$= 2161{\cdot}9 \text{ kW}$$

5.

$$R_r \propto L^3$$

$$\therefore \quad R_r = 36 \times \left(\frac{20}{1}\right)^3$$

$$= 288\,000 \text{ N}$$

$$V \propto \sqrt{L}$$

$$\therefore \quad V = 3 \times \sqrt{\frac{20}{1}}$$

$$= 13{\cdot}417 \text{ knots}$$

$$\text{Power} = 288\,000 \times 13{\cdot}417 \times \frac{1852}{3600}$$

$$= 1\,987\,800 \text{ W}$$
$$= 1987{\cdot}8 \text{ kW}$$

6.
$$V \propto \sqrt{L}$$
$$\propto \triangle^{\frac{1}{6}}$$

$$\therefore \quad V = 16 \times \left(\frac{24\,000}{14\,000}\right)^{\frac{1}{6}}$$
$$= 17{\cdot}503 \text{ knots}$$

$$R_r \propto \triangle$$
$$\therefore \quad R_r = 113 \times \left(\frac{24\,000}{14\,000}\right)$$
$$= 193{\cdot}7 \text{ kN}$$

7. Let v = speed in m/s

At 3 m/s
$$R_f = 11 \times 1{\cdot}025$$
$$= 11{\cdot}275 \text{ N/m}^2 \text{ in sea water}$$

$$\therefore \quad R_f = 11{\cdot}275 \times 2500$$
$$= 28\,190 \text{ N}$$

At v m/s
$$R_f = 28{\cdot}19 \left(\frac{v}{3}\right)^{1{\cdot}92} \text{ kN}$$

and
$$R_t = \frac{R_f}{0{\cdot}72}$$

$$\text{effective power} = R_t \times v$$
$$= 1100 \text{ kW}$$

$$\frac{R_f}{0{\cdot}72} \times v = 1100$$

$$\frac{28{\cdot}19}{0{\cdot}72}\left(\frac{v}{3}\right)^{1{\cdot}92} \times v = 1100$$

$$v^{2{\cdot}92} = 1100 \times \frac{0{\cdot}72}{28{\cdot}19} \times 3^{1{\cdot}92}$$

$$v = 6{\cdot}454 \text{ m/s}$$

$$V = 6{\cdot}454 \times \frac{3600}{1852}$$

$$= 12{\cdot}545 \text{ knots}$$

8. Model:

$$R_t = 35 \text{ N in fresh water}$$
$$= 35 \times 1{\cdot}025$$
$$= 35{\cdot}875 \text{ N}$$

$$f = 0{\cdot}417 + \frac{0{\cdot}773}{6 + 2{\cdot}862}$$

$$= 0{\cdot}417 + 0{\cdot}0872$$
$$= 0{\cdot}5042$$

$$R_f = 0{\cdot}5042 \times 7 \times 3^{1{\cdot}825}$$
$$= 26{\cdot}208 \text{ N}$$

$$R_r = R_t - R_f$$
$$= 35{\cdot}875 - 26{\cdot}208$$
$$= 9{\cdot}667 \text{ N}$$

Ship:

$$R_r \propto L^3$$

$$\therefore \quad R_r = 9{\cdot}667 \times \left(\frac{120}{6}\right)^3$$

$$= 77\,336 \text{ N}$$

$$S \propto L^2$$

$$\therefore \quad S = 7 \times \left(\frac{120}{6}\right)^2$$

$$= 2800 \text{ m}^2$$

$$V \propto \sqrt{L}$$

$$\therefore \quad V = 3\sqrt{\frac{120}{6}}$$

$$= 13{\cdot}416 \text{ knots}$$

$$f = 0{\cdot}417 + \frac{0{\cdot}773}{120 + 2{\cdot}862}$$

$$= 0{\cdot}417 + 0{\cdot}0063$$
$$= 0{\cdot}4233$$

$$R_f = 0{\cdot}4233 \times 2800 \times 13{\cdot}416^{1{\cdot}825}$$
$$= 135\,400 \text{ N}$$

$$R_t = 77\,336 + 135\,400$$
$$= 212\,736 \text{ N}$$

$$\text{Effective power (naked)} = 212\,736 \times 13{\cdot}416 \times \frac{1852}{3600}$$

$$= 1468{\cdot}2 \text{ kW}$$

$$\begin{aligned}\text{Effective power} &= \text{ep}_n \times \text{SCF}\\ &= 1468{\cdot}2 \times 1{\cdot}15\\ &= 1688{\cdot}4 \text{ kW}\end{aligned}$$

9.

$$\text{sp} = \frac{\Delta^{\frac{2}{3}}\, V^3}{C}$$

$$= \frac{12\,000^{\frac{2}{3}} \times 16^3}{550}$$

$$= 3903 \text{ kW}$$

10. (a)

$$2800 = \frac{\Delta^{\frac{2}{3}} \times 14^3}{520}$$

$$\Delta^{\frac{2}{3}} = \frac{2800 \times 520}{14^3}$$

$$\Delta = 12\,223 \text{ tonne}$$

(b)

$$\begin{aligned}\text{New speed} &= 0{\cdot}85V\\ &= 0{\cdot}85 \times 14\end{aligned}$$

$$\text{sp} \propto V^3$$

$$\therefore \quad \frac{\text{sp}_1}{\text{sp}_2} = \left(\frac{V_1}{V_2}\right)^3$$

$$\text{sp}_2 = 2800 \times \left(\frac{0{\cdot}85 \times 14}{14}\right)^3$$

$$= 1720 \text{ kW}$$

11.

$$2100 = \frac{8000^{\frac{2}{3}} \times V^3}{470}$$

$$V^3 = \frac{2100 \times 470}{8000^{\frac{2}{3}}}$$

$$V = 13{\cdot}51 \text{ knots}$$

12. (a)
$$\triangle = 150 \times 19 \times 8 \times 0{\cdot}68 \times 1{\cdot}025$$
$$= 15\,890 \text{ tonne}$$

$$\text{sp} = \frac{15\,890^{\frac{2}{3}} \times 18^3}{600}$$

$$= 6143 \text{ kW}$$

(b)
$$\text{sp} \propto R_t \times V$$
$$\propto V^3 \times V$$
$$\propto V^4$$

$$\therefore \quad \text{sp} = 6143 \times \left(\frac{21}{18}\right)^4$$

$$= 11\,382 \text{ kW}$$

Note: In practice there will be a gradual increase in the index of speed.

13. Fuel cons/day
$$= \frac{\triangle^{\frac{2}{3}} V^3}{\text{fuel coefficient}} \text{ tonne}$$

$$= \frac{15\,000^{\frac{2}{3}} \times 14{\cdot}5^3}{62\,500}$$

$$= 29{\cdot}67 \text{ tonne}$$

14.
$$25 = \frac{9000^{\frac{2}{3}} V^3}{53\,500}$$

$$V^3 = \frac{25 \times 53\,500}{9000^{\frac{2}{3}}}$$

$$V = 14{\cdot}57 \text{ knots}$$

15. At 16 knots: time taken
$$= \frac{2000}{16 \times 24}$$

$$= 5{\cdot}209 \text{ days}$$

$$\text{total consumption} = 28 \times 5{\cdot}209$$
$$= 145{\cdot}8 \text{ tonne}$$

But
$$\text{total consumption} \propto V^2$$

$\therefore$ at 14 knots, total consumption

$$= 145 \cdot 8 \times \left(\frac{14}{16}\right)^2$$

$$= 111 \cdot 6 \text{ tonne}$$

$$\therefore \text{Saving in fuel} = 145 \cdot 8 - 111 \cdot 6$$
$$= 34 \cdot 2 \text{ tonne}$$

16. $$\text{Total fuel used} = 115 - 20$$
$$= 95 \text{ tonne}$$

At 15 knots; total consumption for 1100 nautical miles

$$= 40 \times \frac{1100}{15 \times 24}$$

$$= 122 \cdot 3 \text{ tonne}$$

But total consumption $\propto V^2$

$$\frac{95}{122 \cdot 3} = \frac{V^2}{15^2}$$

$$V = 15 \sqrt{\frac{95}{122 \cdot 3}}$$

$$= 13 \cdot 22 \text{ knots}$$

$$\text{Time taken} = \frac{1100}{13 \cdot 22}$$

$$= 83 \cdot 18 \text{ hours}$$

17 (a) $$\frac{\text{cons}_1}{\text{cons}_2} = \left(\frac{V_1}{V_2}\right)^3$$

$$V_2^3 = V_1^3 \left(\frac{\text{cons}_2}{\text{cons}_1}\right)$$

$$V_2 = V_1 \sqrt[3]{\frac{\text{cons}_2}{\text{cons}_1}}$$

$$= 14 \sqrt[3]{\frac{1 \cdot 15 \times 23}{23}}$$

$$= 14 \cdot 67 \text{ knots}$$

(b) $$V_2 = 14 \sqrt[3]{\frac{0{\cdot}88 \times 23}{23}}$$

$$= 13{\cdot}42 \text{ knots}$$

(c) $$V_2 = 14 \sqrt[3]{\frac{18}{23}}$$

$$= 12{\cdot}90 \text{ knots}$$

18. Fuel consumption/hour $= 0{\cdot}12 + 0{\cdot}001 V^3$ tonne
$= 0{\cdot}12 + 0{\cdot}001 \times 14^3$
$= 2{\cdot}864$ tonne

(a) Over 1700 nautical miles;

$$\text{total fuel consumption} = 2{\cdot}864 \times \frac{1700}{14}$$

$$= 347{\cdot}7 \text{ tonne}$$

(b) Saving in fuel $= 10$ t/day

$$= \frac{10}{24} \text{ t/h}$$

$$\therefore \text{ new fuel consumption} = 2{\cdot}864 - \frac{10}{24}$$

$$= 2{\cdot}447 \text{ t/h}$$

$$2{\cdot}447 = 0{\cdot}12 + 0{\cdot}001\, V^3$$

$$0{\cdot}001\, V^3 = 2{\cdot}447 - 0{\cdot}12$$

$$V^3 = 2327$$

$$V = 13{\cdot}25 \text{ knots}$$

19. Let C = normal consumption per hour
V = normal speed.

For first 8 hours:

$$\text{speed} = 1{\cdot}2V$$

$$\text{cons/h} = C \times \left(\frac{1{\cdot}2V}{V}\right)^3$$

$$= 1{\cdot}728C$$

$$\text{cons for 8 hours} = 8 \times 1{\cdot}728C$$

$$= 13{\cdot}824C$$

For next 10 hours:

$$\text{speed} = 0{\cdot}9V$$

$$\text{cons/h} = C \times \left(\frac{0{\cdot}9V}{V}\right)^3$$

$$= 0{\cdot}729C$$

$$\text{cons for 10 hours} = 10 \times 0{\cdot}729C$$
$$= 7{\cdot}29C$$

For remaining 6 hours:

$$\text{speed} = V$$

$$\text{cons for 6 hours} = 6C$$

$$\text{Total cons for 24 hours} = 13{\cdot}824C + 7{\cdot}29C + 6C$$
$$= 27{\cdot}114C$$

$$\text{Normal cons for 24 hours} = 24C$$

$$\therefore \quad \text{Increase in cons} = 27{\cdot}114C - 24C$$
$$= 3{\cdot}114C$$

$$\%\ \text{increase in cons} = \frac{3{\cdot}114C}{24C} \times 100$$

$$= 12{\cdot}97\%$$

20. Let C = consumption per day at 18 knots

Then $C - 22$ = consumption per day at 14·5 knots

$$\frac{C}{C-22} = \left(\frac{18}{14{\cdot}5}\right)^3$$

$$= 1{\cdot}913$$

$$C = 1{\cdot}913C - 1{\cdot}913 \times 22$$

$$0{\cdot}913C = 1{\cdot}913 \times 22$$

$$C = \frac{1{\cdot}913 \times 22}{0{\cdot}913}$$

$$= 46{\cdot}09 \text{ tonne/day}$$

21. At 17 knots: cons/day = 42 tonne

At V knots: cons/day = 28 tonne

But at V knots: cons/day = $1{\cdot}18C$

Where $$C = 42 \times \left(\frac{V}{17}\right)^3$$

$$1{\cdot}18C = 28$$

$$\therefore \quad 28 = 1{\cdot}18 \times 42 \times \left(\frac{V}{17}\right)^3$$

$$V^3 = \frac{17^3 \times 28}{1{\cdot}18 \times 42}$$

$$V = 17\sqrt[3]{\frac{28}{1{\cdot}18 \times 42}}$$

$$= 14{\cdot}06 \text{ knots}$$

SOLUTIONS TO TEST EXAMPLES 8

1. Theoretical speed $V_T = \dfrac{P \times N \times 60}{1852}$

$$= \frac{5 \times 105 \times 60}{1852}$$

$$= 17{\cdot}01 \text{ knots}$$

Apparent slip $= \dfrac{V_T - V}{V_T} \times 100$

$$= \frac{17{\cdot}01 - 14}{17{\cdot}01} \times 100$$

$$= 17{\cdot}70\%$$

Speed of advance $V_a = V(1 - w)$

$$= 14(1 - 0{\cdot}35)$$

$$= 9{\cdot}10 \text{ knots}$$

Real slip $= \dfrac{V_T - V_a}{V_T} \times 100$

$$= \frac{17{\cdot}01 - 9{\cdot}10}{17{\cdot}01} \times 100$$

$$= 46{\cdot}50\%$$

2. $$p = \frac{P}{D}$$

$\therefore$ Pitch $P = p \times D$

$$= 0{\cdot}8 \times 5{\cdot}5$$

$$= 4{\cdot}4 \text{ m}$$

Theoretical speed $V_T = \dfrac{4{\cdot}4 \times 120 \times 60}{1852}$

$$= 17{\cdot}11 \text{ knots}$$

$$\text{Real slip} = \frac{V_T - V_a}{V_T} \times 100$$

$$0{\cdot}35 = \frac{17{\cdot}11 - V_a}{17{\cdot}11}$$

$$\therefore \text{Speed of advance } V_a = 17{\cdot}11\,(1 - 0{\cdot}35)$$
$$= 11{\cdot}12 \text{ knots}$$

But $$V_a = V(1 - w)$$

$$\therefore \quad \text{Ship speed } V = \frac{V_a}{1 - w}$$

$$= \frac{11{\cdot}12}{1 - 0{\cdot}32}$$

$$= 16{\cdot}35 \text{ knots}$$

$$\text{Apparent slip} = \frac{V_T - V}{V_T} \times 100$$

$$= \frac{17{\cdot}11 - 16{\cdot}35}{17{\cdot}11} \times 100$$

$$= 4{\cdot}44\%$$

3. $$C_b = \frac{\triangle}{L \times B \times d \times \rho}$$

$$= \frac{12\,400}{120 \times 17{\cdot}5 \times 7{\cdot}5 \times 1{\cdot}025}$$

$$= 0{\cdot}768$$

$$w = 0{\cdot}5 \times 0{\cdot}768 - 0{\cdot}05$$
$$= 0{\cdot}334$$

$$\text{Speed of advance } V_a = V(1 - w)$$
$$= 12\,(1 - 0{\cdot}334)$$
$$= 7{\cdot}992 \text{ knots}$$

$$\text{Real slip} = \frac{V_T - V_a}{V_T} \times 100$$

$$0{\cdot}30 V_T = V_T - 7{\cdot}992$$

$$V_T = \frac{7{\cdot}992}{0{\cdot}7}$$

$$= 11{\cdot}42 \text{ knots}$$

But $$V_T = \frac{P \times N \times 60}{1852}$$

$\therefore$ Pitch $$P = \frac{11{\cdot}42 \times 1852}{100 \times 60}$$

$$= 3{\cdot}52 \text{ m}$$

$$\text{Diameter } D = \frac{P}{p}$$

$$= \frac{3{\cdot}52}{0{\cdot}75}$$

$$= 4{\cdot}70 \text{ m}$$

$$\text{Apparent slip} = \frac{11{\cdot}42 - 12}{11{\cdot}42} \times 100$$

$$= -5{\cdot}08\%$$

4. $$\text{Theoretical speed } V_T = \frac{4{\cdot}8 \times 110 \times 60}{1852}$$

$$= 17{\cdot}11 \text{ knots}$$

Apparent slip $$-s = \frac{V_T - V}{V_T} \times 100 \quad \ldots (1)$$

Real slip $$+1{\cdot}5s = \frac{V_T - V_a}{V_T} \times 100$$

$$V_a = V(1 - w)$$
$$= 0{\cdot}75V$$

$$\therefore \text{ Real slip} \qquad + 1{\cdot}5s = \frac{V_T - 0{\cdot}75V}{V_T} \times 100 \qquad \ldots(2)$$

Multiply (1) by 1·5

$$- 1{\cdot}5s = 1{\cdot}5 \times \frac{V_T - V}{V_T} \times 100 \qquad \ldots(3)$$

Adding (2) and (3)

$$\frac{V_T - 0{\cdot}75V}{V_T} \times 100 + 1{\cdot}5 \times \frac{V_T - V}{V_T} \times 100 = 0$$

Hence

$$V_T - 0{\cdot}75V + 1{\cdot}5V_T - 1{\cdot}5V = 0$$

$$2{\cdot}25V = 2{\cdot}5V_T$$

$$V = \frac{2{\cdot}5}{2{\cdot}25} \times 17{\cdot}11$$

$$= 19{\cdot}01 \text{ knots}$$

Substitute for V in (1)

$$- s = \frac{17{\cdot}11 - 19{\cdot}01}{17{\cdot}11} \times 100$$

$$\text{Apparent slip} = - 11{\cdot}10\%$$

$$\text{Real slip} = - 1{\cdot}5 \times (- 11{\cdot}10)$$
$$= + 16{\cdot}66\%$$

5. $$\text{Theoretical speed } V_T = \frac{4{\cdot}3 \times 95 \times 60}{1852}$$

$$= 13{\cdot}23 \text{ knots}$$

$$\text{Real slip } 0{\cdot}28 = \frac{13{\cdot}23 - V_a}{13{\cdot}23}$$

$$V_a = 13{\cdot}23\,(1 - 0{\cdot}28)$$
$$= 9{\cdot}53 \text{ knots}$$

$$\text{Effective disc area } A = \frac{\pi}{4}(D^2 - d^2)$$

$$= \frac{\pi}{4}(4{\cdot}6^2 - 0{\cdot}75^2)$$

$$= 16{\cdot}18 \text{ m}^2$$

$$\text{Thrust } T = \rho\, A\, s\, P^2\, n^2$$

$$= 1{\cdot}025 \times 16{\cdot}18 \times 0{\cdot}28 \times 4{\cdot}3^2 \times \left(\frac{95}{60}\right)^2$$

$$= 215{\cdot}2 \text{ kN}$$

$$\text{Thrust power} = T \times v_a$$

$$= 215{\cdot}2 \times 9{\cdot}53 \times \frac{1852}{3600}$$

$$= 1055 \text{ kW}$$

6.

$$T_1\, N_1 = T_2\, N_2$$

$$17{\cdot}5 \times 115 = T_2 \times 90$$

$$T_2 = \frac{17{\cdot}5 \times 115}{90}$$

$$\text{Thrust pressure} = 22{\cdot}36 \text{ b}$$

7.

$$\frac{tp_1}{tp_2} = \frac{T_1\, V_1}{T_2\, V_2}$$

$$T_2 = \frac{T_1\, V_1\, tp_2}{V_2\, tp_1}$$

$$= 19{\cdot}5 \times \frac{V_1}{0{\cdot}88 V_1} \times \frac{2900}{3400}$$

$$\text{Thrust pressure} = 18{\cdot}9 \text{ b}$$

8.

$$\text{tp} = 2550 \text{ kW}$$

$$\text{dp} = \frac{\text{tp}}{\text{propeller efficiency}}$$

$$= \frac{2550}{0{\cdot}65}$$

$$\text{sp} = \frac{\text{dp}}{\text{transmission efficiency}}$$

$$= \frac{2550}{0{\cdot}65 \times 0{\cdot}94}$$

$$\text{ip} = \frac{\text{sp}}{\text{mechanical efficiency}}$$

$$= \frac{2550}{0{\cdot}65 \times 0{\cdot}94 \times 0{\cdot}83}$$

$$\text{Indicated power} = 5028 \text{ kW}$$

$$\text{ep} = \text{dp} \times \text{QPC}$$

$$= \frac{2550}{0{\cdot}65} \times 0{\cdot}71$$

$$\text{Effective power} = 2785 \text{ kW}$$

$$\text{sp} = \frac{\Delta^{\frac{2}{3}} V^3}{C}$$

$$V^3 = \frac{\text{sp} \times C}{\Delta^{\frac{2}{3}}}$$

$$= \frac{2550}{0{\cdot}65 \times 0{\cdot}94} \times \frac{420}{15\,000^{\frac{2}{3}}}$$

$$\text{Ship speed } V = 14{\cdot}23 \text{ knots}$$

9. $$\text{Theoretical speed } v_T = 4 \times \frac{125}{60}$$

$$= 8{\cdot}33 \text{ m/s}$$

$$\text{Real slip} = \frac{v_T - v_a}{v_T}$$

$$0{\cdot}36 = \frac{8{\cdot}33 - v_a}{8{\cdot}33}$$

$$v_a = 8 \cdot 33\,(1 - 0 \cdot 36) = 5 \cdot 33 \text{ m/s}$$

$$\text{tp} = \text{dp} \times \text{propeller efficiency} = 2800 \times 0 \cdot 67 = 1876 \text{ kW}$$

But
$$\text{tp} = T \times v_a$$

$$\therefore \quad 1876 = T \times 5 \cdot 33$$

$$T = \frac{1876}{5 \cdot 33}$$

$$\text{Propeller thrust} = 351 \cdot 8 \text{ kN}$$

10.
$$\frac{pm_1}{pm_2} = \left(\frac{N_1}{N_2}\right)^2$$

$$\therefore \quad pm_2 = pm_1 \left(\frac{N_2}{N_1}\right)^2 = 4 \cdot 5 \times \left(\frac{105}{130}\right)^2 = 2 \cdot 94 \text{ b}$$

11. Since the power is reduced, the rev/min may be found from the Admiralty Coefficient formula.

The displacement and Admiralty Coefficient are constant, hence:

$$\text{power} \propto \text{speed}^3$$

$$P \propto \text{revs}^3$$

$$\frac{P_1}{P_2} = \left(\frac{N_1}{N_2}\right)^3$$

$$N_2 = 120 \sqrt[3]{\frac{0 \cdot 8 P_2}{P_1}} = 111 \cdot 4 \text{ rev/min}$$

Also $$P \propto p_m \times N$$

$$\therefore \quad \frac{P_1}{P_2} = \frac{p_{m1}}{p_{m2}} \times \frac{N_1}{N_2}$$

$$p_{m2} = p_{m1} \times \frac{N_1}{N_2} \times \frac{P_2}{P_1}$$

$$= 5{\cdot}5 \times \frac{120}{111{\cdot}4} \times \frac{0{\cdot}8P_1}{P_1}$$

$$= 4{\cdot}74 \text{ b}$$

12.
$$\text{Pitch} = \tan\theta \times 2\pi R$$
$$= \tan 21{\cdot}5^\circ \times 2\pi \times 2$$
$$= 4{\cdot}95 \text{ m}$$

13.
$$\tan\theta = \frac{40}{115}$$

$$= 0{\cdot}3478$$

$$\theta = 19^\circ\ 10'$$

$$\text{Pitch} = 0{\cdot}3478 \times 2\pi \times 2{\cdot}6$$
$$= 5{\cdot}682 \text{ m}$$

$$\sin\theta = \frac{\text{horizontal ordinate}}{\text{blade width}}$$

$$\therefore \quad \text{width} = \frac{40}{\sin\ 19^\circ\ 10'}$$

$$= 121{\cdot}8 \text{ cm}$$
$$= 1{\cdot}218 \text{ m}$$

SOLUTIONS TO TEST EXAMPLES 9

1. $$\text{Ship speed} = 18 \text{ knots}$$

$$= 18 \times \frac{1852}{3600}$$

$$= 9{\cdot}26 \text{ m/s}$$

$$\text{Torque } T = F \sin \alpha \times b$$
$$= 580 A\, v^2 \sin \alpha \times b$$
$$= 580 \times 25 \times 9{\cdot}26^2 \times 0{\cdot}5736 \times 1{\cdot}2$$
$$= 855\ 816 \text{ N m}$$

But $$\frac{T}{J} = \frac{q}{r}$$

and $$J = \frac{\pi r^4}{2}$$

$\therefore$ $$T = \frac{\pi r^4 q}{2r}$$

$$r^3 = \frac{T \times 2}{\pi \times q}$$

$$= \frac{855\ 816 \times 2}{\pi \times 85 \times 10^6}$$

$$r = 0{\cdot}1858 \text{ m}$$

$$\text{Diameter} = 0{\cdot}3716 \text{ m}$$
$$= 372 \text{ mm}$$

2. $$\text{Torque } T = 580\ A\ v^2 \sin \alpha \times b$$

$$= 580 \times 13 \times \left(\frac{14 \times 1852}{3600}\right)^2 \times 1{\cdot}1 \sin \alpha$$

$$= 430\ 226 \sin \alpha$$

Angle α	$\sin\alpha$	T	SM	product
0°	0	0	1	0
10°	0·1736	74 690	4	298 760
20°	0·3420	147 140	2	294 280
30°	0·5000	215 110	4	860 440
40°	0·6428	276 550	1	276 550
				1 730 030

$$\text{Common interval} = \frac{10}{57 \cdot 3}$$

$$\text{Work done} = \frac{1}{3} \times \frac{10}{57 \cdot 3} \times 1\,730\,030$$

$$= 100\,640 \text{ J}$$
$$= 100 \cdot 64 \text{ kJ}$$

3. $$\text{Rudder area} = \frac{L \times d}{60}$$

$$= \frac{150 \times 8 \cdot 5}{60}$$

$$= 21 \cdot 25 \text{ m}^2$$

$$\text{Torque } T = \frac{Jq}{r}$$

$$= \frac{\pi r^4 q}{2r}$$

$$= \frac{\pi r^3 q}{2}$$

$$= \frac{\pi}{2} \times 0 \cdot 16^3 \times 70 \times 10^6$$

$$= 0 \cdot 4504 \times 10^6 \text{ N m}$$

$$\therefore \quad 0 \cdot 4504 \times 10^6 = 580 \times 21 \cdot 25 \times v^2 \times 0 \cdot 9 \times \sin 35°$$

J

$$v^2 = \frac{0.4504 \times 10^6}{580 \times 21.25 \times 0.9 \times 0.5736}$$

$$v = 8.414 \text{ m/s}$$

$$\text{Ship speed } V = 8.414 \times \frac{3600}{1852}$$

$$= 16.35 \text{ knots}$$

4. Normal force on rudder F_n
$= 580\ A\ v^2 \sin \alpha$

Transverse force on rudder F_t
$= 580\ A\ v^2 \sin \alpha \cos \alpha$

$$= 580 \times 12 \times \left(\frac{16 \times 1852}{3600}\right)^2 \times \sin 35^\circ \cos 35^\circ$$

$= 221\ 600$ N
$= 221.6$ kN

$$\tan \theta = \frac{F_t \times NL}{\triangle \times g \times GM}$$

$$= \frac{221.6 \times 1.6}{5000 \times 9.81 \times 0.24}$$

$$= 0.0301$$

Angle of heel $\theta = 1^\circ\ 43'$

5. Speed of ship $V = 17$ knots

$$v = 17 \times \frac{1852}{3600}$$

$$= 8.746 \text{ m/s}$$

$$\tan \theta = \frac{v^2 \times GL}{g \times \rho \times GM}$$

$$= \frac{8.746^2 \times (7.00 - 4.00)}{9.81 \times 450 \times (7.45 - 7.00)}$$

$= 0.1155$

Angle of heel $\theta = 6^\circ\ 35'$

SELECTION OF EXAMINATION QUESTIONS
SECOND CLASS

** Questions marked with an asterisk have been selected from Department of Trade papers and are reproduced by kind permission of The Controller of Her Majesty's Stationary Office.*

1. A box barge is 15 m long, 6 m wide and floats in water of 1.016 t/m^3 at a draught of 3 m. 150 tonne cargo is now added. Calculate the load exerted by the water on the sides, ends and bottom.

2. A ship has a load displacement of 6000 tonne and centre of gravity 5.30 m above the keel. 1000 tonne are then removed 2.1 m above the keel, 300 tonne moved down 2.4 m, 180 tonne placed on board 3 m above the keel and 460 tonne placed on board 2.2 m above the keel. Find the new position of the centre of gravity.

3. A ship of 8000 tonne displacement floats in sea water of 1.025 t/m^3 and has a TPC of 14. The vessel moves into fresh water of 1.000 t/m^3 and loads 300 tonne of oil fuel. Calculate the change in mean draught.

4. The wetted surface area of one ship is 40% that of a similar ship. The displacement of the latter is 4750 tonne more than the former. Calculate the displacement of the smaller ship.

5. A ship of 9000 tonne displacement floats in fresh water of 1.000 t/m^3 at a draught 50 mm below the sea water line. The waterplane area is 1650 m^2. Calculate the mass of cargo which must be added so that when entering sea water of 1.025 t/m^3 it floats at the sea water line.

6. A thin plate drawn through the water at 3 m/s has a frictional resistance of 14 N/m^2 and its resistance varies as $(\text{speed})^{1.97}$.

A ship 90 m long has a breadth of 16.7 m, draught of 7.5 m and block coefficient of 0.87. If the total frictional resistance is 300 kN, calculate the speed of the ship. The wetted surface area may be found from the formula.

$$S = 1.7\,Ld + \frac{\nabla}{d}$$

7. The effective power of ship is 1400 kW at 12 knots, the propulsive efficiency 65% and the fuel consumption 0.3 kg/kW h, based on shaft power. Calculate the fuel required to travel 10 000 nautical miles at 10 knots.

8. A ship is 60 m long, 16 m beam and has a draught of 5 m in sea water, block coefficient 0.7 and waterplane area coefficient 0.8. Calculate the draught at which it will float in fresh water.

9. A tank top manhole 0.50 m wide and 0.65 m long has semi-circular ends. The studs are pitched 30 mm outside the line of hole and 100 mm apart. The cross-sectional area of the studs between the threads is 350 mm^2. The tank is filled with salt water to a height of 7.5 m up the sounding pipe. Calculate the stress in the studs.

10. A small vessel has the following particulars before modifications are carried out. Displacement 150 tonne, *GM* 0.45 m, *KG* 1.98 m, *KB* 0.9 m, TPC 2.0 and draught 1.65 m.

After modification, 20 tonne has been added, *Kg* 3.6 m. Calculate the new *GM* assuming constant waterplane area over the change in draught.

11. A ship displacing 10 000 tonne and travelling at 16 knots has a fuel consumption of 41 tonne per day. Calculate the consumption per day if the displacement is increased to 13 750 tonne and the speed is increased to 17 knots. Within this speed range, fuel consumption per day varies as $(\text{speed})^{3 \cdot 8}$.

12. The ½ ordinates of a waterplane 320 m long are 0, 9, 16, 23, 25, 25, 22, 18 and 0 m respectively. Calculate:

(a) waterplane area
(b) TPC
(c) waterplane area coefficient.

13. A double bottom tank is filled with sea water to the top of the air pipe. The pressure on the outer bottom is found to be 1.20 bar while the pressure on the inner bottom is found to be 1.05 bar. Calculate the height of the air pipe above the inner bottom and the depth of the tank.

14. A ship 125 m long and 17.5 m beam floats in sea water of 1.025 t/m^3 at a draught of 8 m. The waterplane area coefficient is 0.83, block coefficient 0.759 and midship section area coefficient 0.98. Calculate:

(a) prismatic coefficient
(b) TPC
(c) change in mean draught if the vessel moves into river water of 1.016 t/m^3.

15. At 90 rev/min a propeller of 5 m pitch has an apparent slip of 15% and wake fraction 0.10. Calculate the real slip.

16. A hopper barge of box form 50 m long and 10 m wide floats at a draught of 2 m in sea water when the hopper, which is 15 m long and 5 m wide, is loaded with mud having relative density twice that of the sea water, to the level of the waterline.

Doors in the bottom of the hopper are now opened allowing the mud to be discharged. Calculate the new draught.

17. It is found that by reducing the fuel consumption of a vessel 43 tonne/day, the speed is reduced 2.2 knots and the saving in fuel for a voyage of 3500 nautical miles is 23%. Determine:

(a) the original daily fuel consumption, and
(b) the original speed.

18. A ship of 7000 tonne displacement, having *KG* 6 m and TPC 21, floats at a draught of 6 m. 300 tonne of cargo is now added at *Kg* 1.0 m and 130 tonne removed at *Kg* 5 m. The final draught is to be 6.5 m and *KG* 5.8 m. Two holds are available for additional cargo, one having *Kg* 5 m and the other *Kg* 7 m. Calculate the mass of cargo to be added to each hold.

19. A block of wood of uniform density has a constant cross-section in the form of a triangle, apex down. The width is 0.5 m and the depth 0.5 m. It floats at a draught of 0.45 m. Calculate the metacentric height.

20. The waterplane area of a ship at 8.40 m draught is 1670 m^2. The areas of successive waterplanes at 1.40 m intervals below this are 1600, 1540, 1420, 1270, 1080 and 690 m^2 respectively. Calculate the displacement in fresh water at 8.40 m draught and the draught at which the ship would lie in sea water with the same displacement.

21. A floating dock 150 m long, 24 m overall width and 9 m draught consists of a rectangular bottom compartment 3 m deep and rectangular wing compartments 2.5 m wide. A ship with a draught of 5.5 m is floated in. 4000 tonne of ballast are pumped out of the dock to raise the ship 1.2 m. Calculate the mean TPC of the ship.

22. A ship of 7500 tonne displacement has its centre of gravity 6.5 m above the keel. Structural alterations are made, when 300 tonne are added 4.8 m above the keel, 1000 tonne of oil fuel are then added 0.7 m above the keel.

(a) Calculate the new position of the centre of gravity.
(b) Calculate the final centre of gravity when 500 tonne of oil fuel are used.

23. A ship of 15 000 tonne displacement floats at a draught of 7 m in water of 1.000 t/m³. It is required to load the maximum amount of oil to give the ship a draught of 7 m in sea water of 1.025 t/m³. If the waterplane area is 2150 m², calculate the mass of oil required.

24. A bulkhead 12 m wide and 9 m high is secured at the base by an angle bar having 20 mm diameter rivets on a pitch of 80 mm. The bulkhead is loaded on one side only to the top edge with sea water. Calculate the stress in the rivets.

25. The ½ ordinates of a waterplane 96 m long are 1.2, 3.9, 5.4, 6.0, 6.3, 6.3, 6.3, 5.7, 4.5, 2.7 and 0 m respectively. A rectangular double bottom tank with parallel sides is 7.2 m wide, 6 m long and 1.2 m deep. When the tank is completely filled with oil of 1.15 m³/tonne the ship's draught is 4.5 m. Calculate the draught when the sounding in the tank is 0.6 m.

* 26. A ship 120 m in length has a wetted surface area of 2680 m² and operates at a speed of 14.5 knots. If the effective power is 2050 kW, calculate:

(a) the residuary resistance
(b) the ratio of frictional resistance to total resistance.

Frictional coefficient in water of density 1025 kg/m³ is 1.42. Speed in m/s with index for ship 1.83.

27. A ship enters harbour and discharges 6% of its displacement. It then travels upriver to a berth and the total change in draught is found to be 20 cm. The densities of the harbour and berth water are respectively 1.023 t/m³ and 1.006 t/m³ and the TPC in the harbour water is 19. Calculate the original displacement and state whether the draught has been increased or reduced.

28. The fuel consumption of a vessel varies within certain limits as $(\text{speed})^{2.95}$. If, at 1.5 knots above and below the normal speed, the power is 9200 kW and 5710 kW respectively, find the normal speed.

29. The length of a ship is 7.6 times the breadth, while the breadth is 2.85 times the draught. The block coefficient is 0.69, prismatic coefficient 0.735, waterplane area coefficient 0.81 and the wetted surface area 7000 m². The wetted surface area S is given by:

$$S = 1.7\,L\,d + \frac{\nabla}{d}$$

Calculate:

(a) displacement in tonne
(b) area of immersed midship section
(c) waterplane area

30. A ship 120 m long, 17 m beam and 7.2 m draught has a block coefficient of 0.76. A parallel section 6 m long is added to the ship amidships. The midship sectional area coefficient is 0.96. Find the new displacement and block coefficient.

31. State what is meant by the Admiralty Coefficient and what its limitations are.

A ship has an Admiralty Coefficient of 355, a speed of 15 knots and shaft power 7200 kW. Calculate its displacement.

If the speed is now reduced by 16%, calculate the new power required.

* 32. A ship of 8100 tonne displacement, 120 m long and 16 m beam floats in water of density 1025 kg/m³. In this condition the ship has the following hydrostatic data:

Prismatic coefficient	=	0.70
Midship area coefficient	=	0.98
Waterplane area coefficient	=	0.82

A full depth midship compartment, which extends the full breadth of the ship, is flooded. Calculate the length of the compartment if, after flooding, the mean draught is 7.5 m.

33. A double bottom tank 1.15 m deep has transverse floors 0.90 m apart connected to the tank top by rivets spaced 7 diameters apart. When the tank is filled with oil (*rd* 0.81) to the top of the sounding pipe, the pressure on the outer bottom is 1.06 bar, while the stress in the rivets in the tank top is 320 bar. Calculate:

(a) the height of the sounding pipe above the tank top
(b) the diameter of the rivets.

34. A ship of 22 000 tonne displacement has a draught of 9.00 m in river water of 1.008 t/m³. The waterplane area is 3200 m². The vessel then enters sea water of 1.026 t/m³. Calculate the change in displacement as a percentage of the original displacement in order to:

(a) keep the draught the same
(b) give a draught of 8.55 m.

* 35. A ship has a fuel consumption of 60 tonne per 24 hours when the displacement is 15 500 tonne and the ship speed 14 knots. Determine the ship speed during a passage of 640 nautical miles if the displacement is 14 500 tonne and the total fuel consumption is 175 tonne.

36. The TPC values of a ship at 1.2 m intervals of draught, commencing at the load waterline are 19, 18.4, 17.4, 16.0, 13.8, 11.0 and 6.6 respectively. If 6.6 represents the value at the keel, calculate the displacement in tonne and state the load draught.

37. If the density of sea water is 1.025 t/m³ and the density of fresh water is 1.000 t/m³, prove that the Statutory Fresh Water Allowance is $\frac{\Delta}{40\ \text{TPC}}$ cm.

A ship of 12 000 tonne displacement loads in water of 1.012 t/m³. By how much will the Summer Load Line be submerged if it is known that 130 tonne must be removed before sailing? The TPC in sea water is 17.7.

38. A ship of 2543 tonne displacement is 85 m long, 10.5 m beam, 5 m draught and has a speed of 14 knots. A similar ship has a displacement of 3000 tonne. Calculate the dimensions of the larger ship and its corresponding speed.

* 39. A collision bulkhead is in the form of an isosceles triangle and has a depth of 7.0 m and a width at the deck of 6.0 m. The bulkhead is flooded on one side with water of density 1025 kg/m³ and the resultant load on the bulkhead is estimated at 195 kN. Calculate the depth of water to which the bulkhead is flooded.

40. A ship of 8000 tonne displacement has a metacentric height of 0.46 m, centre of gravity 6.6 m above the keel and centre of buoyancy 3.6 m above the keel. Calculate the second moment of area of the waterplane about the centreline of the ship.

41. A ship has a displacement of 9800 tonne. 120 tonne of oil fuel are moved from an after tank to a tank forward. The centre of gravity of the ship moves 0.75 m forward.

The forward tank already contains 320 tonne of oil fuel and after the transfer 420 tonne of fuel are used. The centre of gravity of the ship now moves to a new position 0.45 m aft of the vessel's original centre of gravity. Find the distance from the ship's original centre of gravity to the centre of gravity of each tank.

* 42. A ship of 18 000 tonne displacement carries 500 tonne of fuel. If the fuel coefficient is 47 250 when the ship's speed is 12 knots, calculate:

(a) the range of operation of the ship
(b) the percentage difference in range of the ship's operation if the speed of the ship is increased by 5%.

43. Before bunkering in harbour the draught of a vessel of 12 000 tonne displacement is 8.16 m, the waterplane area being 1625 m^2. After loading 1650 tonne of fuel and entering sea water of 1.024 t/m^3, the draught is 9.08 m. Assuming that the waterplane area remains constant and neglecting any fuel etc. expended in moving the vessel, calculate the density of the harbour water.

44. Describe how an inclining experiment is carried out.

A vessel of 8000 tonne displacement was inclined by moving 5 tonne through 12 m. The recorded deflections of a 6 m pendulum were 73, 80, 78 and 75 mm. If the KM for this displacement was 5.10 m, calculate KG.

45. A rectangular bulkhead 17 m wide and 6 m deep has a head of sea water on one side only, of 2.5 m above the top of the bulkhead. Calculate:

(a) the load on the bulkhead
(b) the pressure at the top and bottom of the bulkhead.

* 46. For a ship of 4600 tonne displacement the metacentric height (GM) is 0.77 m. A 200 tonne container is moved from the hold to the upper deck.

Determine the angle of heel developed if, during this process, the centre of mass of the container is moved 8 m vertically and 1.1 m transversely.

47. The load draught of a ship is 7.5 m in sea water and the corresponding waterplane area is 2100 m^2. The areas of parallel waterplanes at intervals of 1.5 m below the load waterplane are 1930, 1720, 1428 and 966 m^2 respectively.

Draw the TPC curve. Assuming that the displacement of the portion below the lowest given waterplane is 711 tonne, calculate the displacement of the vessel when:

(a) fully loaded, and
(b) floating at a draught of 4.5 m in sea water.

48. Define *centre of buoyancy* and show with the aid of sketches how a vessel which is stable will return to the upright after being heeled by an external force.

A vessel displacing 8000 tonne has its centre of gravity 1 m above the centre of buoyancy when in the upright condition. If the moment tending to right the vessel is 570 tonne m when the vessel is heeled over 7°, calculate the horizontal distance the centre of buoyancy has moved from its original position.

* 49. A propeller rotates at 2 rev/s with a speed of advance of 12 knots and a real slip of 0.30. The torque absorbed by the propeller is 250 kN m and the thrust delivered is 300 kN. Calculate:

(a) the pitch of the propeller
(b) the thrust power
(c) the delivered power.

50. A box barge 40 m long and 7.5 m wide floats in sea water with draughts forward and aft of 1.2 m and 2.4 m respectively. Where should a mass of 90 tonne be added to obtain a level keel draught?

* 51. For a box-shaped barge of 16 m beam floating at an even keel draught of 6 m in water of density 1025 kg/m^3, the tonne per centimetre immersion (TPC) is 17. A full-depth midship compartment 20 m in length and 16 m breadth has a permeability of 0.80.

If the compartment is bilged, determine:

(a) the draught
(b) the position of the metacentre above the keel, if the second moment of area of the intact waterplane about the centreline is 75 000 m^4.

52. A box barge is 7.2 m wide and 6 m deep. Draw the metacentric diagram using 1 m intervals of draught up to the deck line.

* 53. A propeller has a diameter of 4.28 m, pitch ratio of 1.1 and rotates at a speed of 2 rev/s. If the apparent and true slip are 0.7% and 12% respectively, calculate the wake speed.

$$\textit{Pitch ratio} = \frac{\textit{propeller pitch}}{\textit{propeller diameter}}$$

* 54. At a ship speed of 12 knots the shaft power for a vessel is 1710 kW and the fuel consumption is 0.55 kg/kW h.

Determine for a speed of 10 knots:

(a) the quantity of fuel required for a voyage of 7500 miles
(b) the fuel coefficient if the ship's displacement is 6000 tonne.

55. A ship of 12 000 tonne displacement has a metacentric height of 0.6 m and a centre of buoyancy 4.5 m above the keel. The second moment of area of the waterplane about the centreline is 42.5×10^3 m^4. Calculate height of centre of gravity above keel.

* 56. A box-shaped barge 37 m long, 6.4 m beam, floats at an even keel draught of 2.5 m in water of density 1025 kg/m^3. If a mass is added and the vessel moves into water of density 1000 kg/m^3, determine the magnitude of this mass if the forward and aft draughts are 2.4 m and 3.8 m respectively.

* 57. Calculate the effective power of a ship given the following information:

Length of ship	= 145 m
Wetted surface area	= 3400 m^2
Ship speed	= 16 knots

The residuary resistance can be assumed to be 7.1 newtons per tonne displacement.

The frictional coefficient for the ship in water of density 1025 kg/m^3 is 1.42. Speed in m/s with index for ship 1.83.

Note: Wetted surface area $= 2.56\sqrt{\textit{displacement} \times \textit{length}}$

58. A ship of 7200 tonne displacement has two similar bunkers adjacent to each other, the capacity of each being 495 tonne and their depth 9.9 m. If one of the bunkers is completely full and the other completely empty, find how much fuel must be transferred to lower the ship's centre of gravity 120 mm.

59. The ½ ordinates of the waterplane of a ship 75 m long are:

Station	0	½	1	2	3	4	5	5½	6
½ ord	0.06	1.25	2.68	4.55	5.30	5.20	4.20	2.60	0 m

Calculate the tonne per cm immersion.

60. A ship of 7000 tonne displacement has *KM* 7.30 m. Masses of 150 tonne at a centre 3.0 m above and 60 tonne at a centre 5.5 m below the original centre of gravity of the ship are placed on board. A ballast tank containing 76 tonne of water at *Kg* 0.60 m is then discharged.

Calculate the original height of the ship's centre of gravity above the keel if the final metacentric height is 0.50 m and *KM* is assumed to remain constant.

SOLUTIONS TO SECOND CLASS EXAMINATION QUESTIONS

1.
$$\text{TPC} = \frac{A_W \times 1.016}{100}$$
$$= 15 \times 6 \times 0.01016$$
$$= 0.914$$

$$\text{Bodily sinkage} = \frac{150}{0.914}$$
$$= 164 \text{ cm}$$

$$\text{New draught} = 3.0 + 1.64$$
$$= 4.64 \text{ m}$$

$$\text{Load on side} = \rho g A H$$
$$= 1.016 \times 9.81 \times 15 \times 4.64 \times 2.32$$
$$= 1609 \text{ kN}$$

$$\text{Load on end} = 1.016 \times 9.81 \times 6 \times 4.64 \times 2.32$$
$$= 643.8 \text{ kN}$$

$$\text{Load on bottom} = 1.016 \times 9.81 \times 15 \times 6 \times 4.64$$
$$= 4162 \text{ kN}$$

2.

Mass	distance	moment
6000	5.30	31 800
− 1000	2.10	− 2100
(300)	− 2.40	− 720
180	3.00	+ 540
460	2.20	+ 1012
5640		30 532

Note: The 300 tonne remains on board and therefore does not alter the final displacement.

$$\text{Centre of gravity above keel} = \frac{30\,532}{5640}$$
$$= 5.413 \text{ m}$$

3. Change in mean draught

$$= \frac{\Delta \times 100 \times 1.025}{\text{TPC} \times 100}\left(\frac{1.025 - 1.000}{1.000 \times 1.025}\right)$$

$$= \frac{8000 \times 0.025}{14}$$

$$= 14.29 \text{ cm increase}$$

$$\text{TPC in fresh water} = 14 \times \frac{1.000}{1.025}$$

$$\text{Bodily sinkage} = \frac{300}{\text{TPC}}$$

$$= \frac{300 \times 1.025}{14}$$

$$= 21.96 \text{ cm}$$

$$\text{Total increase} = 14.29 + 21.96$$
$$= 36.25 \text{ cm}$$

4. Let S = wetted surface area of small ship

Δ = displacement of small ship

Then $\frac{S}{0.4}$ = wetted surface area of large ship

$\Delta + 4750$ = displacement of large ship

Now $\Delta \propto S^{\frac{3}{2}}$

$$\frac{\Delta}{\Delta_1} = \left(\frac{S}{S_1}\right)^{\frac{3}{2}}$$

$$\frac{\Delta}{\Delta + 4750} = \left(\frac{S \times 0.4}{S}\right)^{\frac{3}{2}}$$

$$= 0.2523$$

$$\Delta = 0.2523(\Delta + 4750)$$

$$\Delta(1 - 0.2523) = 0.2523 \times 4750$$

$$\Delta = \frac{0.2523 \times 4750}{0.7477}$$

$$= 1603 \text{ tonne}$$

5. Change in mean draught $= \dfrac{9000 \times 100}{1650} \left(\dfrac{1.025 - 1.000}{1.000 \times 1.025}\right)$

$= 13.30$ cm

Thus new waterline $= 13.30 + 5.0$
$= 18.30$ cm below SW line

TPC $= 1650 \times 0.01025$
$= 16.91$

Mass of cargo $= 18.3 \times 16.91$
$= 309.5$ tonne

6. Volume of displacement $= 90 \times 16.7 \times 7.5 \times 0.87$
$= 9807 \text{ m}^3$

Wetted surface area $S = 1.7 \times 90 \times 7.5 + \dfrac{9807}{7.5}$

$= 1147.5 + 1307.6$
$= 2455.1 \text{ m}^2$

At 3 m/s $\quad R_f = 14 \times 2455.1$ N
$= 34.371$ kN

At v m/s $\quad R_f = 34.371 \times \left(\dfrac{v}{3}\right)^{1\cdot 97}$

$\therefore \quad 300 = 34.371 \times \left(\dfrac{v}{3}\right)^{1\cdot 97}$

$v = 3\sqrt[1\cdot 97]{\dfrac{300}{34.371}}$

$= 9.010$ m/s

Ship speed $V = \dfrac{9.010 \times 3600}{1852}$

$= 17.51$ knots

7. At 12 knots:

$$ep = 1400 \text{ kW}$$

$$sp = \frac{1400}{0.65}$$

$$= 2154 \text{ kW}$$

$$\text{Cons/h} = 0.3 \times 2154$$
$$= 646.2 \text{ kg}$$

$$\text{Cons/h} \propto \text{speed}^3$$

$$\therefore \text{ at 10 knots: cons/h} = 646.2 \times \left(\frac{10}{12}\right)^3$$

$$= 374.0 \text{ kg}$$

$$\text{Time on voyage} = \frac{10\,000}{10}$$

$$= 1000 \text{ h}$$

$$\therefore \quad \text{voyage consumption} = 374 \times 1000$$
$$= 374 \text{ tonne}$$

8. $$\text{Displacement } \Delta = L\,B\,d \times C_b \times 1.025$$

$$\text{Area of waterplane } A_w = L\,B \times C_w$$

$$\frac{\Delta}{A_w} = 1.025d \times \frac{C_b}{C_w}$$

$$\text{Change in draught} = \frac{\Delta \times 100}{A_w}\left(\frac{\rho_s - \rho_R}{\rho_R \times \rho_s}\right)$$

$$= \frac{1.025 \times 5 \times 0.7 \times 100}{0.8}\left(\frac{1.025 - 1.000}{1.000 \times 1.025}\right)$$

$$= 10.94 \text{ cm}$$

$$\text{New draught} = 5.0 + 0.109$$
$$= 5.109 \text{ m}$$

9.

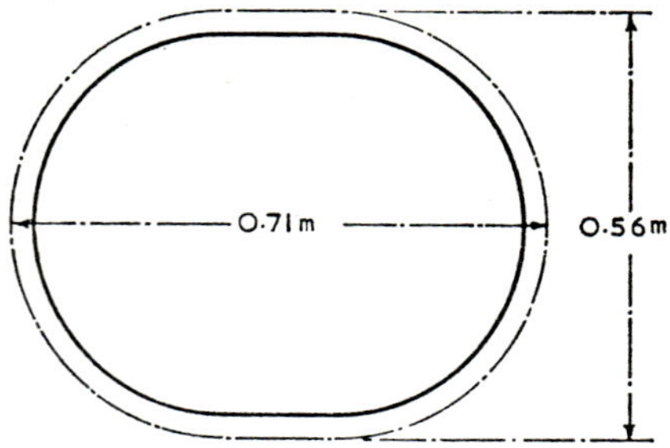

Fig. 109

$$\text{Perimeter of stud line} = \pi \times 0.56 + 2 \times 0.15$$
$$= 2.059 \text{ m}$$

$$\text{Number of studs} = \frac{2.059}{0.10}$$

$$= 21$$

$$\text{Cross-sectional area of studs} = 21 \times 350$$
$$= 7350 \text{ mm}^2$$
$$= 7.35 \times 10^{-3} \text{ m}^2$$

$$\text{Effective area of door} = \frac{\pi}{4} \times 0.56^2 + 0.15 \times 0.56$$

$$= 0.2463 + 0.0840$$
$$= 0.3303 \text{ m}^2$$

$$\text{Load on door} = \rho g A H$$
$$= 1.025 \times 9.81 \times 0.3303 \times 7.5$$
$$= 24.91 \text{ kN}$$

$$\text{Stress} = \frac{\text{load}}{\text{area}}$$

$$= \frac{24.91}{7.35 \times 10^{-3}}$$

$$= 3.389 \times 10^3 \text{ kN/m}^2$$
$$= 3.389 \text{ MN/m}^2$$

10.

$$KM = 1.98 + 0.45$$
$$= 2.43 \text{ m}$$

$$BM = 2.43 - 0.90$$
$$= 1.53 \text{ m}$$

$$= \frac{I}{\nabla}$$

$$I = \frac{1.53 \times 150}{1.025}$$

$$= 223.9 \text{ m}^4$$

$$KG = \frac{150 \times 1.98 + 20 \times 3.6}{150 + 20}$$

$$= \frac{297.0 + 72}{170}$$

$$= 2.170 \text{ m}$$

$$\text{Increase in draught} = \frac{20}{2}$$

$$= 10 \text{ cm}$$

$$KB_1 = \frac{150 \times 0.9 + 20(1.65 + 0.05)}{170}$$

$$= \frac{135 + 34}{170}$$

$$= 0.994 \text{ m}$$

$$BM_1 = \frac{223.9}{170} \times 1.025$$

$$= 1.350 \text{ m}$$

$$KM_1 = 0.994 + 1.350$$
$$= 2.344 \text{ m}$$

$$GM_1 = 2.344 - 2.170$$
$$= 0.174 \text{ m}$$

11. Fuel cons/day $\propto \Delta^{\frac{2}{3}} V^{3.8}$

$$\therefore \quad \frac{C}{41} = \left(\frac{13\,750}{10\,000}\right)^{\frac{2}{3}} \times \left(\frac{17}{16}\right)^{3.8}$$

$$C = 41\left(\frac{13\,750}{10\,000}\right)^{\frac{2}{3}} \times \left(\frac{17}{16}\right)^{3.8}$$

Fuel cons/day = 63.81 tonne

12.

½ ordinate	SM	Product
0	1	—
9	4	36
16	2	32
23	4	92
25	2	50
25	4	100
22	2	44
18	4	72
0	1	—
		426

$$h = \frac{320}{8}$$

$$= 40 \text{ m}$$

(a) Waterplane area $= \frac{2}{3} \times 40 \times 426$
$= 11\,360 \text{ m}^2$

(b) TPC $= A_w \times 0.01025$
$= 116.44$

(c) Waterplane area coefficient

$$= \frac{11\,360}{320 \times 50}$$

$$= 0.710$$

13. Bottom pressure $= 1.2 \times 10^5 \text{ N/m}^2$
$= \rho g h$

$$h = \frac{1.2 \times 10^5}{1025 \times 9.81}$$

Height of air pipe above outer bottom
$= 11.93 \text{ m}$

Top pressure $= 1.05 \times 10^5 \text{ N/m}^2$

$$h_1 = \frac{1.05 \times 10^5}{1025 \times 9.81}$$

Height of air pipe above inner bottom
$= 10.44 \text{ m}$

Depth of tank $= 11.93 - 10.44$
$= 1.49 \text{ m}$

14. (a)

$$C_p = \frac{C_b}{C_m}$$

$$= \frac{0.759}{0.98}$$

$$= 0.7745$$

(b) Waterplane area $= 125 \times 17.5 \times 0.83$
$= 1815.6 \text{ m}^2$

TPC $= 1815.6 \times 0.01025$
$= 18.61$

(c) Displacement $= 125 \times 17.5 \times 8 \times 0.759 \times 1.025$
$= 13\,615$ tonne

$$\text{Change in draught} = \frac{13\,615 \times 100}{1815.6} \left(\frac{1.025 - 1.016}{1.016 \times 1.025}\right)$$

$$= 6.48 \text{ cm}$$

15. Theoretical speed $V_t = \dfrac{5 \times 90 \times 60}{1852}$

$= 14.58$ knots

Apparent slip $0.15 = \dfrac{14.58 - V}{14.58}$

$V = 14.58\,(1 - 0.15)$
$= 12.39$ knots

Wake fraction $0.10 = \dfrac{12.39 - V_a}{12.39}$

$V_a = 12.39\,(1 - 0.10)$
$= 11.15$ knots

Real slip $= \dfrac{14.58 - 11.15}{14.58}$

$= 0.2352$
or 23.52%

16. Mass of mud in hopper $= 2 \times 1.025 \times 15 \times 5 \times 2$
Mass of buoyancy in hopper $= 1.025 \times 15 \times 5 \times 2$

When the doors are opened, the mud drops out, but at the same time the buoyancy of the hopper is lost. Since the reduction in displacement exceeds the reduction in buoyancy, there will be a reduction in draught.

Nett loss of displacement $= 1.025 \times 15 \times 5 \times 2$
Nett volume of lost displacement
$= 15 \times 5 \times 2\ \text{m}^3$

Area of intact waterplane $= 50 \times 10 - 15 \times 5$
$= 425\ \text{m}^2$

Reduction in draught $= \dfrac{15 \times 5 \times 2}{425}$

$= 0.353$ m

New draught $= 2.0 - 0.353$
$= 1.647$ m

17. Let C = original daily consumption
V = original speed
K = original voyage consumption

$$C \propto V^3$$
$$K \propto V^2$$

$$\frac{K}{(1-0.23)K} = \left(\frac{V}{V-2.2}\right)^2$$

$$\frac{1}{0.77} = \left(\frac{V}{V-2.2}\right)^2$$

$$\frac{1}{\sqrt{0.77}} = \frac{V}{V-2.2}$$

$$V-2.2 = \sqrt{0.77} \times V$$
$$= 0.8775V$$
$$V(1-0.8775) = 2.2$$
Original speed V = 17.97 knots

$$\frac{C}{C-43} = \left(\frac{V}{V-2.2}\right)^3$$

$$= \left(\frac{17.97}{15.77}\right)^3$$

$$= 1.480$$
$$C = 1.480(C-43)$$
$$0.48C = 1.48 \times 43$$

$$C = \frac{1.48 \times 43}{0.48}$$

Original consumption = 132.6 tonne/day

18. Total change in draught = $6.5 - 6.0$
= 0.50 m

Total change in displacement = 50×21
= 1050 tonne

Let mass of additional cargo = m

Then $1050 = +300 - 130 + m$
$m = 1050 - 300 + 130$
= 880 tonne

Let mass of cargo in one hold = x

Then mass of cargo in other hold

$$= 880 - x$$

Taking moments about the keel.

Mass	Kg	moment
7000	6.0	42 000
+ 300	1.0	+ 300
− 130	5.0	− 650
+ x	5.0	+ $5x$
+ $(880 - x)$	7.0	+ $(6160 - 7x)$
8050		$47\,810 - 2x$

Final KG $$5.8 = \frac{47\,810 - 2x}{8050}$$

$$5.8 \times 8050 = 47\,810 - 2x$$
$$2x = 47\,810 - 46\,690$$
$$x = 560 \text{ tonne}$$

∴ Mass of cargo in holds Kg 5.0 m and Kg 7.0 m respectively are 560 tonne and 320 tonne.

19.

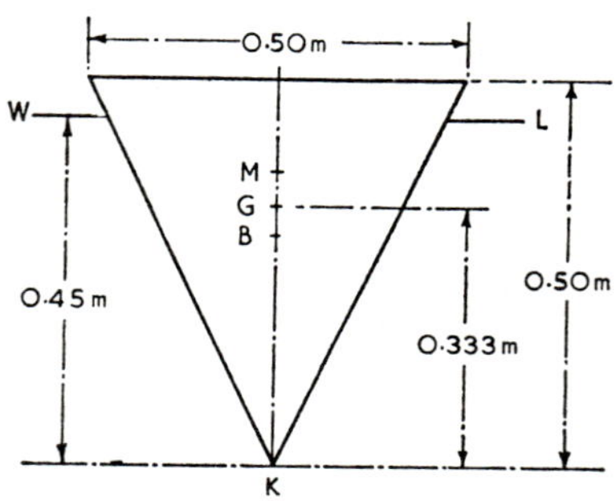

Fig. 110

$$\text{Width at waterline} = 0.45 \text{ m}$$
$$KB = \tfrac{2}{3}d$$
$$= 0.30 \text{ m}$$
$$KG = \tfrac{2}{3}D$$
$$= 0.333 \text{ m}$$

$$BM = \frac{b^2}{6d}$$

$$= \frac{0.45^2}{6 \times 0.45}$$

$$= 0.075 \text{ m}$$

$$KM = KB + BM$$
$$= 0.30 + 0.075$$
$$= 0.375 \text{ m}$$
$$GM = KM - KG$$
$$= 0.375 - 0.333$$
$$\text{Metacentric height} = 0.042 \text{ m}$$

20.

Waterplane area	SM	product
1670	1	1670
1600	4	6400
1540	2	3080
1420	4	5680
1270	2	2540
1080	4	4320
690	1	690
		24 380

$$h = 1.4 \text{ m}$$

$$\text{Displacement in fresh water} = \frac{1.4}{3} \times 24\,380 \times 1.000$$

$$= 11\,377 \text{ tonne}$$

$$\text{Reduction in draught} = \frac{11\,377 \times 100}{1670}\left(\frac{1.025 - 1.000}{1.000 \times 1.025}\right)$$

$$= 16.62 \text{ cm}$$

$$\text{Draught in sea water} = 8.40 - 0.166$$
$$= 8.234 \text{ m}$$

21.

The dock must rise 0.5 m before the ship touches. Thus if the ship rises 1.2 m, the dock must rise 1.70 m.

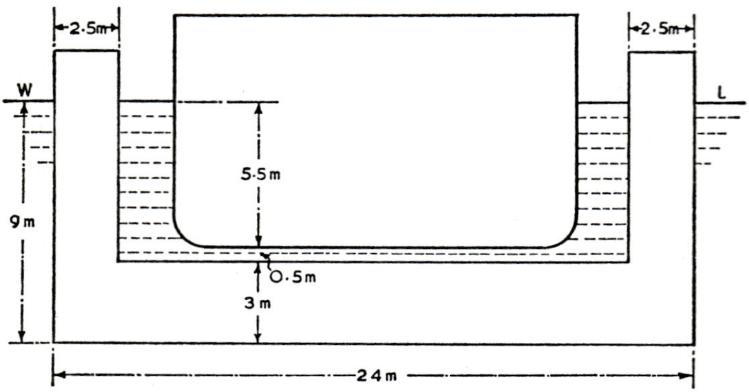

Fig. 111

Mass removed to raise dock 1.70 m

$$= 150 \times (2.5 + 2.5) \times 1.70 \times 1.025$$
$$= 1307 \text{ tonne}$$

Mass removed to raise ship 1.20 m

$$= 4000 - 1307$$
$$= 2693 \text{ tonne}$$
$$\therefore 120 \times \text{TPC} = 2693$$

$$\text{Mean TPC} = \frac{2693}{120}$$

$$= 22.44$$

22. (a)

Mass	Kg	moment
7500	6.5	48 750
300	4.8	1440
1000	0.7	700
8800		50 890

$$\text{New } KG = \frac{50\,890}{8800}$$

$$= 5.783 \text{ m}$$

(b) After burning oil:

$$\text{New } KG = \frac{50\,890 - 500 \times 0.7}{8800 - 500}$$

$$= 6.089 \text{ m}$$

23. With displacement of 15 000 tonne:

$$\text{Reduction in draught} = \frac{15\,000 \times 100}{2150} \left(\frac{1.025 - 1.000}{1.000 \times 1.025}\right)$$

$$= 17.02 \text{ cm}$$

∴ Increase in draught required

$$= 17.02 \text{ cm}$$

$$\text{TPC} = 2150 \times 0.01025$$

$$= 22.04$$

$$\text{Mass of oil required} = 17.02 \times 22.04$$

$$= 375.1 \text{ tonne}$$

24.

$$\text{Load on bulkhead} = \rho g A H$$

$$= 1.025 \times 9.81 \times 12 \times 9 \times 4.5$$

$$= 4887 \text{ kN}$$

$$\text{Shearing force at bottom} = \tfrac{2}{3} \times 4887$$

$$= 3258 \text{ kN}$$

$$\text{Number of rivets} = \frac{12}{0.080}$$

$$= 150$$

Total cross-sectional area of rivets

$$= 150 \times \frac{\pi}{4} \times 20^2 \times 10^{-6}$$

$$= 47.13 \times 10^{-3} \text{m}^2$$

$$\text{Stress in rivets} = \frac{\text{load}}{\text{area}}$$

$$= \frac{3258}{47.13 \times 10^{-3}}$$

$$= 69.13 \times 10^3 \text{ kN/m}^2$$

$$= 69.13 \text{ MN/m}^2$$

25.

½ ordinate	SM	product
1.2	1	1.2
3.9	4	15.6
5.4	2	10.8
6.0	4	24.0
6.3	2	12.6
6.3	4	25.2
6.3	2	12.6
5.7	4	22.8
4.5	2	9.0
2.7	4	10.8
0	1	—
		144.6

$$h = 9.6 \text{ m}$$

$$\text{Waterplane area} = \tfrac{2}{3} \times 9.6 \times 144.6$$

$$= 925.44 \text{ m}^2$$

$$\text{TPC} = 925.44 \times 0.01025$$

$$= 9.486$$

$$\text{Mass of oil in tank} = \frac{7.2 \times 6.0 \times 1.2}{1.15}$$

$$= 45.08 \text{ tonne}$$

$$\therefore \text{mass removed} = 22.54 \text{ tonne}$$

$$\text{Reduction in draught} = \frac{22.54}{9.486}$$

$$= 2.38 \text{ cm}$$

$$\text{Final draught} = 4.50 - 0.024$$

$$= 4.476 \text{ m}$$

26.

$$\text{Effective power} = R_t \times v$$

$$v = 14.5 \times \frac{1852}{3600}$$

$$= 7.459 \text{ m/s}$$

$$2050 = R_t \times 7.459$$

$$R_t = 274.8 \text{ kN}$$

$$R_f = fSv^n$$

$$= 1.42 \times 2680 \times 7.459^{1.83}$$
$$= 150.0 \text{ kN}$$
$$R_r = R_t - R_f$$
$$= 274.8 - 150.0$$
$$\text{Residuary resistance} = 124.8 \text{ kN}$$
$$\frac{R_f}{R_t} = \frac{150.0}{274.8}$$
$$= 0.546$$

i.e. Frictional resistance 54.6% of total resistance.

27. Let Δ = original displacement

Change in draught due to removal of mass

$$a = \frac{0.06\Delta}{\text{TPC}} \text{ cm}$$
$$= \frac{0.06\Delta}{19}$$
$$= 0.003158\,\Delta \text{ cm reduction}$$

Change in draught due to change in density

$$b = \frac{0.94\,\Delta \times 100 \times 1.023}{\text{TPC} \times 100} \left(\frac{1.023 - 1.006}{1.006 \times 1.023}\right)$$
$$= \frac{0.94\,\Delta \times 0.017}{19 \times 1.006}$$
$$= 0.000836\,\Delta \text{ cm increase}$$

Assuming a to be greater than b

$$20 = a - b$$
$$= 0.003158\Delta - 0.000836\Delta$$
$$= 0.002322\Delta$$

Original displacement

$$\Delta = \frac{20}{0.002322}$$
$$= 8613 \text{ tonne}$$

The draught will be *reduced* by 20 cm.

If the above assumption were wrong, the displacement would work out as a *negative* value.

28. Let V = normal speed in knots

$$\frac{9200}{5710} = \left(\frac{V+1.5}{V-1.5}\right)^{2.95}$$

$$\frac{V+1.5}{V-1.5} = \sqrt[2.95]{\frac{9200}{5710}}$$

$$= 1.176$$

$$V + 1.5 = 1.176V - 1.176 \times 1.5$$

$$0.176V = (1 + 1.176)\,1.5$$

$$V = \frac{2.176 \times 1.5}{0.176}$$

Normal speed V = 18.55 knots

29.

$$L = 7.6 \times B$$

$$B = 2.85 \times d$$

$$\nabla = L \times B \times d \times C_b$$

$$= 7.6B \times B \times \frac{B}{2.85} \times 0.69$$

$$= \frac{7.6 \times 0.69\,B^3}{2.85}$$

$$S = 1.7\,L\,d + \frac{\nabla}{d}$$

$$= 1.7 \times 7.6B \times \frac{B}{2.85} + \frac{7.6 \times 0.69}{2.85}\,B^3 \times \frac{2.85}{B}$$

$$7000 = 4.533B^2 + 5.244B^2$$

$$= 9.777B^2$$

$$B^2 = \frac{7000}{9.777}$$

$$B = 26.76\text{ m}$$

$$L = 7.6 \times 26.76$$

$$= 203.38\text{ m}$$

$$d = \frac{26.76}{2.85}$$

$$= 9.389\text{ m}$$

(a) Displacement Δ
$= 203.38 \times 26.76 \times 9.389 \times 0.69 \times 1.025$
$= 36\,140$ tonne

(b) Area of immersed midship section

$$A_m = B \times d \times \frac{C_b}{C_p}$$

$$= 26.76 \times 9.389 \times \frac{0.69}{0.735}$$

$$= 235.9 \text{ m}^2$$

(c) Waterplane area

$$A_w = L \times B \times C_w$$
$$= 203.38 \times 26.76 \times 0.81$$
$$= 4410 \text{ m}^2$$

30. Original displacement $= 120 \times 17 \times 7.2 \times 0.76 \times 1.025$
$= 11\,442$ tonne
Additional displacement $= 6 \times 17 \times 7.2 \times 0.96 \times 1.025$
$= 723$ tonne
New displacement $= 11\,442 + 723$
$= 12\,165$ tonne
New length $= 126$ m

$$\text{New block coefficient} = \frac{12\,165}{126 \times 17 \times 7.2 \times 1.025}$$

$$= 0.770$$

31.

$$\text{sp} = \frac{\Delta^{\frac{2}{3}} V^3}{C}$$

$$\Delta^{\frac{2}{3}} = \frac{7200 \times 355}{15^3}$$

Displacement $\Delta = 20\,840$ tonne
$\text{sp} \propto V^3$

$$\therefore \quad \text{sp}_1 = 7200 \times \left(\frac{0.84V}{V}\right)^3$$

$$= 4267 \text{ kW}$$

32.

$$\Delta = L \times B \times d \times C_b \times \rho$$

$$C_b = C_p \times C_m$$

$$\therefore \quad 8100 = 120 \times 16 \times d \times 0.70 \times 0.98 \times 1.025$$

$$d = \frac{8100}{120 \times 16 \times 0.70 \times 0.98 \times 1.025}$$

$$= 6.00 \text{ m}$$

Let l = length of compartment

Volume of lost buoyancy = $l \times 16 \times 6.0 \times 0.98$

Area of intact waterplane = $120 \times 16 \times 0.82 - l \times 16$

Increase in draught = $7.5 - 6.0$

= 1.5 m

$$\therefore \quad 1.5 = \frac{l \times 16 \times 6 \times 0.98}{120 \times 16 \times 0.82 - l \times 16}$$

$$2361.6 - 24l = 94.08\,l$$

$$94.08\,l + 24\,l = 2361.6$$

$$l = \frac{2361.6}{118.08}$$

Length of compartment = 20 m

33.

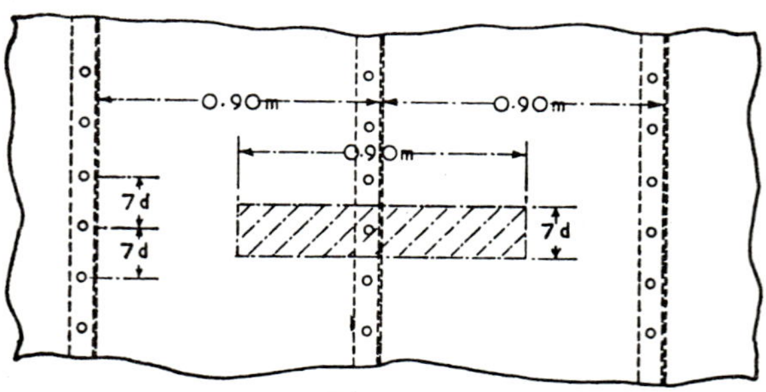

Fig. 112

(a) Pressure on outer bottom = $\rho g h$

$$1.06 \times 10^5 = 810 \times 9.81 \times h$$

$$h = \frac{1.06 \times 10^5}{810 \times 9.81}$$

$$= 13.34 \text{ m}$$

∴ Sounding pipe above tank top

$$= 13.34 - 1.15$$

$$= 12.19 \text{ m}$$

(b) Area supported by one rivet

$$= 0.90 \times 7d$$
$$= 6.3d \text{ m}^2$$
$$\text{Load on one rivet} = \rho g A H$$
$$= 810 \times 9.81 \times 6.3d \times 12.19$$
$$= 610\,240d \text{ N}$$

$$\text{But} \qquad \text{stress} = \frac{\text{load}}{\text{area}}$$

$$\therefore \quad \text{load on one rivet} = \text{stress} \times \text{area of one rivet}$$

$$610\,240d = 320 \times 10^5 \times \frac{\pi}{4} d^2$$

$$d = \frac{610\,240 \times 4}{320 \times 10^5 \times \pi}$$

$$= 0.0243 \text{ m}$$
$$= 24 \text{ mm}$$

34 (a) Change in draught due to density

$$= \frac{22\,000 \times 100}{3200} \left(\frac{1.026 - 1.008}{1.008 \times 1.026}\right)$$

$$= 11.97 \text{ cm reduction}$$
$$\text{Change in displacement} = 11.97 \times 3200 \times 1.026 \times 10^{-2}$$
$$= 393 \text{ tonne}$$

$$\% \text{ difference} = \frac{393}{22\,000} \times 100$$

$$= 1.79\% \text{ increase}$$

$$\text{(b)} \qquad \text{New draught} = 9.00 - 0.1197$$
$$= 8.8803 \text{ m}$$
$$\text{Final draught} = 8.55 \text{ m}$$
$$\text{Change in draught} = 0.3303 \text{ m}$$
$$\text{Change in displacement} = 0.3303 \times 3200 \times 1.026$$
$$= 1084 \text{ tonne}$$

$$\% \text{ difference} = \frac{1084}{22\,000} \times 100$$

$$= 4.93\% \text{ reduction}$$

35. Fuel cons/day $\propto \Delta^{\frac{2}{3}} V^3$

Let V = new ship speed in knots

Then new cons/day $= \left(\frac{14\,500}{15\,500}\right)^{\frac{2}{3}} \times \left(\frac{V}{14}\right)^3 \times 60$

$= 0.02092\, V^3$ tonne

Number of days $= \frac{640}{24V}$

$\therefore$ Voyage cons $= \frac{640}{24V} \times 0.02092 V^3$

$175 = \frac{640}{24} \times 0.02092\, V^2$

$V^2 = 313.7$

Ship speed $V = 17.71$ knot

36.

TPC	SM	product
19.0	1	19.0
18.4	4	73.6
17.4	2	34.8
16.0	4	64.0
13.8	2	27.6
11.0	4	44.0
6.6	1	6.6
		269.6

$h = 1.2$ m

Displacement $= \frac{1.2}{3} \times 269.6 \times 100$

$= 10\,784$ tonne

Load draught $= 1.2 \times 6$

$= 7.2$ m

37. Change in draught due to density

$$= \frac{12\,000 \times 100 \times 1.025}{17.7 \times 100} \left(\frac{1.025 - 1.012}{1.012 \times 1.025}\right)$$

$= 8.71$ cm reduction

Change in draught due to removal of 130 tonne

$$= \frac{130}{17.7}$$

$= 7.344$ cm reduction

Total change in draught

$= 8.71 + 7.344$

$= 16.054$ cm

i.e. Summer Load Line would have been submerged 16.05 cm.

38. For similar ships

$$\Delta \propto L^3$$
$$L \propto \Delta^{\frac{1}{3}}$$

$$\therefore \quad \text{Length } L = 85\sqrt[3]{\frac{3000}{2543}}$$

$$= 85 \times 1.057$$
$$= 89.81 \text{ m}$$

$$\text{Breadth } B = 10.5 \times 1.057$$
$$= 11.09 \text{ m}$$

$$\text{Draught } d = 5 \times 1.057$$
$$= 5.28 \text{ m}$$

$$\text{Corresponding speed} \propto \Delta^{\frac{1}{6}}$$

$$\therefore \quad V = 14\sqrt[6]{\frac{3000}{2543}}$$

$$= 14.39 \text{ knots}$$

39.

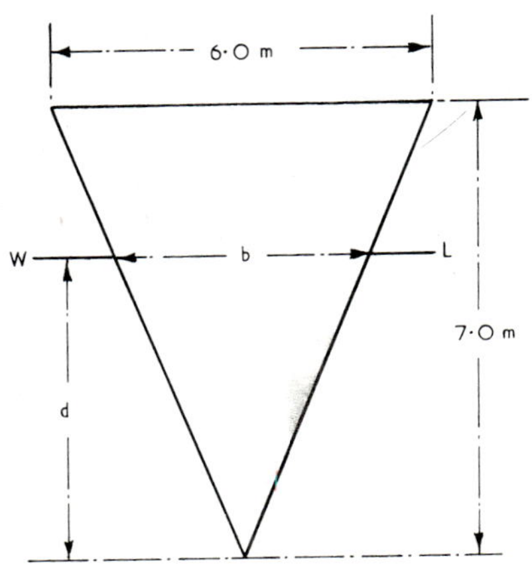

Fig. 113

The effective area of the bulkhead will depend upon whether the water level is above or below the top. Assume that the water is *at* the top.

$$\text{Load on bulkhead} = \rho g A H$$
$$= 1.025 \times 9.81 \times \frac{6 \times 7}{2} \times \frac{7}{3}$$
$$= 492.7 \text{ kN}$$

Thus the water must lie below the top.

Let d = depth of water;
b = width of bulkhead at water level.

$$b = \frac{6}{7} \times d$$

$$\text{Load on bulkhead} = 1.025 \times 9.81 \times \frac{b \times d}{2} \times \frac{d}{3}$$

$$195 = 1.025 \times 9.81 \times \frac{6}{7} d \times \frac{d}{2} \times \frac{d}{3}$$

$$d^3 = \frac{195 \times 7 \times 2 \times 3}{1.025 \times 9.81 \times 6}$$

$$\text{Depth of water} \quad d = 5.14 \text{ m}$$

40.
$$GM = KB + BM - KG$$
$$\therefore \quad BM = KG + GM - KB$$
$$= 6.60 + 0.46 - 3.60$$
$$= 3.46 \text{ m}$$

But
$$BM = \frac{I}{\nabla}$$

$$\therefore \quad I = BM \times \nabla$$
$$= 3.46 \times \frac{8000}{1.025}$$
$$\text{Second moment of area} = 27\,005 \text{ m}^4$$

41. When 120 tonne transferred:

$$\text{Shift in CG} = \frac{m \times d}{\Delta}$$
$$d = \frac{\Delta \times GG_1}{m}$$
$$= \frac{9800 \times 0.75}{120}$$
$$\text{Distance between tanks} = 61.25 \text{ m}$$

When 420 tonne used:

Let x = distance from original CG to cg of forward tank.

Taking moments about the original CG:

$$
\begin{aligned}
(9800-420)(-0.45) &= 9800 \times 0.75 - 420 \times x \\
-4221 &= 7350 - 420x \\
420x &= 7350 + 4221 \\
x &= \frac{11\,571}{420} \\
&= 27.55\ \text{m} \\
d - x &= 61.25 - 27.55 \\
&= 33.70\ \text{m}
\end{aligned}
$$

Thus the forward tank is 27.55 m from the vessel's original CG and the after tank is 33.70 m from the vessel's original CG.

42.

$$
\begin{aligned}
\text{Fuel cons day} &= \frac{18\,000^{\frac{2}{3}} \times 12^3}{47250} \\
&= 25.12\ \text{tonne} \\
\text{Number of days} &= \frac{500}{25.12} \\
&= 19.91
\end{aligned}
$$

(a)

$$
\begin{aligned}
\text{Range of operation} &= 19.91 \times 12 \times 24 \\
&= 5734\ \text{nautical miles}
\end{aligned}
$$

(b)

$$
\begin{aligned}
\text{new speed} &= 12 \times 1.05 \\
&= 12.6\ \text{knot} \\
\text{new cons/day} &= 25.12 \times \left(\frac{12.6}{12}\right)^3 \\
&= 29.08\ \text{tonne} \\
\text{number of days} &= \frac{500}{29.08} \\
&= 17.19 \\
\text{Range of operation} &= 17.19 \times 12.6 \times 24 \\
&= 5200\ \text{nautical miles} \\
\text{Percentage difference} &= \frac{5734 - 5200}{5734} \times 100 \\
&= 9.31\%\ \text{reduction}
\end{aligned}
$$

43. TPC in sea water $= 1625 \times 0.01024$
$= 16.64$

$$\text{Bodily sinkage} = \frac{1650}{16.64}$$

$$= 99.16 \text{ cm}$$

Thus at 12 000 tonne displacement:

Mean draught in sea water $= 9.08 - 0.992$
$= 8.088 \text{ m}$

Thus change in draught due to density
$= 8.16 - 8.088$
$= 0.072 \text{ m}$

Let ρ_R = density of harbour water in t/m³

Then
$$7.2 = \frac{12\,000 \times 100}{1650}\left(\frac{1.024 - \rho_R}{1.024\rho_R}\right)$$

$$1.024\ \rho_R \times 7.2 = 727.3(1.024 - \rho_R)$$

$$\rho_R(7.37 + 727.3) = 727.3 \times 1.024$$

$$\rho_R = \frac{727.3 \times 1.024}{734.67}$$

Density of harbour water $= 1.014 \text{ t/m}^3$

44. Mean deflection $= \frac{1}{4}(73 + 80 + 78 + 75)$
$= 76.5 \text{ mm}$

$$GM = \frac{m \times d}{\Delta \tan\theta}$$

$$= \frac{5 \times 12 \times 6}{8000 \times 0.0765}$$

$$= 0.588 \text{ m}$$

$KG = KM - GM$
$= 5.10 - 0.588$
$= 4.512 \text{ m}$

45. (a) Load on bulkhead $= \rho g A H$
$= 1.025 \times 9.81 \times 17 \times 6 \times (3 + 2.5)$
$= 5641 \text{ kN}$

(b) Pressure at top of bulkhead

$$= \rho g h$$
$$= 1.025 \times 9.81 \times 2.5$$
$$= 25.14 \text{ kN/m}^2$$

Pressure at bottom of bulkhead

$$= 1.025 \times 9.81 \times (6 + 2.5)$$
$$= 85.47 \text{ kN/m}^2$$

46.

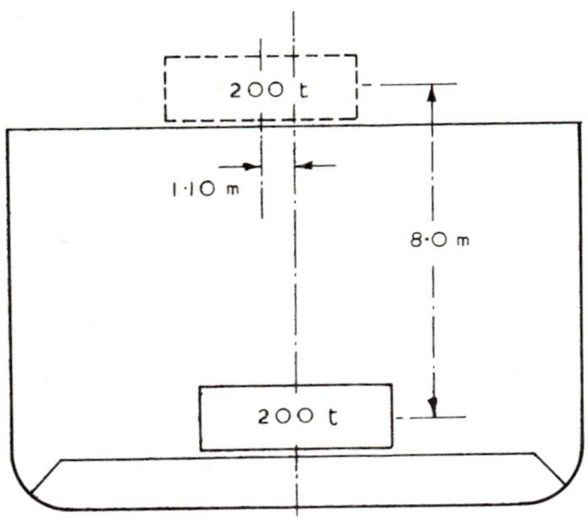

Fig. 114

$$\text{Rise in } G = \frac{200 \times 8}{4600}$$

$$= 0.348 \text{ m}$$

$$\text{New } GM = 0.77 - 0.348$$

$$= 0.422 \text{ m}$$

$$\tan\theta = \frac{200 \times 1.1}{4600 \times 0.422}$$

$$= 0.1133$$

$$\text{Angle of heel } \theta = 6° \ 27'$$

47. $$TPC = A_w \times 0.01025$$

Draught	A_w	TPC	SM_1	$product_1$	SM_2	$product_2$
7.5	2100	21.52	1	21.52		
6.0	1930	19.78	4	79.12		
4.5	1720	17.63	2	35.26	1	17.63
3.0	1428	14.64	4	58.56	4	58.56
1.5	966	9.90	1	9.90	1	9.90
				204.36		86.09

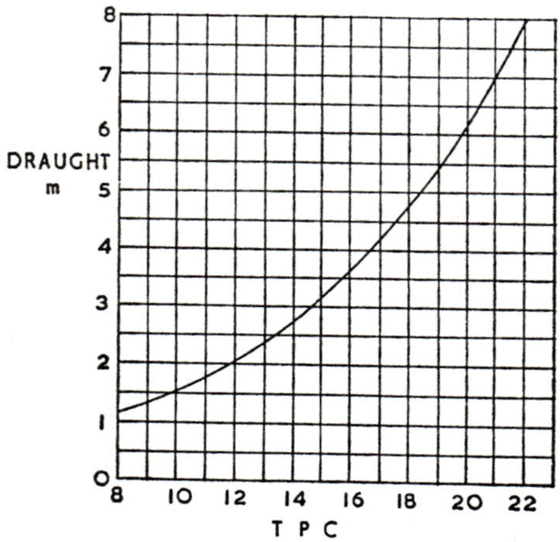

Fig. 115

(a) Displacement at 7.5 m draught

$$= 711 + \frac{1.5}{3} \times 100 \times 204.36$$

$$= 711 + 10\,218$$
$$= 10\,929 \text{ tonne}$$

(b) Displacement at 4.5 m draught

$$= 711 + \frac{1.5}{3} \times 100 \times 86.09$$

$$= 711 + 4305$$
$$= 5016 \text{ tonne}$$

48.

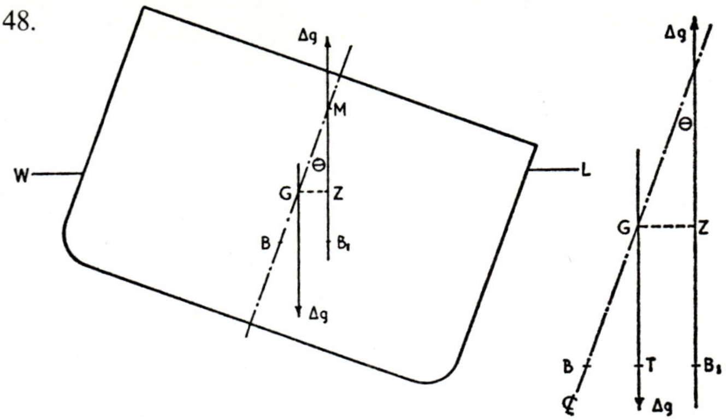

Fig. 116

$$\text{Righting moment} = \Delta \times GZ$$
$$570 = 8000 \times GZ$$

$$GZ = \frac{570}{8000}$$
$$= 0.0713 \text{ m}$$

$$BT = BG \sin\theta$$
$$= 1.0 \times 0.1219$$
$$= 0.1219 \text{ m}$$

$$\text{Horizontal movement of B} = BB_1$$
$$= BT + TB_1$$
$$= BT + GZ$$
$$= 0.1219 + 0.0713$$
$$= 0.1932 \text{ m}$$

49.

$$\text{Speed of advance} = 12 \times \frac{1852}{3600}$$
$$= 6.173 \text{ m/s}$$

$$\text{Real slip} = \frac{vt - va}{vt}$$

$$0.30 = \frac{vt - 6.173}{vt}$$

$$vt = \frac{6.173}{1 - 0.30}$$

$$= 8.819 \text{ m/s}$$

$$= P \times n$$

(a) $$P = \frac{8.819}{2}$$

$$= 4.410 \text{ m}$$

(b) $$\text{Thrust power} = 300 \times 6.173$$
$$= 1852 \text{ kW}$$

(c) $$\text{Delivered power} = 250 \times 2\pi \times 2$$
$$= 3142 \text{ kW}$$

50.

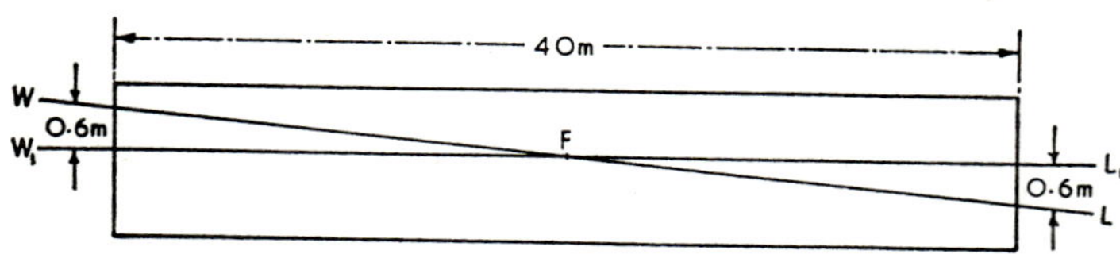

Fig. 117

If it is assumed that the mass is first added amidships, there will be a bodily increase in draught without change of trim.

To obtain a level keel draught the wedge of buoyancy WFW_1 must be transferred to L_1FL.

$$\text{Mass of wedge} = 20 \times 7.5 \times 0.6 \times \tfrac{1}{2} \times 1.025$$
$$= 46.125 \text{ tonne}$$

$$\text{Distance moved} = \tfrac{2}{3} \times 40$$

Let x = distance moved forward by 90 tonne, then

$$90x = 46.125 \times \tfrac{2}{3} \times 40$$

$$x = \frac{46.125 \times 2 \times 40}{90 \times 3}$$

$$= 13.67 \text{ m}$$

51.

$$\text{TPC} = \frac{L \times B \times 1.025}{100}$$

$$L = \frac{17 \times 100}{16 \times 1.025}$$

$$= 103.6 \text{ m}$$

Volume of lost buoyancy = $20 \times 16 \times 6 \times 0.80$
Area of intact waterplane = $103.6 \times 16 - 20 \times 16 \times 0.8$

$$\text{Increase in draught} = \frac{20 \times 16 \times 6 \times 0.8}{103.6 \times 16 - 20 \times 16 \times 0.8}$$

$$= 1.096 \text{ m}$$

(a) New draught = $6 + 1.096$

$$= 7.096 \text{ m}$$

(b)

$$KB = \frac{7.096}{2}$$

$$= 3.548 \text{ m}$$

$$BM = \frac{75\,000}{103.6 \times 16 \times 6}$$

$$= 7.541 \text{ m}$$

$$KM = 3.548 + 7.541$$

$$= 11.089 \text{ m}$$

52.

$$KB = \frac{d}{2}$$

$$BM = \frac{B^2}{12d}$$

$$= \frac{7.2^2}{12d}$$

$$= \frac{4.32}{d}$$

Draught	*KB*	*BM*	*KM*
0	0	∞	∞
1	0.5	4.32	4.82
2	1.0	2.16	3.16
3	1.5	1.44	2.94
4	2.0	1.08	3.08
5	2.5	0.86	3.36
6	3.0	0.72	3.72

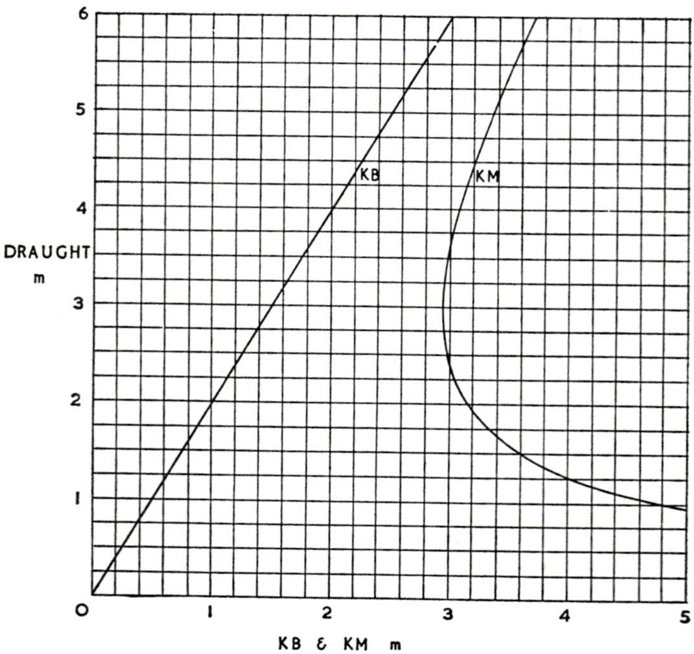

Fig. 118

53.

$$\text{Propeller pitch} = 1.1 \times 4.28$$
$$= 4.708 \text{ m}$$

$$vt = 4.708 \times 2$$

$$= 9.416 \text{ m/s}$$

$$\text{Apparent slip } 0.7 = \frac{9.416 - v}{9.416} \times 100$$

$$v = 9.416(1 - 0.007)$$

$$= 9.34 \text{ m/s}$$

$$\text{Real slip } 12 = \frac{9.416 - va}{9.416} \times 100$$

$$va = 9.416(1 - 0.12)$$

$$= 8.286 \text{ m/s}$$

$$\text{Wake speed} = 9.34 - 8.286$$
$$= 1.054 \text{ m/s}$$
$$= 2.05 \text{ knot}$$

54.

$$\text{At 12 knot; Cons/day} = \frac{0.55 \times 1710 \times 24}{1000}$$

$$= 22.572 \text{ tonne}$$

(a)

$$\text{At 10 knot: Cons/day} = 22.572 \times \left(\frac{10}{12}\right)^3$$

$$= 13.062 \text{ tonne}$$

$$\text{Number of days} = \frac{7500}{10 \times 24}$$

$$= 31.25$$
$$\text{Fuel required} = 13.062 \times 31.25$$
$$= 408.2 \text{ tonne}$$

(b)

$$\text{Fuel coefficient} = \frac{6000^{\frac{2}{3}} \times 12^3}{22.572}$$

$$= 25\,278$$

55.

$$KG = KB + BM - GM$$

$$KB = 4.5\text{ m}$$

$$BM = \frac{I}{\nabla}$$

$$= \frac{42.5 \times 10^3 \times 1.025}{12\,000}$$

$$= 3.630\text{ m}$$

$$GM = 0.6\text{ m}$$

$$\therefore \quad KG = 4.50 + 3.63 - 0.60$$
$$= 7.53\text{ m}$$

56.

$$\text{Original displacement} = 37 \times 6.4 \times 2.5 \times 1.025$$

$$= 606.8\text{ tonne}$$

$$\text{New mean draught} = \frac{2.4 + 3.8}{2}$$

$$= 3.10\text{ m}$$

$$\text{New displacement} = 37 \times 6.4 \times 3.1 \times 1.00$$

$$= 734.08\text{ tonne}$$

$$\text{Mass added} = 734.08 - 606.8$$

$$= 127.28\text{ tonne}$$

57.

$$\text{Ship speed} = 16 \times \frac{1852}{3600}$$

$$= 8.231\text{ m/s}$$

$$R_f = fSv^n$$

$$= 1.42 \times 3400 \times 8.231^{1.83}$$

$$= 228.58 \text{ kN}$$

$$S = 2.56\sqrt{\Delta L}$$

$$\Delta = \left(\frac{S}{2.56}\right)^2 \times \frac{1}{L}$$

$$= \left(\frac{3400}{2.56}\right)^2 \times \frac{1}{145}$$

$$= 12\,165 \text{ tonne}$$

$$R_r = 7.1 \times 12\,165$$

$$= 86.37 \text{ kN}$$

$$R_t = 228.58 + 86.37$$

$$= 314.95 \text{ kN}$$

Effective power $\quad \text{ep} = 314.95 \times 8.231$

$$= 2592 \text{ kW}$$

58.

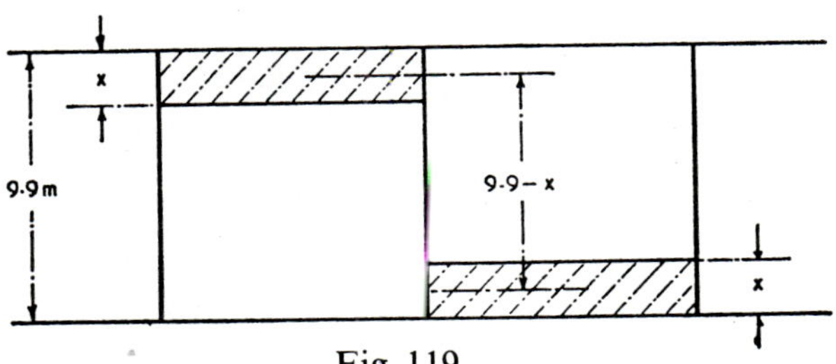

Fig. 119

$$\text{Mass of fuel/ m depth} = \frac{495}{9.9}$$

$$= 50 \text{ tonne}$$

Let x = depth of fuel which must be transferred

Then $50x$ = mass of fuel transferred

$9.9 - x$ = distance which fuel is transferred

$$\text{Shift in CG} = \frac{m \times d}{\Delta}$$

$$0.12 = \frac{50x \times (9.9 - x)}{7200}$$

$$= \frac{x \times (9.9 - x)}{144}$$

$$0.12 \times 144 = 9.9x - x^2$$

$$x^2 - 9.9x + 17.28 = 0$$

$$x = \frac{9.9 \pm \sqrt{9.9^2 - 4 \times 17.28}}{2}$$

$$= \frac{9.9 \pm 5.375}{2}$$

$$= 7.638 \text{ m or } 2.262 \text{ m}$$

Mass of oil transferred = 50 × 7.638
= 381.9 tonne
or 50 × 2.262
= 113.1 tonne

59.

Station	½ ordinate	SM	product
0	0.06	½	0.03
½	1.25	2	2.50
1	2.68	1½	4.02
2	4.55	4	18.20
3	5.30	2	10.60
4	5.20	4	20.80
5	4.20	1½	6.30
5½	2.60	2	5.20
6	0	½	0
			67.65

$$h = \frac{75}{6}$$

$$= 12.5 \text{ m}$$

$$\text{Waterplane area} = \tfrac{2}{3} \times 12.5 \times 67.65$$
$$= 563.75 \text{ m}^2$$

$$\text{TPC} = A_w \times 0.01025$$
$$= 563.75 \times 0.01025$$
$$= 5.778$$

60.

$$\text{Final } KG = 7.30 - 0.50$$
$$= 6.80 \text{ m}$$

Let x = distance of original CG above the keel

mass	Kg	moment
7000	x	$7000x$
150	$x + 3$	$150x + 450$
60	$x - 5.5$	$60x - 330$
− 76	0.6	−45.6
7134		$7210x + 74.4$

$$7134 \times 6.8 = 7210x + 74.4$$

$$7210x = 48\,511 - 74$$
$$= 48\,437$$

$$x = \frac{48\,437}{7210}$$

$$\text{Original } KG = 6.73 \text{ m}$$

SELECTION OF EXAMINATION QUESTIONS

FIRST CLASS

** Questions marked with an asterisk have been selected from Department of Trade papers and are reproduced by kind permission of The Controller of Her Majesty's Stationary Office.*

1. A ship 120 m long has draughts of 6.6 m forward and 6.9 m aft. The TPC is 20, MCTI cm 101 tonne m and the centre of flotation 3.5 m aft of midships. Calculate the maximum position aft at which 240 tonne may be added so that the after draught does not exceed 7.2 m.

2. A vessel, when floating at a draught of 3.6 m has a displacement of 8172 tonne, KB 1.91 m and LCB 0.15 m aft of midships. From the following information, calculate the displacement, KB and position of the LCB for the vessel when floating at a draught of 1.2 m.

Draught (m)	TPC	LCF from midships
1.2	23.0	1.37 F
2.4	24.2	0.76 A
3.6	25.0	0.92 A

3. An oil tanker has LBP 142 m, beam 18.8 m and draught 8 m. It displaces 17 000 tonne in sea water of 1.025 t/m³. The face pitch ratio of the propeller is 0.673 and the diameter 4.8 m. The results of the speed trial show that the true slip may be regarded as constant over a range of speeds of 9 to 12 knots and is 35%. The wake fraction may be calculated from the equation:

$$w = 0.5C_b - 0.05$$

If the vessel uses 20 tonne of fuel per day at 12 knots, and the consumption varies as $(\text{speed})^3$, find the consumption per day at 100 rev/min.

4. A ship of 5000 tonne displacement has a KM of 6.4 m. When 5 tonne are moved 15 m across the ship a pendulum 6 m long has a deflection of 12 cm. A double bottom tank 7.5 m long, 9 m wide and 1.2 m deep is half-full of sea water.

Calculate the KG of the light ship.

5. A propeller has a pitch ratio of 0.95. When turning at 120 rev/min the real slip is 30%, the wake fraction 0.28 and the ship speed 16 knots. The thrust is found to be 400 kN, the torque 270 kN m and the QPC 0.67. Calculate:

(a) the propeller diameter
(b) the shaft power
(c) the propeller efficiency
(d) the thrust deduction factor.

6. A pontoon has a constant cross-section as shown in Fig. 120. The metacentric height is 2.5 m. Find the height of the centre of gravity above the keel.

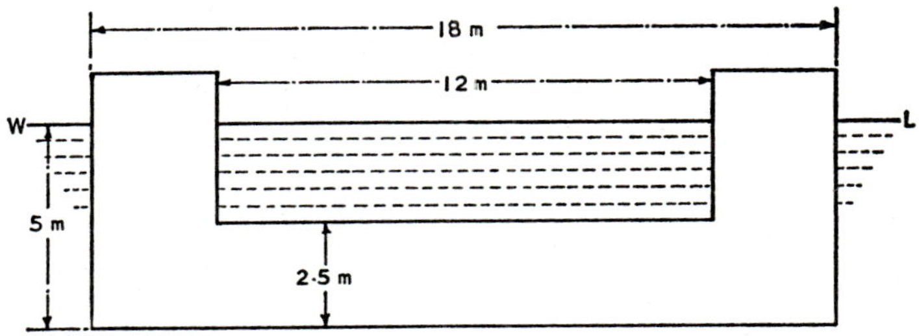

Fig. 120

7. (a) Derive the Admiralty Coefficient formula and show how this may be modified to suit a fast ship.

(b) A ship of 14 000 tonne displacement requires 23 000 kW shaft power to drive it at 24 knots. Using the modified Admiralty Coefficient formula, calculate the shaft power required for a similar ship of 12 000 tonne displacement at 21 knots.

8. A ship 80 m long has equally-spaced immersed cross-sectional areas of 0, 11.5, 27, 38.5, 44, 45, 44.5, 39, 26.5, 14.5 and 0 m^2 respectively. Calculate:

(a) displacement
(b) distance of centre of buoyancy from midships
(c) prismatic coefficient.

9. The following data refer to two similar ships

	L	S	V	ep_n	f_{sw}
Ship A	160	4000	18	6400	0.420
Ship B	140				0.425

Calculate ep_n for ship B at the corresponding speed.

10. A ship of 11 200 tonne displacement has a double bottom tank containing oil, whose centre of gravity is 16.5 m forward and 6.6 m below the centre of gravity of the ship. When the oil is used the ship's centre of gravity moves 380 mm. Calculate:

(a) the mass of oil used
(b) the angle which the centre of gravity moves relative to the horizontal.

11. A watertight door is 1.2 m high and 0.75 m wide, with a 0.6 m sill. The bulkhead is flooded with sea water to a depth of 3 m on one side and 1.5 m on the other side. Draw the load diagram and from it determine the resultant load and position of the centre of pressure on the door.

12. A ship of 10 000 tonne displacement has *KM* 8 m and *GM* 0.6 m. A rectangular double bottom tank is 1.5 m deep, 18 m long and 15 m wide. Assuming that *KM* remains constant, determine the new *GM* when the tank is now:

(a) filled with sea water
(b) half-filled with sea water.

13. A propeller 6 m diameter has a pitch ratio of 0.9, BAR 0.48 and, when turning at 110 rev/min, has a real slip of 25% and wake fraction 0.30. If the propeller delivers a thrust of 300 kN and the propeller efficiency is 0.65, calculate:

(a) blade area
(b) ship speed
(c) thrust power
(d) shaft power
(e) torque.

14. When a ship is 800 nautical miles from port its speed is reduced by 20%, thereby reducing the daily fuel consumption by 42 tonne and arriving in port with 50 tonne on board. If the fuel consumption in t/h is given by the expression $(0.136 + 0.001\ V^3)$ where V is the speed in knots, estimate:

(a) the reduced consumption per day
(b) the amount of fuel on board when the speed was reduced

(c) the percentage decrease in consumption for the latter part of the voyage
(d) the percentage increase in time for this latter period.

15. An oil tanker 160 m long and 22 m beam floats at a draught of 9 m in sea water. C_w is 0.865.

The midship section is in the form of a rectangle with 1.2 m radius at the bilges. A midship tank 10.5 m long has twin longitudinal bulkheads and contains oil of 1.4 m^3/t to a depth of 11.5 m. The tank is holed to the sea for the whole of its transverse section. Find the new draught.

16. A ship 160 m long and 8700 tonne displacement floats at a waterline with half-ordinates of: 0, 2.4, 5.0, 7.3, 7.9, 8.0, 8.0, 7.7, 5.5, 2.8 and 0 m respectively. While floating at this waterline, the ship develops a list of 10° due to instability. Calculate the negative metacentric height when the vessel is upright in this condition.

17. The speed of a ship is increased to 18% above normal for 7.5 hours, then reduced to 9% below normal for 10 hours. The speed is then reduced for the remainder of the day so that the consumption for the day is the normal amount. Find the percentage difference between the distance travelled in that day and the normal distance travelled per day.

18. A double bottom tank containing sea water is 6 m long, 12 m wide and 1 m deep. The inlet pipe from the pump has its centre 75 mm above the outer bottom. The pump has a pressure of 70 kN/m^2 and is left running indefinitely. Calculate the load on the tank top:

(a) if there is no outlet
(b) if the overflow pipe extends 5 m above the tank top.

19. A ship 120 m long displaces 12 000 tonne. The following data are available from trial results:

V (knots	10	11	12	13	14	15
sp (kW)	880	1155	1520	2010	2670	3600

(a) Draw the curve of Admiralty Coefficients on a base of speed
(b) Estimate the shaft power required for a similar ship 140 m long at 14 knots.

20. A box barge 60 m long and 10 m wide floats at a level keel draught of 3 m. Its centre of gravity is 2.5 m above the keel. Determine the end draughts if an empty, fore end compartment 9 m long is laid open to the sea.

21. A vessel of constant triangular cross-section floats apex down at a draught of 4 m, the width of the waterplane being 8 m, when its keel just touches a layer of mud having relative density twice that of the water. The tide now falls 2 m. Calculate the depth to which the vessel sinks in the mud.

22. The following information relates to a model propeller of 400 mm pitch:

Rev/min		400	450	500	550	600
Thrust	N	175	260	365	480	610
Torque	Nm	16.8	22.4	28.2	34.3	40.5

(a) Plot curves of thrust and torque against rev/min
(b) When the speed of advance of the model is 150 m/min and slip 0.20, calculate the efficiency.

23. A ship of 8100 tonne displacement floats upright in sea water. KG = 7.5 m and GM = 0.45 m. A tank, whose centre of gravity is 0.5 m above the keel and 4 m from the centreline, contains 100 tonne of water ballast. Neglecting free surface effect, calculate the angle of heel when the ballast is pumped out.

24. The ½ ordinates of a waterplane 120 m long are 0.7, 3.3, 5.5, 7.2, 7.5, 7.5, 7.5, 6.8, 4.6, 2.2 and 0 m respectively. The ship displaces 11 000 tonne. Calculate the transverse BM.

25. A box barge 85 m long, 18 m beam and 6 m draught floats in sea water of 1.025 t/m³. A midship compartment 18 m long contains cargo stowing at 1.8 m³/t and having a density of 1.600 t/m³. There is a watertight flat 6 m above the keel. Calculate the new draught if this compartment is bilged below the flat.

26. The ½ ordinates of a waterplane 90 m long are as follows:

Station	AP	½	1	2	3	4	5	6	7	7½	FP
½ ordinate	0.6	2.7	4.6	6.0	6.3	6.3	6.3	5.7	4.8	2.0	0 m

Calculate the area of the waterplane and the distance of the centre of flotation from midships.

27. A ship of 6600 tonne displacement has KG 3.6 m and KM 4.3 m. A mass of 50 tonne is now lifted from the quay by one of the ship's derricks whose head is 18 m above the keel. The ship heels to a maximum of 9.5° while the mass is being transferred. Calculate the outreach of the derrick from the ship's centreline.

28. A ship 120 m long displaces 10 500 tonne and has a wetted surface area of 3000 m². At 15 knots the shaft power is 4100 kW, propulsive coefficient 0.6 and 55% of the thrust is available to overcome frictional resistance.

Calculate the shaft power required for a similar ship 140 m long at the corresponding speed. $f = 0.42$ and $n = 1.825$.

29. A box-shaped vessel 30 m long and 9 m wide floats in water of 1.025 t/m³ at a draught of 0.75 m when empty. The vessel moves from water of 1.000 t/m³ to water of 1.025 t/m³ in a partially laden state and, on reaching the sea water, it is found that the mean draught is reduced by 3.2 cm. Calculate the mass of cargo on board.

30. A ship of 5000 tonne displacement has a double bottom tank 12 m long. The ½ breadths of the top of the tank are 5, 4 and 2 m respectively. The tank has a watertight centreline division. Calculate the free surface effect if the tank is partially full of fresh water *on one side only*.

* 31. The following data apply to a ship operating at a speed of 15 knots:

Shaft power	=	3050 kW
Propeller speed	=	1.58 rev/s
Propeller thrust	=	360 kN
Apparent slip	=	0

Calculate the propeller pitch, real slip and the propulsive coefficient if the Taylor wake fraction and thrust deduction factor are 0.31 and 0.20 respectively.

32. The force acting normal to the plane of a rudder at angle α is given by:

$$F_n = 577\, A\, v^2 \sin\alpha \quad \text{N}$$

where A = area of rudder = 22 m^2

and v = water speed in m/s.

When the rudder is turned to 35°, the centre of effort is 1.1 m from the centreline of stock. Allowing 20% for race effect, calculate the diameter of the stock if the maximum ship speed is 15 knots and the maximum allowable stress is 70 MN/m^2.

If the effective diameter is reduced by corrosion and wear to 330 mm, calculate the speed at which the vessel must travel so that the above stress is not exceeded.

33. A ship 100 m long and 15 m beam floats at a mean draught of 3.5 m. The semi-ordinates of the waterplane at equal intervals are: 0, 3.0, 5.5, 7.3, 7.5, 7.5, 7.5, 7.05, 6.10, 3.25 and 0 m respectively. The section amidships is constant and parallel for 20 m and the submerged cross-sectional area is 50 m^2 at this section.

Calculate the new mean draught when a midship compartment 15 m long is opened to the sea. Assume the vessel to be wall-sided in the region of the waterplane.

34. A ship 85 m long displaces 8100 tonne when floating in sea water at draughts of 5.25 m forward and 5.55 m aft. TPC 9.0, GM_L 96 m, LCF 2 m aft of midships. It is decided to introduce water ballast to completely submerge the propeller and a draught aft of 5.85 m is required. A ballast tank 33 m aft of midships is available. Find the least amount of water required and the final draught forward.

35. A solid block of wood has a square cross-section of side S and length L greater than S. Calculate the relative density of the wood if it floats with its sides vertical in fresh water.

36. A ship travelling at 15.5 knots has a propeller of 5.5 m pitch turning at 95 rev/min. The thrust of the propeller is 380 kN and the delivered power 3540 kW. If the real slip is 20% and the thrust deduction factor 0.198, calculate the QPC and the wake fraction.

37. (a) Describe briefly the inclining experiment and explain how the results are used.

(b) A ship of 8500 tonne displacement has a double bottom tank 11 m wide extending for the full breadth of the ship, having a free surface of sea water. If the apparent loss in metacentric height due to slack water is 14 cm, find the length of the tank.

38. (a) Derive an expression for the change in draught of a vessel moving from sea water into river water.

(b) A ship of 8000 tonne displacement has TPC 17 when at a level keel draught of 7 m in sea water of 1.024 t/m³. The vessel then moves into water of 1.008 t/m³. The maximum draught at which the vessel may enter dock is 6.85 m. Calculate how much ballast must be discharged.

39 (a). What is meant by the Admiralty Coefficient and the Fuel Coefficient?

(b) A ship of 14 900 tonne displacement has a shaft power of 4460 kW at 14.55 knots. The shaft power is reduced to 4120 kW and the fuel consumption at the same displacement is 541 kg/h. Calculate the fuel coefficient for the ship.

40. A ship of 12 000 tonne displacement has a rudder 15 m² in area, whose centre is 5 m below the waterline. The metacentric height of the ship is 0.3 m and the centre of buoyancy is 3.3 m below the waterline. When travelling at 20 knots the rudder is turned through 30°. Find the initial angle of heel if the force F_n perpendicular to the plane of the rudder is given by:

$$F_n = 577\,A\,v^2 \sin\alpha \quad \text{N}$$

Allow 20% for the race effect.

41. A ship 120 m long has a light displacement of 4000 tonne and LCG in this condition 2.5 m aft of midships.

The following items are then added:

Cargo	10 000 tonne	Lcg 3.0 m	forward of midships
Fuel	1500 tonne	Lcg 2.0 m	aft of midships
Water	400 tonne	Lcg 8.0 m	aft of midships
Stores	100 tonne	Lcg 10.0 m	forward of midships

Using the following hydrostatic data, calculate the final draughts:

Draught (m)	Displacement (t)	MCT1 cm (t m)	LCB from midships (m)	LCF from midships (m)
8.50	16 650	183	1.94 F	1.20 A
8.00	15 350	175	2.10 F	0.06 F

42. A box barge 30 m long and 9 m beam floats at a draught of 3 m. The centre of gravity lies on the centreline and KG is 3.50 m. A mass of 10 tonne, which is already on board, is now moved 6 m across the ship.

(a) Estimate the angle to which the vessel will heel, using the formula:

$$GZ = \sin\theta \left(GM + \tfrac{1}{2} BM \tan^2\theta \right)$$

(b) Compare the above result with the angle of heel obtained by the metacentric formula.

43. The fuel consumption of a ship at 17 knots is 47 tonne/day. The speed is reduced and the consumption is reduced to 22 tonne/day. At the lower speed, however, the consumption per unit power is 13.2% greater than at 17 knots. Find the reduced speed and the percentage saving on a voyage of 3000 nautical miles.

44. A ship of 14 000 tonne displacement is 135 m long and floats at draughts of 7.30 m forward and 8.05 m aft. GM_L is 127 m, TPC 18 and LCF 3.0 m aft of midships. Calculate the new draughts when 180 tonne of cargo are added 40 m forward of midships.

45. A propeller has a pitch of 5.5 m. When turning at 93 rev/min the apparent slip is found to be – S% and the real slip + S%, the wake speed being 10% of the ship's speed. Calculate the speed of the ship, the apparent slip and the real slip.

46. The ½ ordinates of a waterplane at 15 m intervals, commencing from aft, are 1, 7, 10.5, 11, 11, 10.5, 8, 4 and 0 m. Calculate:

(a) TPC

(b) distance of the centre of flotation from midships

(c) second moment of area of the waterplane about a transverse axis through the centre of flotation.

Note:—Second moment of area about any axis y – y which is parallel to an axis N – A through the centroid and distance x from it, is given by:

$$I_{yy} = I_{NA} + Ax^2$$

47. A ship travelling at 12 knots has a metacentric height of 0.25 m. The distance between the centre of gravity and the centre of lateral resistance is 2.7 m. If the vessel turns in a circle of 600 m radius, calculate the angle to which it will heel.

48. The following data are available for a twin screw vessel:

V	(knots)	15	16	17	18
ep_n	(kW)	3000	3750	4700	5650
QPC		0.73	0.73	0.72	0.71

Calculate the service speed if the brake power for each engine is 3500 kW. The transmission losses are 3% and the allowances for weather and appendages 30%.

49. A ship 120 m long displaces 8000 tonne, GM_L is 102 m, TPC 17.5 and LCF 2 m aft of midships. It arrives in port with draughts of 6.3 m forward and 6.6 m aft.

During the voyage the following changes in loading have taken place:

Fuel used	200 tonne	18 m forward of midships
Water used	100 tonne	3 m aft of midships
Stores used	10 tonne	9 m aft of midships
Ballast added	300 tonne	24 m forward of midships

Calculate the *original* draughts.

50. A propeller has a pitch of 5.5 m. When turning at 80 rev/min the ship speed is 13.2 knots, speed of advance 11 knots, propeller efficiency 70% and delivered power 3000 kW. Calculate:

(a) real slip
(b) wake fraction
(c) propeller thrust.

* 51. A watertight bulkhead 6.0 m deep is supported by vertical inverted angle stiffeners 255 mm x 100 mm x 12.5 mm, spaced 0.6 m apart. The ends of the stiffeners in contact with the tank top are welded all round, and the thickness of weld at its throat is 5 mm.

Calculate the shear stress in the weld metal at the tank top when the bulkhead is covered on one side, by water of density 1025 kg/m^3, to a depth of 4.85 m.

52. A ship of 5000 tonne displacement has three rectangular double bottom tanks; A 12 m long and 16 m wide; B 14 m long and 15 m wide; C 14 m long and 16 m wide.

Calculate the free surface effect for any one tank and state in which order the tanks should be filled when making use of them for stability correction.

53. A box barge 75 m long and 8.5 m beam floats at draughts of 2.13 m forward and 3.05 m aft.

An empty compartment is now flooded and the vessel finally lies at a draught of 3.00 m level keel. Calculate the length and Lcg of the flooded compartment.

54. A vessel has a maximum allowable draught of 8.5 m in fresh water and 8.25 m in sea water of 1.026 t/m^3, the TPC in the sea water being 27.5. The vessel is loaded in river water of 1.012 t/m^3 to a draught of 8.44 m. If it now moves into sea water, is it necessary to pump out any ballast, and if so, how much?

55. A bulkead is in the form of a trapezoid 13 m wide at the deck, 10 m wide at the tank top and 7.5 m deep.

Calculate the load on the bulkead and the position of the centre of pressure if it is flooded to a depth of 5 m with sea water on one side only.

56. A double bottom tank is 23 m long. The half breadths of the top of the tank are 5.5, 4.6, 4.3, 3.7 and 3.0 m respectively. When the ship displaces 5350 tonne, the loss in metacentric height due to free surface is 0.2 m. Calculate the density of the liquid in the tank.

57. A vessel of constant rectangular cross-section is 100 m long and floats at a draught of 5 m. It has a mid-length compartment 10 m long extending right across the vessel, but sub-divided by a horizontal watertight flat 3 m above the keel. GM is 0.8 m.

Calculate the new draught and metacentric height if the compartment is bilged below the flat.

58 (a). If resistance $\propto S V^2$ and $S \propto \Delta^{\frac{2}{3}}$, derive the Admiralty Coefficient formula.

(b). A ship 160 m long, 22 m beam and 9.2 m draught has a block coefficient of 0.765. The pitch of the propeller is 4 m and when it turns at 96 rev/min the true slip is 33%, the wake fraction 0.335 and shaft power 2900 kW. Calculate the Admiralty Coefficient and the shaft power at 15 knots.

59. A ship model 6 m long has a total resistance of 40 N when towed at 3.6 knots in fresh water.

The ship itself is 180 m long and displaces 20 400 tonne. The wetted surface area may be calculated from the formula

$$S = 2.57 \sqrt{\Delta L.}$$

Calculate ep_n for the ship at its corresponding speed in sea water. $f(\text{model})_{FW} = 0.492$; $f(\text{ship})_{SW} = 0.421$; $n = 1.825$.

60 (a). Why is an inclining experiment carried out? Write a short account of the method adopted.

(b) An inclining experiment was carried out on a ship of 8000 tonne displacement. The inclining ballast was moved transversely through 12 m and the deflections of a pendulum 5.5 m long, measured from the centreline, were as follows:

3 tonne port to starboard	64 mm S
3 tonne port to starboard	116 mm S
Ballast restored	3 mm S
3 tonne starboard to port	54 mm P
3 tonne starboard to port	113 mm P

Calculate the metacentric height of the vessel.

61. A ship of 8000 tonne displacement, 110 m long, floats in sea water of 1.024 t/m³ at draughts of 6 m forward and 6.3 m aft. The TPC is 16, LCB 0.6 m aft of midships, LCF 3 m aft of midships and MCT1 cm 65 tonne m. The vessel now moves into fresh water of 1.000 t/m³. Calculate the distance a mass of 50 tonne must be moved to bring the vessel to an even keel and determine the final draught.

62. A rectangular watertight bulkhead 9 m high and 14.5 m wide has sea water on both sides, the height of water on one side being four times that on the other side. The resultant centre of pressure is 7 m from the top of the bulkead. Calculate:

(a) the depths of water

(b) the resultant load on the bulkhead.

* 63. The following values refer to a vessel 143 m in length which is to have a service speed of 14 knots:

Service speed (knots)	13.0	14.1	15.2	16.3
Effective power naked ep_n (kW)	1690	2060	2670	3400

If allowances for the above ep_n for trial and service conditions are 13 per cent and 33 per cent respectively, and the ratio of service indicated power to maximum available indicated power is 0.9, calculate using the data below:

(a) the indicated power (i p) of the engine to be fitted.
(b) the service and trial speed of the vessel if the total available ep were used.

The vessel has the following data:

Quasi-propulsive coefficient (QPC) = 0.72
Shaft losses = 3.5 per cent
Mechanical efficiency of the engine to be fitted = 87 per cent.

64. A ship of 15 000 tonne displacement has righting levers of 0, 0.38, 1.0, 1.41 and 1.2 m at angles of heel of 0°, 15°, 30°, 45° and 60° respectively and an assumed *KG* of 7.0 m. The vessel is loaded to this displacement but the *KG* is found to be 6.80 m and *GM* 1.50 m.

(a) Draw the amended stability curve

(b) Estimate the dynamical stability at 60°.

65. On increasing the speed of a vessel by 1.5 knots it is found that the daily consumption of fuel is increased by 25 tonne and the percentage increase in fuel consumption for a voyage of 2250 nautical miles is 20. Estimate:

(a) the original daily fuel consumption
(b) the original speed of the ship.

66. The end bulkhead of the wing tank of an oil tanker has the following widths at 3 m intervals, commencing at the deck: 6.0, 6.0, 5.3, 3.6 and 0.6 m. Calculate the load on the bulkhead and the position of the centre of pressure if the tank is full of oil *rd* 0.8.

* 67. The following data for a ship has been produced from propulsion experiments on a model:

Ship speed (knot)	12.50	13.25	14.00
Effective power (kW)	1440	1800	2230
QPC	0.705	0.713	0.708
Propeller efficiency	0.565	0.584	0.585
Taylor wake fraction	0.391	0.362	0.356

Determine the speed of the ship and propeller thrust when the delivered power is 2385 kW.

68. A box barge 45 m long and 15 m wide floats at a level keel draught of 2 m in sea water, the load being uniformly distributed over the full length. Two masses, each of 30 tonne, are added at 10 m from each end and 50 tonne is evenly distributed between them. Sketch the shear force diagram and give the maximum shear force.

* 69. The power delivered to a propeller is 3540 kW at a ship speed of 15.5 knots. The propeller rotates at 1.58 rev/s, develops a thrust of 378 kN and has a pitch of 4.87 m.

If the thrust deduction fraction is 0.24, real slip 30 per cent and transmission losses are 3 per cent, calculate:

(a) the effective power,
(b) the Taylor wake fraction,
(c) the propulsive coefficient,
(d) the quasi-propulsive coefficient, assuming the appendage and weather allowance is 15 per cent.

70. A ship of 14 000 tonne displacement is 125 m long and floats at draughts of 7.9 m forward and 8.5 m aft. The TPC is 19, GM_L 120 m and LCF 3 m forward of midships. It is required to bring the vessel to an even keel draught of 8.5 m. Calculate the mass which should be added and the distance of the centre of the mass from midships.

71. A ship of 4000 tonne displacement has a mass of 50 tonne on board, on the centreline of the tank top. A derrick, whose head is 18 m above the cg of the mass, is used to lift it. Find the shift in the ship's centre of gravity from its original position when the mass is:

(a) lifted just clear of the tank top
(b) raised to the derrick head
(c) placed on the deck 12 m above the tank top
(d) swung outboard 14 m.

* 72. A rectangular bulkhead 8 m wide has water of density 1000 kg/m³ to a depth of 7 m on one side and on the other side oil of density 850 kg/m³ to a depth of 4 m. Calculate:

(a) the resultant pressure on the bulkhead
(b) the position of the resultant centre of pressure.

* 73. A ship of 91.5 m length between perpendiculars contains ballast water in a forward compartment and has the following equidistant half areas of immersed sections commencing at the after perpendicular (AP).

Section	0(AP)	1	2	3	4	5	6	7	8	9	10(FP)
Half-area of immersed sections (m^2)	0.4	7.6	21.4	33.5	40.8	45.5	48.4	52.0	51.1	34.4	0

If, prior to ballasting, the ship's displacement was 5750 tonne and the position of the longitudinal centre of buoyancy (LCB) was 4.6 m forward of midships, calculate:

(a) the mass of water of density 1025 kg/m³ added as ballast,
(b) the distance of the centre of gravity of the ballast water contained in the forward compartment from midships.

74. A ship of 7500 tonne displacement has a double bottom tank 14 m long, 12 m wide and 1.2 m deep full of sea water. The centre of gravity is 6.7 m above the keel and the metacentric height 0.45 m.

Calculate the new *GM* if half of the water is pumped out of the tank. Assume that *KM* remains constant.

75. A ship 120 m long displaces 9100 tonne. It loads in fresh water of 1.000 t/m³ to a level keel draught of 6.70 m. It then moves into sea water of 1.024 t/m³. TPC in sea water 16.8, MCT1 cm 122 tonne m, LCF 0.6 aft of midships, LCB 2.25 m forward of midships. Calculate the end draughts in the sea water.

76. A box-shaped vessel is 20 m long and 10 m wide. The weight of the vessel is uniformly distributed throughout the length and the draught is 2.5 m. The vessel contains ten evenly-spaced double bottom tanks, each having a depth of 1 m.

Draw the shear force diagrams:

(a) with No. 1 and No. 10 tanks filled
(b) with No. 3 and No. 8 tanks filled
(c) with No. 5 and No. 6 tanks filled.

Which ballast condition is to be preferred from the strength point of view?

* 77. For a box-shaped barge of 216 tonne displacement, 32 m in length, 5.5 m breadth and floating in water of density 1025 kg/m³, the *KG* is 1.8 m. An item of machinery of mass 81 tonne is loaded amidships and to maintain a positive metacentric height, 54 tonne of solid ballast is taken aboard and evenly distributed over the bottom of the barge so that the average *Kg* of the ballast is 0.15 m. If in the final condition the *GM* is 0.13 m, calculate the *Kg* of the machinery.

78. The maximum allowable draught of a ship in fresh water of 1.000 t/m³ is 9.50 m and in sea water of 1.025 t/m³ is 9.27 m.

The vessel is loaded to a draught of 9.50 m in a river, but when it proceeds to sea it is found that 202 tonne of water ballast must be pumped out to prevent the maximum draught being exceeded. If the TPC in the sea water is 23, calculate the density of the river water.

* 79. The following data are recorded from tests carried out on a model propeller 0.3 m diameter rotating at 8 rev/s in water of density 1000 kg/m³.

Speed of advance *va* (m/s)	1.22	1.46	1.70	1.94
Thrust (N)	93.7	72.3	49.7	24.3
Torque (Nm)	3.90	3.23	2.50	1.61

Draw graphs of thrust and delivered power against speed of advance *va*.

A geometrically similar propeller 4.8 m diameter operates in water of 1025 kg/m³. If the propeller absorbs 3000 kW delivered power and satisfies the law of comparison, determine for the propeller:

(a) the thrust power,

(b) the efficiency.

Note: For geometrically similar propellers the thrust power and delivered power vary directly as $(\text{diameter})^{3\cdot5}$.

* 80. A ship 128 m in length, 16.75 m in breadth, has the following hydrostatic data:

Draught (m)	1.22	2.44	3.66	4.88	6.10
Waterplane area coefficient	0.78	0.82	0.85	0.88	0.90
Position of longitudinal centre of flotation (LCF) from midships (m)	1.30 for'd	1.21 for'd	0.93 for'd	0.50 for'd	0.06 aft

Calculate:

(a) the displacement in water of density 1025 kg/m^3 of a layer of shipbody between the waterplanes at 1.22 m and 6.10 m draught,

(b) for the layer

(i) the position of the longitudinal centre of buoyancy,

(ii) the position of the vertical centre of buoyancy.

SOLUTIONS TO FIRST CLASS EXAMINATION QUESTIONS

1. $$\text{Bodily sinkage} = \frac{240}{20}$$

$$= 12\,\text{cm}$$

$$\text{New draught aft} = 6.9 + 0.12$$
$$= 7.02\,\text{m}$$

i.e. the after draught may be increased by a further 0.18 m and this becomes the change in trim aft.

$$\text{Change in trim aft} = \text{total change in trim} \times \frac{56.5}{120}$$

$$18 = t \times \frac{56.5}{120}$$

$$t = 18 \times \frac{120}{56.5}$$

$$= 38.23\,\text{cm}$$

But $$t = \frac{m \times d}{\text{MCT1 cm}}$$

$$d = \frac{38.23 \times 101}{240}$$

$$= 16.09\,\text{m aft of the centre of flotation}$$
$$= 19.59\,\text{m aft of midships}$$

2.

draught	TPC	SM	product for displacement	lever*	product for vertical moment
1.2	23.0	1	23.0	1	23.0
2.4	24.2	4	96.8	2	193.6
3.6	25.0	1	25.0	3	75.0
			144.8		291.6

*Using a lever of 1 at 1.2 m draught produces a vertical moment about the *keel*.

$$\text{Displacement } 1.2\,\text{m to } 3.6\,\text{m} = \frac{1.2}{3} \times 144.8 \times 100$$

$$= 5792 \text{ tonne}$$

$$\text{Displacement } 0\,\text{m to } 3.6\,\text{m} = 8172 \text{ tonne}$$

$$\therefore \text{Displacement } 0\,\text{m to } 1.2\,\text{m} = 2380 \text{ tonne}$$

Vertical moment 1.2 m to 3.6 m

$$= \frac{1.2}{3} \times 1.2 \times 291.6 \times 100$$

$$= 14\,000 \text{ tonne m}$$

$$\text{Vertical moment } 0\,\text{m to } 3.6\,\text{m} = 8172 \times 1.91$$

$$= 15\,610 \text{ tonne m}$$

∴ Vertical moment 0 m to 1.2 m

$$= 1610 \text{ tonne m}$$

$$KB \text{ at } 1.2\,\text{m draught} = \frac{1610}{2380}$$

$$= 0.676\,\text{m}$$

draught	TPC	LCF*	TPC × LCF	SM	product for longitudinal moment
1.2	23.0	− 1.37	− 31.5	1	− 31.5
2.4	24.2	+ 0.76	+ 18.39	4	+ 73.56
3.6	25.0	+ 0.92	+ 23.0	1	+ 23.0
					+ 65.06

* Taking forward as negative and aft as positive.

Longitudinal moment 1.2 m to 3.6 m

$$= \frac{1.2}{3} \times 65.06 \times 100$$

$$= +2602.4 \text{ tonne m}$$

Longitudinal moment 0 m to 3.6 m

$$= 8172 \times 0.15$$

$$= +1225.8 \text{ tonne m}$$

∴ Longitudinal moment 0 m to 1.2 m

$$= -1376.6 \text{ tonne m}$$

$$\text{LCB at 1.2 m draught} = -\frac{1376.6}{2380}$$

$$= 0.578 \text{ m forward of midships}$$

i.e. at 1.2 m draught, the displacement is 2380 t, *KB* 0.676 m and the LCB 0.578 m forward of midships.

3. $$\text{Block coefficient } C_b = \frac{17\,000}{1.025 \times 142 \times 18.8 \times 8}$$

$$= 0.776$$

$$w = 0.5 \times 0.776 - 0.05$$
$$= 0.338$$

$$\text{Pitch} = 4.8 \times 0.673$$
$$= 3.23 \text{ m}$$

$$\text{Theoretical speed} = \frac{3.23 \times 100 \times 60}{1852}$$

$$= 10.46 \text{ knots}$$

$$\text{Speed of advance} = 10.46 \times 0.65$$
$$= 6.80 \text{ knots}$$

$$\text{Ship speed} = \frac{6.80}{1 - 0.338}$$

$$= 10.27 \text{ knots}$$

$$\text{New fuel consumption} = 20 \times \left(\frac{10.27}{12}\right)^3$$

$$= 12.54 \text{ tonne per day}$$

4. $$\text{Effective } GM = \frac{5 \times 15 \times 6}{5000 \times 0.12}$$

$$= 0.75 \text{ m}$$

$$\text{Free surface effect} = \frac{1.025}{1.025} \times \frac{7.5 \times 9^3 \times 1.025}{12 \times 5000}$$

$$= 0.093 \text{ m}$$

Hence, actual $GM = 0.7 + 0.093$

$$= 0.843 \text{ m}$$
$$KM = 6.40 \text{ m}$$
$$\therefore \quad KG = 5.557 \text{ m}$$

$$\text{Mass of water in tank} = 1.025 \times 7.5 \times 9 \times 0.6$$
$$= 41.51 \text{ tonne}$$

Taking moments about the keel:

$$\text{Light ship } KG_1 = \frac{5000 \times 5.557 - 41.51 \times 0.3}{500 - 41.51}$$
$$= 5.601 \text{ m}$$

5.

$$w = \frac{V - Va}{V}$$

$$\text{Speed of advance} \quad Va = 16(1 - 0.28)$$
$$= 11.52 \text{ knots}$$

$$\text{Real slip } s = \frac{V_t - Va}{V_t}$$

$$\text{Theoretical speed } V_t = \frac{11.52}{1 - 0.30}$$
$$= 16.46 \text{ knots}$$

But

$$V_t = \frac{P \times N \times 60}{1852}$$

$$\text{Pitch } P = \frac{16.46 \times 1852}{120 \times 60}$$
$$= 4.23 \text{ m}$$

$$\text{Pitch ratio } p = \frac{P}{D}$$

(a)

$$\therefore \text{Diameter} \quad D = \frac{4.23}{0.95}$$
$$= 4.45 \text{ m}$$

(b)

$$\text{Shaft power sp} = 2\pi \times \frac{120}{60} \times 270$$
$$= 3393 \text{ kW}$$

(*Note.* This is the power at the after end of the shaft and hence is strictly the delivered power.)

(c)

$$\text{Thrust power tp} = 400 \times 11.52 \times \frac{1852}{3600}$$
$$= 2370 \text{ kW}$$

$$\text{Propeller efficiency} = \frac{2370}{3393} \times 100$$

$$= 69.84\%$$

(d)

$$ep = 3393 \times 0.67$$
$$= 2273.3$$

But

$$ep = R_t \times v$$

$$R_t = \frac{2273.3 \times 3600}{16 \times 1852}$$

$$= 276.2 \text{ kN}$$

$$R_t = T(1-t)$$

$$1 - t = \frac{276.2}{400}$$

$$= 0.69$$

$$\text{Thrust deduction factor } t = 0.31$$

6.

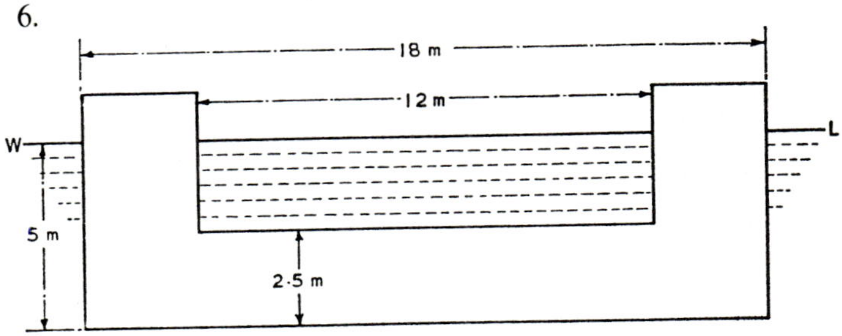

Fig. 121

$$KB = \frac{12 \times 2.5 \times 1.25 + 2 \times 3 \times 5 \times 2.5}{12 \times 2.5 + 2 \times 3 \times 5}$$

$$= 1.875 \text{ m}$$

$$I = \frac{1}{12} L(18^3 - 12^3)$$

$$= \frac{1}{12} \times 4104L$$

$$\nabla = L(12 \times 2.5 + 2 \times 3 \times 5)$$
$$= 60L$$

$$BM = \frac{4104\,L}{12 \times 60\,L}$$

$$= 5.70 \text{ m}$$

$$KM = 1.875 + 5.70$$
$$= 7.575 \text{ m}$$

$$GM = 2.50 \text{ m}$$

$$\therefore \quad KG = 5.075 \text{ m}$$

7(a). Derive $\text{sp} = \frac{\Delta^{\frac{2}{3}} V^3}{C}$

and hence $\frac{\text{sp}_1}{\text{sp}_2} = \left(\frac{\Delta_1}{\Delta_2}\right)^{\frac{2}{3}} \left(\frac{V_1}{V_2}\right)^4$

(b) $\text{sp}_2 = 23\,000 \times \left(\frac{12\,000}{14\,000}\right)^{\frac{2}{3}} \times \left(\frac{21}{24}\right)^4$

$$= 12\,166 \text{ kW}$$

8.

CSA	SM	f of ▽	lever	f of M
0	1	—	+5	—
11.5	4	46.0	+4	+184.0
27.0	2	54.0	+3	+162.0
38.5	4	154.0	+2	+308.0
44.0	2	88.0	+1	+ 88.0
45.0	4	180.0	0	+742.0
44.5	2	89.0	−1	− 89.0
39.0	4	156.0	−2	−312.0
26.5	2	53.0	−3	−159.0
14.5	4	58.0	−4	−232.0
0	1	—	−5	—
		878.0		−792.0

$$h = 8 \text{ m}$$

(a) $\text{Displacement} = \frac{8}{3} \times 878 \times 1.025$

$$= 2400 \text{ tonne}$$

(b) LCB from midships $= \dfrac{8\,(742 - 792)}{878}$

$= -0.455$ m

$= 0.455$ m forward

(c) Prismatic coefficient $= \dfrac{8 \times 878}{3 \times 80 \times 45}$

$= 0.650$

9. Ship A

$$ep_n = R_t \times v$$

$$R_t = \frac{6400 \times 3600}{18 \times 1852}$$

$$= 691.1 \text{ kN}$$

$$R_f = f S V^n$$
$$= 0.42 \times 4000 \times 18^{1.825}$$
$$= 328.2 \text{ kN}$$

$$R_r = 691.1 - 328.2$$
$$= 362.9 \text{ kN}$$

Ship B

$$R_r \propto L^3$$

$$\therefore \quad R_r = 362.9 \times \left(\frac{140}{160}\right)^3$$

$$= 243.1 \text{ kN}$$

$$S \propto L^2$$

$$\therefore \quad S = 4000 \times \left(\frac{140}{160}\right)^2$$

$$= 3062$$

$$V \propto \sqrt{L}$$

$$\therefore \quad V = 18\sqrt{\frac{140}{160}}$$

$$= 16.84$$

$$R_f = 0.425 \times 3062 \times 16.84^{1.825}$$
$$= 225.1 \text{ kN}$$

$$R_t = 225.1 + 243.1$$
$$= 468.2 \text{ kN}$$

$$ep_n = \frac{468.2 \times 16.84 \times 1852}{3600}$$

$$= 4055 \text{ kN}$$

10.

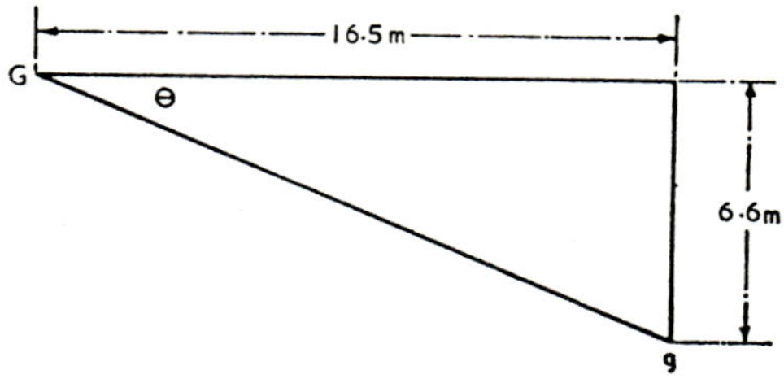

Fig. 122

$$Gg = \sqrt{16.5^2 + 6.6^2}$$

$$= \sqrt{315.81}$$
$$= 17.77 \text{ m}$$

This is the distance from the centre of gravity of the tank to the original centre of gravity of the ship.

(a) Let m = mass of oil used

Then shift in centre of gravity $= \dfrac{m \times Gg}{\text{final displacement}}$

$$0.380 = \frac{m \times 17.77}{11\,200 - m}$$

$$m = \frac{11\,200 \times 0.38}{17.77 + 0.38}$$

$$= 234.5 \text{ tonne}$$

(b) $$\tan\theta = \frac{6.6}{16.5}$$

$$= 0.40$$

Angle of shift $\theta = 21° 48'$

11.

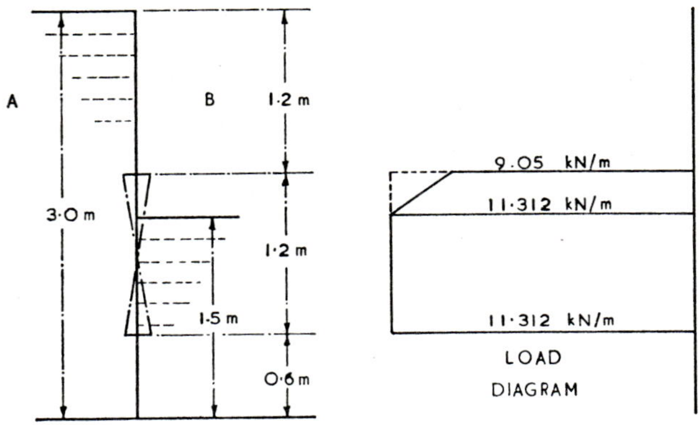

Fig. 123

Load/m at top of door from side A

$$= 1.025 \times 9.81 \times 1.2 \times 0.75$$
$$= 9.050 \text{ kN}$$

Load/m 0.3 m from top of door side A

$$= 1.025 \times 9.81 \times 1.5 \times 0.75$$
$$= 11.312 \text{ kN}$$

Load/m at bottom of door from side A

$$= 1.025 \times 9.81 \times 2.4 \times 0.75$$
$$= 18.099 \text{ kN}$$

Load/m at bottom of door from side B

$$= 1.025 \times 9.81 \times 0.9 \times 0.75$$
$$= 6.787 \text{ kN}$$

∴ nett load/m at bottom of door

$$= 18.099 - 6.787$$
$$= 11.312 \text{ kN}$$

Thus the load diagram is in the form shown by Fig. 123.

The area of this diagram represents the load, while the centroid represents the position of the centre of pressure.

Taking moments about the top of the door:

Centre of pressure from top

$$= \frac{11.312 \times 1.2 \times 0.6 - (11.312 - 9.05) \times 0.3 \times \tfrac{1}{2} \times 0.10}{11.312 \times 1.2 - (11.312 - 9.05) \times 0.3 \times \tfrac{1}{2}}$$

$$= \frac{8.145 - 0.0339}{13.574 - 0.339}$$

$$= \frac{8.111}{13.235}$$

$$= 0.613 \text{ m}$$

Resultant load

$$= 13.235 \text{ kN}$$

12(a). Mass of water in tank $= 18 \times 15 \times 1.5 \times 1.025$

$= 415.1$ tonne

$$\text{New } KG = \frac{10\,000 \times 7.4 + 415.1 \times 0.75}{10\,000 + 415.1}$$

$$= \frac{74\,000 + 311}{10\,415.1}$$

$$= 7.135 \text{ m}$$

New GM $= 8.00 - 7.135$ m

$= 0.865$ m

(b) Mass of water in tank $= \dfrac{415.1}{2}$

$= 207.55$ tonne

$$\text{New } KG = \frac{74\,000 + 207.55 \times 0.375}{10\,000 + 207.55}$$

$$= 7.257 \text{ m}$$

$$\text{Free surface effect} = \frac{1.025 \times 18 \times 15^3 \times 1.025}{1.025 \times 12 \times 10\,207.55}$$

$$= 0.508 \text{ m}$$

New GM $= 8.00 - 7.257 - 0.508$

$= 0.235$ m

13(a). Blade area $= 0.48 \times \dfrac{\pi}{4} \times 6^2$

$= 13.57 \text{ m}^2$

(b) Theoretical speed $v_t = \dfrac{6 \times 0.9 \times 110}{60}$

$= 9.9 \text{ m/s}$

Real slip $0.25 = \dfrac{9.9 - v_a}{9.9}$

Speed of advance $v_a = 9.9(1 - 0.25)$
$= 7.425 \text{ m/s}$

Wake fraction $0.30 = \dfrac{v - 7.425}{v}$

Ship speed $v = \dfrac{7.425}{1 - 0.30}$
$= 10.61 \text{ m/s}$
$= 10.61 \times \dfrac{3600}{1852}$

$= 20.62 \text{ knots}$

(c) Thrust power tp $= 300 \times 7.425$
$= 2227.5 \text{ kW}$

(d) Shaft power sp $= \dfrac{2227.5}{0.65}$

$= 3427 \text{ kW}$

(e) sp $= 2\pi n Q$

Torque $Q = \dfrac{3427 \times 60}{2\pi \times 110}$

$= 297 \text{ kN m}$

14. Let C = normal cons/h at V knots

C_1 = cons/h at reduced speed of 0.8 V knots

Then $C_1 = C - \dfrac{42}{24}$ tonne/h

Now $C = 0.136 + 0.001\,V^3$

$C - \dfrac{42}{24} = 0.136 + 0.001\,(0.8V)^3$

Subtracting: $\frac{42}{24} = 0.001V^3 - 0.001(0.512V^3)$

$$= 0.001V^3(1 - 0.512)$$

$$V^3 = \frac{42}{24 \times 0.001 \times 0.488}$$

$$V = 15.31 \text{ knots}$$

$$\text{Reduced speed} = 0.8 \times 15.31$$
$$= 12.25 \text{ knots}$$

$$C = 0.136 + 0.001 \times 15.31^3$$
$$= 3.722 \text{ tonne/h}$$

$$\text{Normal cons/day} = 3.722 \times 24$$
$$= 89.33 \text{ tonne}$$

(a) $$\text{Reduced cons/day} = 89.33 - 42$$
$$= 47.33 \text{ tonne}$$

(b) Time taken to travel 800 nautical miles at normal speed

$$= \frac{800}{15.31}$$

$$= 52.26 \text{ h}$$

$$\text{Time taken at reduced speed} = \frac{800}{12.25}$$

$$= 65.30 \text{ h}$$

Fuel consumption for 800 nautical miles at reduced speed

$$= 47.33 \times \frac{65.30}{24}$$

$$= 128.8 \text{ tonne}$$

Fuel on board when speed reduced

$$= 128.8 + 50$$
$$= 178.8 \text{ tonne}$$

(c) Normal cons for 800 nm $= 89.33 \times \dfrac{52.26}{24}$

$= 194.5$ tonne

% reduction in consumption $= \dfrac{194.5 - 128.8}{194.5} \times 100$

$= \dfrac{65.7}{194.5} \times 100$

$= 33.78\%$

(d) % increase in time $= \dfrac{65.30 - 52.26}{52.26} \times 100$

$= \dfrac{13.04}{52.26} \times 100$

$= 24.95\%$

15. Complete waterplane area

$= 160 \times 22 \times 0.865$
$= 3045.8\ m^2$

Intact waterplane area $= 3045.8 - 10.5 \times 22$
$= 2814.8\ m^2$

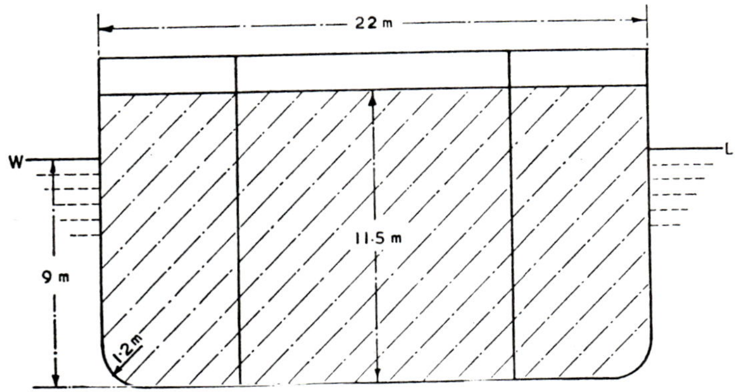

Fig. 124

It may be assumed that the whole of the mass of the oil is taken from the ship and that all the buoyancy of the compartment is lost.

Cross-sectional area of oil

$$= \tfrac{1}{2}\pi \times 1.2^2 + (22 - 2.4) \times 1.2 + 22\,(11.5 - 1.2)$$
$$= 2.26 + 23.52 + 226.6$$
$$= 252.38\ \text{m}^2$$

Immersed cross-sectional area

$$= 252.38 - 22 \times 2.5$$
$$= 197.38\ \text{m}^2$$

Mass of oil in compartment

$$= \frac{252.38 \times 10.5}{1.4}$$

$$= 1892.85\ \text{tonne}$$

Mass of buoyancy lost

$$= 197.38 \times 10.5 \times 1.025$$
$$= 2124.30\ \text{tonne}$$

Nett loss in buoyancy

$$= 2124.30 - 1892.85$$
$$= 231.55\ \text{tonne}$$

Equivalent volume

$$= \frac{231.55}{1.025}$$

$$= 225.9\ \text{m}^3$$

Increase in draught

$$= \frac{\text{nett volume of lost buoyancy}}{\text{area of intact waterplane}}$$

$$= \frac{225.9}{2814.8}$$

$$= 0.0802\ \text{m}$$

New draught

$$= 9.08\ \text{m}$$

16.

½ ord	½ ord³	SM	product
0	0	1	—
2.4	13.82	4	55.28
5.0	125.00	2	250.00
7.3	389.02	4	1556.08
7.9	493.04	2	986.08
8.0	512.00	4	2048.00
8.0	512.00	2	1024.00
7.7	456.53	4	1826.12
5.5	166.38	2	332.76
2.8	21.95	4	87.80
0	0	1	—
			8166.12

$$h = 16\,\text{m}$$

$$I = \frac{2}{9} \times 16 \times 8166.12$$

$$= 29\,035\,\text{m}^4$$

$$BM = \frac{29\,035}{8700} \times 1.025$$

$$= 3.421\,\text{m}$$

At angle of loll

$$\tan\theta = \sqrt{\frac{-2GM}{BM}}$$

$$GM = -\tfrac{1}{2}\,BM \tan^2\theta$$
$$= -\tfrac{1}{2} \times 3.421 \times 0.1763^2$$
$$= -0.053\,\text{m}$$

17. Let V = normal speed

C = normal consumption per *hour*

Then $24C$ = normal consumption per day.

For first 7.5 hours:

$$\text{Cons/h} = C \times \left(\frac{V_1}{V}\right)^3$$

$$= C \times \left(\frac{1.18V}{V}\right)^3$$

$$= 1.643C$$

$$\text{Cons for 7.5 hours} = 7.5 \times 1.643C = 12.32C$$

For next 10 hours:

$$\text{Cons/h} = C \times \left(\frac{0.91V}{V}\right)^3 = 0.7536C$$

$$\text{Cons for 10 hours} = 7.536C$$

i.e.

$$\text{cons for 17.5 hours} = 12.32C + 7.536C = 19.856C$$

$$\text{Cons for remaining 6.5 hours} = 24C - 19.856C = 4.144C$$

$$\text{Cons/h} = \frac{4.144C}{6.5} = 0.637C$$

$$\text{Reduced speed } V_3 = V\sqrt[3]{\frac{0.637C}{C}} = 0.86V$$

$$\text{Normal distance travelled/day} = 24V$$

$$\text{New distance travelled/day} = 1.18V \times 7.5 + 0.91V \times 10 + 0.86V \times 6.5 = 23.54V$$

$$\text{\% reduction in distance/day} = \frac{24V - 23.54V}{24V} \times 100 = 1.92\%$$

18.

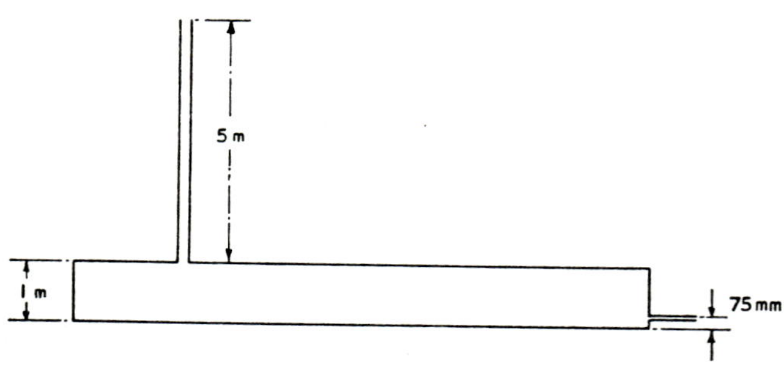

Fig. 125

(a) Pressure at tank top = pressure exerted by pump — pressure due to head of water

$= 70 - 1.025 \times 9.81 \times (1 - 0.075)$

$= 60.70\ \text{kN/m}^2$

Load on tank top $= 60.70 \times 6 \times 12$

$= 4370\ \text{kN}$

$= 4.37\ \text{MN}$

(b) With 70 kN/m² pressure:

maximum head above inlet $= \dfrac{70}{1.025 \times 9.81}$

$= 6.968\ \text{m}$

Maximum head above tank top $= 6.968 - 0.925$

$= 6.043\ \text{m}$

Hence the water will overflow and the maximum head above the tank top is therefore 5 m.

Load on tank top $= 1.025 \times 9.81 \times 6 \times 12 \times 5$

$= 3616\ \text{kN}$

$= 3.616\ \text{MN}$

19(a).

V	V^3	$\Delta^{\frac{2}{3}}$	sp	Ad. Coeff.
10	1000	524.1	880	595.5
11	1331	524.1	1155	604.0
12	1728	524.1	1520	595.8
13	2197	524.1	2010	572.9
14	2744	524.1	2670	538.6
15	3375	524.1	3600	491.3

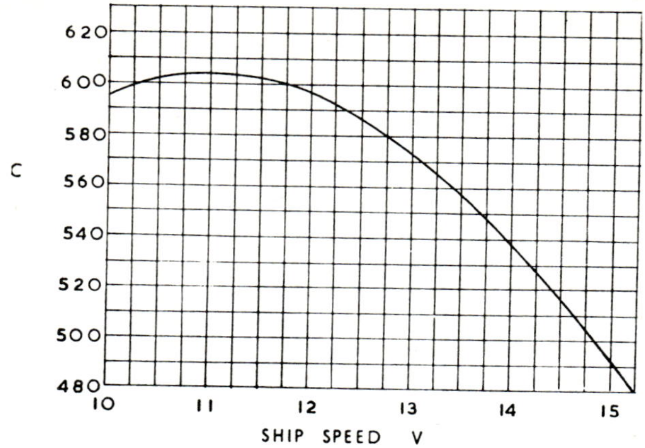

Fig. 126

(b) Corresponding speed of 120 m ship to 14 knots for 140 m ship.

$$= 14\sqrt{\frac{120}{140}}$$

$$= 12.96 \text{ knots}$$

From graph at 12.96 knots, the Admiralty Coefficient is 574.6

$$\Delta \propto L^3$$

$$\therefore \quad \Delta = 12\,000 \times \left(\frac{140}{120}\right)^3$$

$$= 19\,056 \text{ tonne}$$

Hence

$$\text{shaft power} = \frac{19\,056^{\frac{2}{3}} \times 12.96^3}{574.6}$$

$$= 2703 \text{ kW}$$

20.

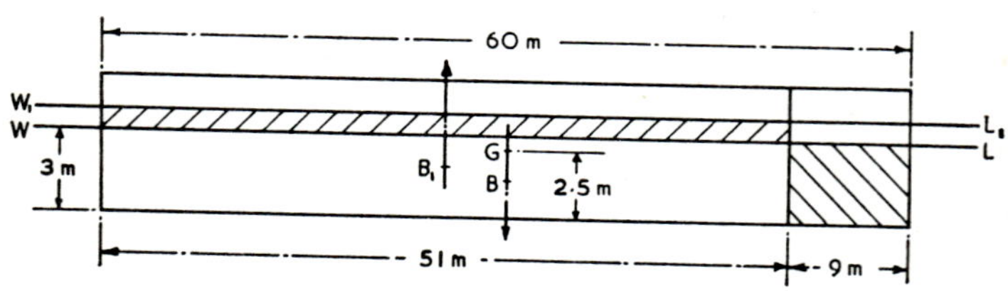

Fig. 127

$$\text{Increase in mean draught} = \frac{9 \times 10 \times 3}{51 \times 10}$$

$$= 0.529$$

$$\text{New mean draught} \quad d_1 = 3 + 0.529$$
$$= 3.529$$

$$KB_1 = \frac{3.529}{2}$$

$$= 1.765$$

$$I_F = \tfrac{1}{12} \times 51^3 \times 10$$

$$\nabla = 60 \times 10 \times 3$$

$$BM_L = \frac{51^3 \times 10}{12 \times 60 \times 10 \times 3}$$

$$= 61.41 \text{ m}$$

$$GM_L = 1.765 + 61.41 - 2.50$$
$$= 60.675 \text{ m}$$

$$BB_1 = \frac{9}{2}$$

$$= 4.5 \text{ m}$$

$$\text{Change in trim} = \frac{\Delta \times 4.5}{\Delta \times 60.675} \times 60$$

$$= 4.45 \text{ m by the head}$$

$$\text{Change forward} = + \frac{4.45}{60} \times \left(\frac{60}{2} + 4.5\right)$$

$$= +2.559 \text{ m}$$

$$\text{Change aft} = -\frac{4.45}{60}\left(\frac{60}{2} - 4.5\right)$$

$$= -1.891 \text{ m}$$

$$\text{New draught forward} = 3.259 + 2.559$$
$$= 5.818 \text{ m}$$

$$\text{New draught aft} = 3.259 - 1.891$$
$$= 1.368 \text{ m}$$

21.

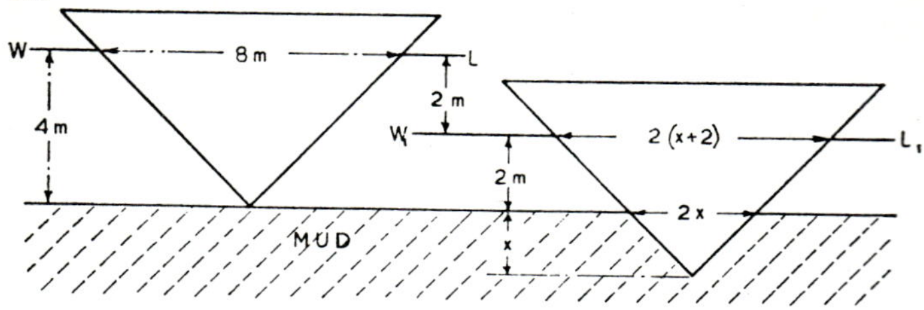

Fig. 128

Let L = length of vessel in m

ρ = density of water in tonne/m³

2ρ = density of mud in tonne/m³

x = depth to which vessel sinks in mud

$$\text{Original displacement} = \rho \times L \times \frac{8 \times 4}{2}$$

$$= 16\rho L \text{ tonne}$$

This displacement remains constant.

$$\text{Displacement of part in mud} = 2\rho \times L \times \frac{2x \times x}{2}$$

$$= 2\rho L x^2$$

$$\text{Displacement of part in water} = \rho \times L \times 2\left(\frac{2x + 2x + 4}{2}\right)$$

$$= \rho L(4x + 4)$$

Hence
$$16\rho L = 2\rho L x^2 + 4\rho L x + 4\rho L$$

$$16 = 2x^2 + 4x + 4$$

$$2x^2 + 4x - 12 = 0$$

from which
$$x = 1.646 \text{ m}$$

i.e. vessel sinks 1.646 m into the mud.

22(a).

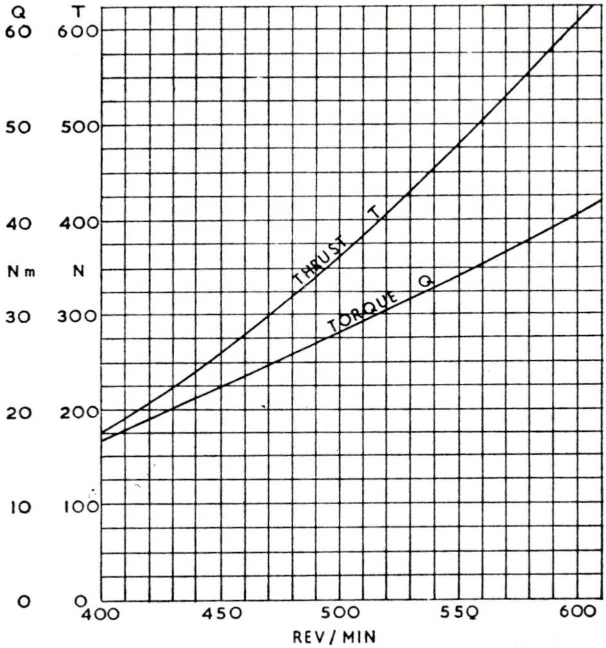

Fig. 129

(b)

$$v_t = P \times N$$

$$v_a = 0.8\, v_t$$

$$150 = 0.8 \times 0.4 \times N$$

$$\text{Rev/min } N = 469$$

$$\text{At 469 rev/min, } T = 298\text{ N and } Q = 24.6\text{ N m}$$

$$\text{Thrust power tp} = 298 \times \frac{150}{60}$$

$$= 745\text{ W}$$

$$\text{Delivered power dp} = 24.6 \times 2\pi \times \frac{469}{60}$$

$$= 1208\text{ W}$$

$$\text{Propeller efficiency} = \frac{\text{tp}}{\text{dp}}$$

$$= \frac{745}{1208}$$

$$= 61.67\%$$

23.

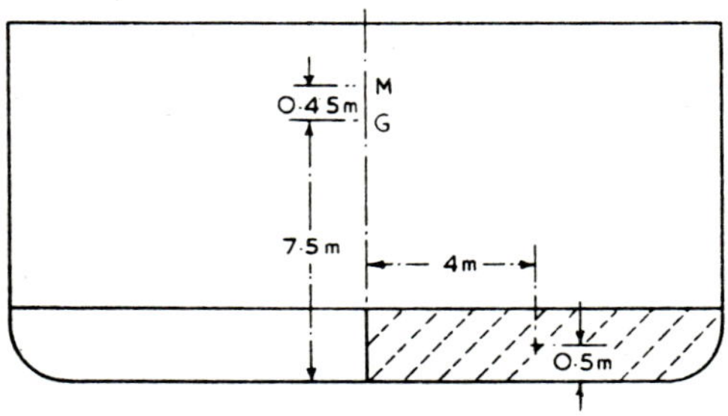

Fig. 130

$$\text{New } KG = \frac{8100 \times 7.5 - 100 \times 0.5}{8100 - 100}$$

$$= \frac{60\,750 - 50}{8000}$$

$$= 7.588 \text{ m}$$

$$\text{New } GM = 7.5 + 0.45 - 7.588$$

$$= 0.362 \text{ m}$$

$$\text{Heeling moment} = 100 \times 4$$

$$\tan\theta = \frac{100 \times 4}{8000 \times 0.362}$$

$$= 0.1381$$

$$\text{Angle of heel } \theta = 7^\circ\ 52'$$

24.

½ ord	½ ord³	SM	product
0.7	0.34	1	0.34
3.3	35.94	4	143.76
5.5	166.38	2	332.76
7.2	373.25	4	1493.00
7.5	421.88	2	843.76
7.5	421.88	4	1687.52
7.5	421.88	2	843.76
6.8	314.43	4	1257.72
4.6	97.34	2	194.68
2.2	10.65	4	42.60
0	—	1	—
			6839.90

$$h = 12\,\text{m}$$

$$I = \frac{2}{9} \times 12 \times 6839.9$$

$$= 18\,240\ \text{m}^4$$

$$BM = \frac{18\,240}{11\,000} \times 1.025$$

$$= 1.70\,\text{m}$$

25. 1 tonne of stowed cargo occupies 1.8 m^3

1 tonne of solid cargo occupies $\frac{1}{1.6}$ or 0.625 m^3

Hence in every 1.8 m^3 of space 0.625 m^3 is occupied by cargo and the remaining 1.175 m^3 is available for water.

$$\text{Permeability} = \frac{1.175}{1.8}$$

$$= 0.653$$

$$\text{Volume of lost buoyancy} = 18 \times 18 \times 6 \times 0.653$$

$$\text{Area of intact waterplane} = 85 \times 18$$

$$\text{Increase in draught} = \frac{18 \times 18 \times 6 \times 0.653}{85 \times 18}$$

$$= 0.830\,\text{m}$$

$$\text{New draught} = 6 + 0.830$$
$$= 6.830\,\text{m}$$

26.

Station	½ ord	SM	product	lever	product
AP	0.6	½	0.3	+ 4	+ 1.2
½	2.7	2	5.4	+ 3½	+ 18.9
1	4.6	1½	6.9	+ 3	+ 20.7
2	6.0	4	24.0	+ 2	+ 48.0
3	6.3	2	12.6	+ 1	+ 12.6
4	6.3	4	25.2	0	+ 101.4
5	6.3	2	12.6	− 1	− 12.6
6	5.7	4	22.8	− 2	− 45.6
7	4.8	1½	7.2	− 3	− 21.6
7½	2.0	2	4.0	− 3½	− 14.0
FP	0	½	—	− 4	—
			121.0		− 93.8

$$\text{Common interval} = \frac{90}{8}$$

$$\text{Area} = \frac{2}{3} \times \frac{90}{8} \times 121.0$$

$$= 907.5 \text{ m}^2$$

$$\text{LCF from midships} = \frac{90}{8} \times \left(\frac{101.4 - 93.8}{121.0}\right)$$

$$= 0.707 \text{ m aft}$$

27.

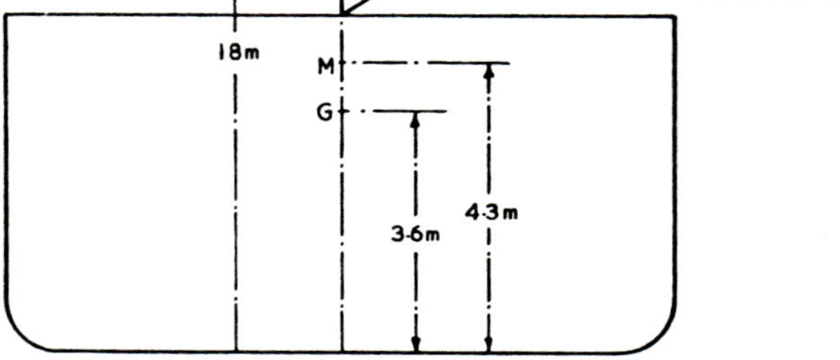

Fig. 131

$$\text{New } KG = \frac{6600 \times 3.6 + 50 \times 18}{6600 + 50}$$

$$= \frac{23\,760 + 900}{6650}$$

$$= 3.708 \text{ m}$$

$$\text{New } GM = 4.30 - 3.708$$

$$= 0.592 \text{ m}$$

$$\tan\theta = \frac{m \times d}{\Delta \times GM}$$

$$d = \frac{6650 \times 0.592 \times \tan 9.5°}{50}$$

$$\text{Outreach of derrick} = 13.18 \text{ m}$$

28. 120 m ship:

$$\text{Shaft power} = 4100 \text{ kW}$$

$$\text{Effective power} = 4100 \times 0.6$$

$$= 2460 \text{ kW}$$

$$= R_t \times v$$

$$R_t = \frac{2460 \times 3600}{15 \times 1852}$$

$$= 318.8 \text{ kN}$$

$$R_f = 0.55\, R_t$$

$$\therefore \quad R_r = 0.45\, R_t$$

$$= 143.46 \text{ kN}$$

140 m ship:

$$R_r \propto L^3$$

$$\therefore \quad R_r = 143.46 \times \left(\frac{140}{120}\right)^3$$

$$= 227.81 \text{ kN}$$

$$S \propto L^2$$

$$\therefore \quad S = 3000 \times \left(\frac{140}{120}\right)^2$$

$$= 4083 \text{ m}^2$$

$$V \propto \sqrt{L}$$

$$\therefore \quad V = 15 \times \sqrt{\frac{140}{120}}$$

$$= 16.20 \text{ knots}$$

$$R_f = 0.42 \times 4083 \times 16.20^{1.825}$$

$$= 276.44 \text{ kN}$$

$$R_t = 276.44 + 227.81$$

$$= 504.25 \text{ kN}$$

$$\text{ep} = \frac{504.25 \times 16.2 \times 1852}{3600}$$

$$= 4202 \text{ kW}$$

$$\text{sp} = \frac{\text{ep}}{\text{pc}}$$

$$= \frac{4202}{0.6}$$

$$= 7003 \text{ kW}$$

29. Original displacement $= 30 \times 9 \times 0.75 \times 1.025$

$$= 207.5 \text{ tonne}$$

$$\text{Change in mean draught} = \frac{\Delta \times 100}{A_w} \left(\frac{\rho_S - \rho_R}{\rho_S \times \rho_R}\right) \text{cm}$$

$$3.2 = \frac{\Delta \times 100}{30 \times 9} \left(\frac{1.025 - 1.000}{1.025 \times 1.000}\right)$$

$$\Delta = \frac{3.2 \times 30 \times 9 \times 1.025}{100 \times 0.025}$$

$$= 354.2 \text{ tonne}$$

$$\text{Cargo added} = 354.2 - 207.5$$

$$= 146.7 \text{ tonne}$$

30.

½ breadth	SM	f of a	$(\frac{1}{2}b)^2$	SM	f of m	$(\frac{1}{2}b)^3$	SM	f of i
5	1	5	25	1	25	125	1	125
4	4	16	16	4	64	64	4	256
2	1	2	4	1	4	8	1	8
		23			93			389

$$\text{Area of tank surface } a = \frac{6}{3} \times 23$$

$$= 46 \text{ m}^2$$

$$\text{Centroid from centreline} = \frac{93}{2 \times 23}$$

$$= 2.022 \text{ m}$$

Second moment of area about centreline

$$= \frac{6}{9} \times 389$$

$$= 259.33 \text{ m}^4$$

Second moment of area about centroid

$$i = 259.33 - 46 \times 2.022^2$$

$$= 71.26 \text{ m}^4$$

$$\text{Free surface effect} = \frac{\rho_1 i}{\rho \nabla}$$

$$= \frac{1.000 \times 71.26}{1.025 \times 5000} \times 1.025$$

$$= 0.0142 \text{ m}$$

31. $$\text{Ship speed} = 15 \times \frac{1852}{3600}$$

$$= 7.717 \text{ m/s}$$

$$\text{Apparent slip} = \frac{v_t - v}{v_t}$$

$$O = v_t - v$$

$$\therefore \qquad v_t = v$$
$$= 7.717 \text{ m/s}$$

$$\text{Propeller pitch} \quad P = \frac{7.717}{1.58}$$
$$= 4.884 \text{ m}$$

$$w_t = \frac{v - v_a}{v}$$

$$v_a = 7.717\,(1 - 0.31)$$
$$= 5.325 \text{ m/s}$$

$$\text{Real slip} = \frac{7.717 - 5.325}{7.717}$$
$$= 0.31$$

$$R_t = T(1 - t)$$
$$= 360\,(1 - 0.20)$$
$$= 288 \text{ kN}$$

$$\text{ep} = 288 \times 7.717$$
$$= 2222.5 \text{ kW}$$

$$\text{Propulsive coefficient} = \frac{2222.5}{3050}$$
$$= 0.729$$

32. Torque = force × lever

$$= F_n \times b$$

$$= 577 \times 22 \times \left(1.2 \times 15 \times \frac{1852}{3600}\right)^2 \times \sin 35° \times 1.1$$

$$= 686\,780 \text{ N m}$$

But $$\frac{T}{J} = \frac{q}{r}$$

And $$J = \frac{\pi}{2} r^4$$

$$\therefore \qquad \frac{686\,780}{\frac{\pi}{2} r^4} = \frac{70 \times 10^6}{r}$$

$$r^3 = \frac{686\,780 \times 2}{70 \times 10^6 \pi}$$

$$r = 0.184 \text{ m}$$

Diameter of stock

$$= 368 \text{ mm}$$

If the diameter is reduced to 330 mm:

$$\frac{T}{\frac{\pi}{2} \times 0.165^4} = \frac{70 \times 10^6}{0.165}$$

$$T = \frac{70 \times 10^6 \times \pi \times 0.165^3}{2}$$

$$= 493\,920 \text{ N m}$$

$$493\,920 = 577 \times 22 \times \left(1.2 \times V \times \frac{1852}{3600}\right) \times 0.5736 \times 1.1$$

$$V^2 = \frac{493\,920 \times 3600^2}{577 \times 22 \times 1.2^2 \times 1852^2 \times 0.5736 \times 1.1}$$

Ship speed V = 12.72 knots

33.

½ ord	SM	product
0	1	—
3.0	4	12.0
5.5	2	11.0
7.3	4	29.2
7.5	2	15.0
7.5	4	30.0
7.5	2	15.0
7.05	4	28.2
6.10	2	12.2
3.25	4	13.0
0	1	—
		165.6

$$h = 10 \text{ m}$$

Waterplane area $= \frac{2}{3} \times 10 \times 165.6$
$= 1104 \text{ m}^2$

Intact waterplane area $= 1104 - 15 \times 15$

$= 879 \text{ m}^2$

$$\text{Immersed cross-sectional area} = 50\ \text{m}^2$$

$$\text{Volume of lost buoyancy} = 15 \times 50$$

$$= 750\ \text{m}^2$$

$$\text{Increase in draught} = \frac{750}{879}$$

$$= 0.853\ \text{m}$$

$$\text{New draught} = 4.353\ \text{m}$$

34. Let m = mass of ballast required

$$\text{MCT 1 cm} = \frac{\Delta \times GM_L}{100\,L}$$

$$= \frac{8100 \times 96}{100 \times 85}$$

$$= 91.48\ \text{tonne m}$$

$$\text{Trimming moment} = m(33 - 2)$$
$$= 31\,m$$

$$\text{Change of trim } t = \frac{31\,m}{91.48}\ \text{cm by the stern}$$

$$\text{Change aft} = +\frac{t}{85}\left(\frac{85}{2} - 2\right)$$

$$= 0.476t\ \text{cm}$$

$$\text{Bodily sinkage} = \frac{m}{9.0}\ \text{cm}$$

$$\text{New draught aft} = \text{old draught aft} + \frac{m}{900} + \frac{0.476t}{100}$$

$$5.85 = 5.55 + 0.00111m + \frac{0.476}{100} \times \frac{31m}{91.48}$$

$$0.30 = 0.002726\,m$$

$$\text{Ballast required } m = 110\ \text{tonne}$$

$$\text{Bodily sinkage} = \frac{110}{9}$$

$$= 12.22\ \text{cm}$$

$$\text{Change in trim} = \frac{31 \times 110}{91.48}$$

$$= 37.28 \text{ cm by the stern}$$

$$\text{Change forward} = -\frac{37.28}{85}\left(\frac{85}{2} + 2\right)$$

$$= -19.51 \text{ cm}$$

$$\text{New draught forward} = 5.25 + 0.122 - 0.195$$
$$= 5.177 \text{ m}$$

35.

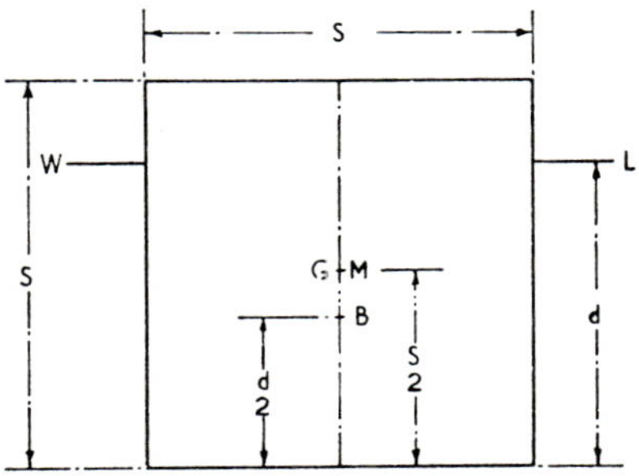

Fig. 132

Let x = relative density of wood

Then draught $d = Sx$

The limit of stability occurs when G and M coincide.

$$KG = \frac{S}{2}$$

$$KB = \frac{d}{2}$$

$$BM = \frac{S^2}{12d}$$

Since $KG = KM$

$$\frac{S}{2} = \frac{d}{2} + \frac{S^2}{12d}$$

$$\frac{S}{2} = \frac{Sx}{2} + \frac{S^2}{12Sx}$$

$$\frac{1}{2} = \frac{x}{2} + \frac{1}{12x}$$

Multiplying by 12:

$$6x = 6x^2 + 1$$

$$6x^2 - 6x + 1 = 0$$

$$x = \frac{6 \pm \sqrt{36 - 24}}{12}$$

$$\text{Relative density } x = 0.212 \text{ or } 0.788$$

It may be seen on referring to the metacentric diagram that the block will be *unstable* between these limits. Thus the relative density must be below 0.212 or between 0.788 and 1.0.

36. $$\text{Theoretical speed } v_t = 5.5 \times \frac{95}{60}$$

$$= 8.708 \text{ m/s}$$

$$\text{Real slip } s = \frac{v_t - v_a}{v_t}$$

$$v_a = 8.708 \times 0.8$$

$$= 6.966 \text{ m/s}$$

$$\text{Thrust power tp} = T \times v_a$$

$$= 380 \times 6.966$$

$$= 2647 \text{ kW}$$

$$R_t = T(1 - t)$$

$$= 380\,(1 - 0.198)$$

$$= 304.8 \text{ kN}$$

$$\text{ep} = R_t \times v$$

$$= 304.8 \times 15.5 \times \frac{1852}{3600}$$

$$= 2430 \text{ kW}$$

$$\text{QPC} = \frac{\text{ep}}{\text{dp}}$$

$$= \frac{2430}{3540}$$

$$= 0.686$$

$$\text{Wake fraction} = \frac{V - V_a}{V}$$

$$= \frac{15.5 - \left(6.966 \times \dfrac{3600}{1852}\right)}{15.5}$$

$$= 0.126$$

37.(a) Description

(b) Free surface effect $= \dfrac{\rho_1 \ i}{\rho \nabla}$

$$= \frac{l b^3}{12 \nabla}$$

$$0.14 = \frac{l \times 11^3}{12 \times 8500} \times 1.025$$

$$l = \frac{0.14 \times 12 \times 8500}{1331 \times 1.025}$$

$$= 10.47 \text{ m}$$

38.(a) Derivation of formula.

(b) Change in draught due to density

$$= \frac{8000 \times 100}{\text{TPC} \times 100} \times 1.024 \left(\frac{1.024 - 1.008}{1.008 \times 1.024}\right)$$

$$= 7.47 \text{ cm increase}$$

New mean draught = 7.075 m

Max. allowable draught

= 6.85 m

Required reduction in draught

= 0.225 m

Mass of ballast discharged

$$= 22.5 \times 17 \times \frac{1.008}{1.024}$$

$$= 376.5 \text{ tonne}$$

39.(a) $\text{Admiralty Coefficient} = \dfrac{\Delta^{\frac{2}{3}} V^3}{\text{sp}}$

$$\text{Fuel Coefficient} = \frac{\Delta^{\frac{2}{3}} V^3}{\text{fuel cons/day}}$$

(b) With constant displacement and Admiralty Coefficient:

$$\text{sp} \propto V^3$$

$$\frac{4460}{4120} = \left(\frac{14.55}{V_1}\right)^3$$

$$V_1 = 14.55\sqrt[3]{\frac{4120}{4460}}$$

$$= 14.17 \text{ knots}$$

At 14.17 knots $\quad \text{fuel cons} = 541 \text{ kg/h}$

$$= 541 \times 24 \times 10^{-3}$$

$$= 12.98 \text{ tonne/day}$$

$$\text{Fuel coefficient} = \frac{14\,900^{\frac{2}{3}} \times 14.17^3}{12.98}$$

$$= 132\,700$$

40. Normal rudder force F_n

$$= 577\, A\, v^2 \sin\alpha \quad \text{N}$$

Transverse force F_t

$$= 577\, A\, v^2 \sin\alpha \cos\alpha \quad \text{N}$$

$$= 577 \times 15 \left(1.2 \times 20 \times \frac{1852}{3600}\right)^2 \times 0.5 \times 0.866$$

$$= 571.29 \text{ kN}$$

$$\text{Heeling moment} = 571.29 \times (5 - 3.3)\cos\theta$$
$$= 971.2\cos\theta \quad \text{kN m}$$

Righting moment

$$= \Delta g\, GZ$$
$$= \Delta g\, GM \sin\theta$$

Steady heel will be produced when the heeling moment is equal to the righting moment.

$$12\,000 \times 9.81 \times 0.3 \sin\theta = 971.2\cos\theta$$

$$\tan\theta = \frac{971.2}{12\,000 \times 9.81 \times 0.3}$$

$$= 0.0275$$

Angle of heel $\theta = 1° 36'$

41.

Item	mass	Lcg	moment forward	moment aft
Cargo	10 000	3.0	30 000	3 000
Fuel	1 500	2.0		3 200
Water	400	8.0		
Stores	100	10.0	1 000	10 000
Lightship	4 000	2.5		
Displacement	16 000		31 000	16 200

$$\text{Excess moment forward} = 31\,000 - 16\,200$$
$$= 14\,800 \text{ tonne m}$$

$$\text{LCG from midships} = \frac{14\,800}{16\,000}$$

$$= 0.925 \text{ m forward}$$

From hydrostatic data at 16 000 tonne displacement:
d = 8.25 m; MCT1 cm = 179 t m; LCB = 2.02 m F; LCF = 0.57 m A.

$$\text{Trimming lever} = 2.02 - 0.925$$

$$= 1.095 \text{ m aft}$$

$$\text{Trim} = \frac{16\,000 \times 1.095}{179}$$

$$= 97.88 \text{ cm by the stern}$$

$$\text{Change forward} = -\frac{97.88}{120}\left(\frac{120}{2} + 0.57\right)$$

$$= -49.40 \text{ cm}$$

$$\text{Change aft} = +\frac{97.88}{120}\left(\frac{120}{2} - 0.57\right)$$

$$= 48.48 \text{ cm}$$

$$\text{Draught forward} = 8.250 - 0.494$$

$$= 7.756 \text{ m}$$

$$\text{Draught aft} = 8.250 + 0.485$$
$$= 8.735 \text{ m}$$

42. $$KB = \frac{3.0}{2}$$
$$= 1.5 \text{ m}$$
$$BM = \frac{9^2}{12 \times 3}$$
$$= 2.25 \text{ m}$$
$$GM = 1.5 + 2.25 - 3.5$$
$$= 0.25 \text{ m}$$
$$\text{Displacement } \Delta = 30 \times 9 \times 3 \times 1.025$$
$$= 830.25 \text{ tonne}$$
$$\text{Righting moment} = \Delta \times GZ$$
$$\text{Heeling moment} = 10 \times 6$$

(a) $$\therefore \quad 10 \times 6 = 830.25\, GZ$$
$$= 830.25 \sin\theta\,(GM + \tfrac{1}{2} BM \tan^2\theta)$$
$$= 830.25 \sin\theta\,(0.25 + \frac{2.25}{2} \tan^2\theta)$$
$$0.07227 = \sin\theta\,(0.25 + 1.125 \tan^2\theta)$$

This expression may be solved graphically.

θ	$\tan\theta$	$\tan^2\theta$	$1.125\tan^2\theta$	$\sin\theta$	GZ
5°	0.0875	0.00766	0.00861	0.0872	0.0226
10°	0.1763	0.03108	0.03497	0.1736	0.0495
15°	0.2680	0.07182	0.08080	0.2588	0.0856
20°	0.3640	0.13250	0.14906	0.3420	0.1365

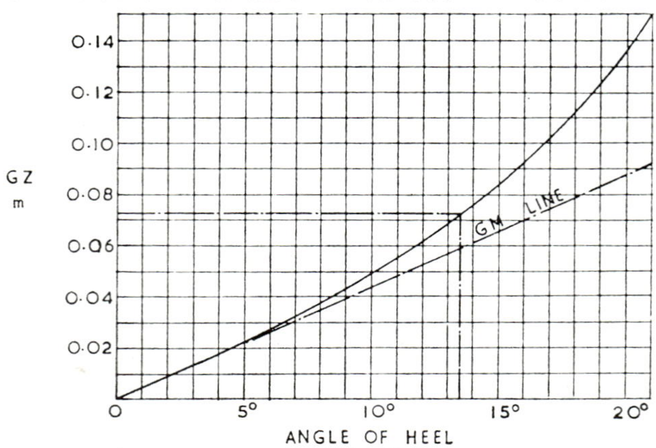

Fig. 133

$$\text{From graph when } GZ = 0.07227$$

$$\text{Angle of heel } \theta = 13° \, 30'$$

$$GZ = GM \sin\theta$$

$$0.07227 = 0.25 \sin\theta$$

$$\sin\theta = \frac{0.07227}{0.25}$$

$$= 0.28908$$

$$\text{Angle of heel } \theta = 16° \, 48'$$

43. Let V = reduced speed in knots
Normally at V knots the consumption per day would be:

$$47 \times \left(\frac{V}{17}\right)^3 \text{ tonne}$$

$$\text{Actual cons/day:} \quad 22 = 1.132 \times 47 \times \left(\frac{V}{17}\right)^3$$

$$V^3 = \frac{22 \times 17^3}{1.132 \times 47}$$

$$\text{Reduced speed } V = 12.66 \text{ knots}$$

$$\text{At 17 knots, time taken} = \frac{3000}{17 \times 24}$$

$$= 7.353 \text{ days}$$

$$\text{and voyage consumption} = 7.353 \times 47$$

$$= 345.6 \text{ tonne}$$

$$\text{At 12.66 knots, time taken} = \frac{3000}{12.66 \times 24}$$

$$= 9.872 \text{ days}$$

$$\text{and voyage consumption} = 9.872 \times 22$$

$$= 217.2 \text{ tonne}$$

$$\text{Difference in consumption} = \frac{345.6 - 217.2}{345.6} \times 100$$

$$= 37.15\%$$

44.
$$\text{Bodily sinkage} = \frac{180}{18}$$
$$= 10 \text{ cm}$$
$$\text{Trimming moment} = 180 \times (40 + 3)$$
$$\text{MCT 1 cm} = \frac{14\,000 \times 127}{100 \times 135}$$
$$\text{Change in trim} = \frac{180 \times 43 \times 100 \times 135}{14\,000 \times 127}$$
$$= 58.76 \text{ cm by the head}$$
$$\text{Change forward} = + \frac{58.76}{135}\left(\frac{135}{2} + 3\right)$$
$$= + 30.68 \text{ cm}$$
$$\text{Change aft} = - \frac{58.76}{135}\left(\frac{135}{2} - 3\right)$$
$$= - 28.08$$
$$\text{New draught forward} = 7.30 + 0.10 + 0.307$$
$$= 7.707 \text{ m}$$
$$\text{New draught aft} = 8.05 + 0.10 - 0.281$$
$$= 7.869 \text{ m}$$

45.
$$\text{Theoretical speed } V_t = \frac{5.5 \times 93 \times 60}{1852}$$
$$= 16.57 \text{ knots}$$

Let
$$\text{ship speed} = V$$

Then
$$-S = \frac{V_t - V}{V_t}$$
$$-S\,V_t = V_t - V \qquad \ldots(1)$$

and
$$+S = \frac{V_t - 0.9V}{V_t}$$
$$+S\,V_t = V_t - 0.9V \qquad \ldots(2)$$

Adding (1) and (2)

$$0 = 2V_t - 1.9V$$

$$V = \frac{2 \times 16.57}{1.9}$$

$$\text{Ship speed } V = 17.44 \text{ knots}$$

Substituting for V:

$$-S = \frac{16.57 - 17.44}{16.57}$$

$$= -0.0525$$

i.e. apparent slip = − 5.25%

and real slip = + 5.25%

46.

½ ord	SM	product	lever	product	lever	product
1	1	1	+4	+ 4	+4	+ 16
7	4	28	+3	+ 84	+3	+252
10.5	2	21	+2	+ 42	+2	+ 84
11	4	44	+1	+ 44	+1	+ 44
				+174		
11	2	22	0		0	0
10.5	4	42	−1	− 42	−1	+ 42
8	2	16	−2	− 32	−2	+ 64
4	4	16	−3	− 48	−3	+144
0	1	—	−4	—	−4	—
		190		−122		+646

$$h = 15 \text{ m}$$

(a) Waterplane area $A = \frac{2}{3} \times 15 \times 190$

$$= 1900 \text{ m}^2$$

$$\text{TPC} = 1900 \times 0.01025$$

$$= 19.475$$

(b) LCF from midships $\bar{x} = \frac{15(174 - 122)}{190}$

$$= 4.11 \text{ m aft}$$

(c) Second moment about midships

$$= \frac{2}{3} \times 15^3 \times 646$$

$$= 1\,453\,500 \text{ m}^4$$

$$A\bar{x}^2 = 1900 \times 4.11^2$$

$$= 32\,095 \text{ m}^4$$

Second moment about centroid

$$= 1\,453\,500 - 32\,095$$

$$= 1\,421\,405 \text{ m}^4$$

47.

$$\text{Centrifugal force} = \frac{\Delta v^2}{r}$$

$$= \frac{\Delta}{600}\left(\frac{12 \times 1852}{3600}\right)^2$$

$$= 0.06\,352\,\Delta$$

$$\text{Heeling moment} = \text{CF} \times GL \cos\theta$$

$$= 0.06\,352\,\Delta \times 2.7 \cos\theta$$

$$\text{Righting moment} = \Delta g\, GM \sin\theta$$

$$= 0.25 \times 9.81\,\Delta \sin\theta$$

$$0.25 \times 9.81\Delta\ \sin\theta = 0.06\,352 \times 2.7\,\Delta \cos\theta$$

$$\tan\theta = \frac{0.06\,352 \times 2.7}{0.25 \times 9.81}$$

$$= 0.06\,993$$

$$\text{Angle of heel } \theta = 4°$$

48.

V	ep_n	$ep = ep_n \times 1.3$	QPC	dp
15	3000	3900	0.73	5342
16	3750	4875	0.73	6678
17	4700	6110	0.72	8486
18	5650	7345	0.71	10 345

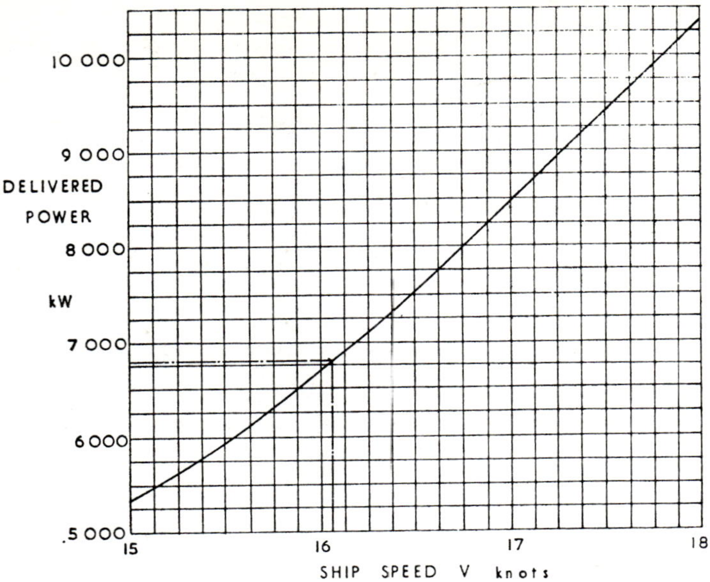

Fig. 134

$$\text{Total brake power} = 2 \times 3500 = 7000\,\text{kW}$$

$$\text{Total delivered power} = 7000(1-0.03) = 6790\,\text{kW}$$

From graph: service speed = 16.06 knots

49.
$$\text{MCT1 cm} = \frac{8000 \times 102}{100 \times 120} = 68 \text{ tonne m}$$

Note: The distances must be measured from the LCF.

	mass	distance from F	moment forward	moment aft
Fuel used	+200	20 F	+4000	
Water used	+100	1 A		+100
Stores used	+ 10	7 A		+ 70
Ballast added	−300	26 F	−7800	
	+ 10		−3800	+170

$$\text{Bodily sinkage} = \frac{10}{17.5}$$

$$= 0.57 \text{ cm}$$

$$\text{Nett moment aft} = 170 - (-3800)$$

$$= +3970 \text{ tonne m}$$

$$\text{Change in trim} = \frac{3970}{68}$$

$$= 58.38 \text{ cm by the stern}$$

$$\text{Change forward} = -\frac{58.38}{120}\left(\frac{120}{2} + 2\right)$$

$$= -30.16 \text{ cm}$$

$$\text{Change aft} = +\frac{58.38}{120}\left(\frac{120}{2} - 2\right)$$

$$= +28.22 \text{ cm}$$

$$\text{Original draught forward} = 6.30 + 0.006 - 0.302$$

$$= 6.004 \text{ m}$$

$$\text{Original draught aft} = 6.60 + 0.006 + 0.282$$

$$= 6.888 \text{ m}$$

50.

$$V_t = \frac{5.5 \times 80 \times 60}{1852}$$

$$= 14.25 \text{ knots}$$

(a)

$$\text{Real slip} = \frac{14.25 - 11}{14.25}$$

$$= 0.2281$$

$$\text{or } 22.81\%$$

(b)

$$\text{Wake fraction} = \frac{13.2 - 11}{13.2}$$

$$= 0.167$$

(c) Thrust power = 3000×0.7

$= 2100$ kW

But thrust power = $T \times v_a$

$\therefore$ Thrust $T = \dfrac{2100 \times 3600}{11 \times 1852}$

$= 371.1$ kN

51.

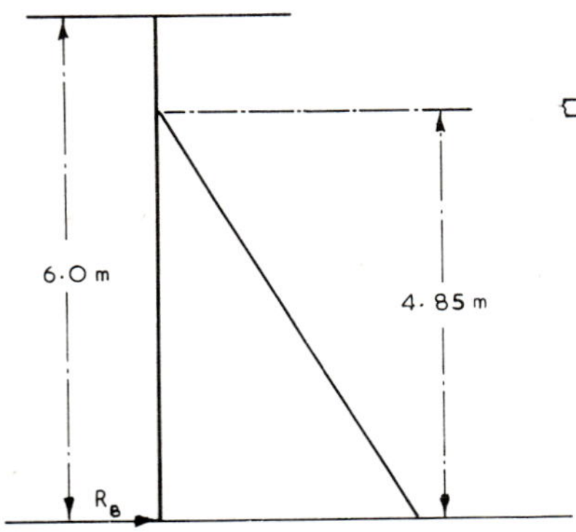

Fig. 135(a)

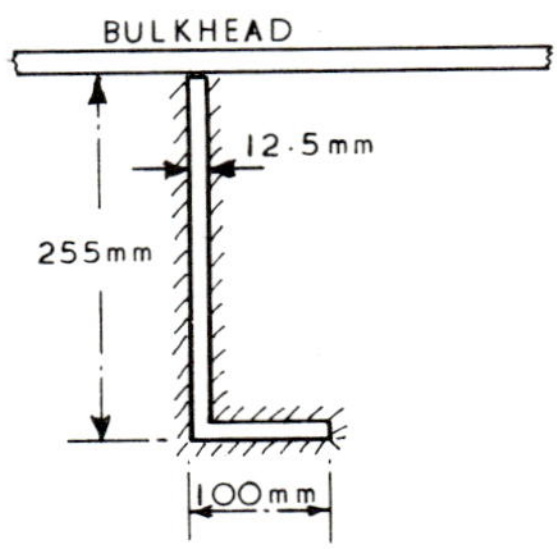

Fig. 135(b)

Load on stiffener = $\rho g A H$

$= 1.025 \times 9.81 \times 4.85 \times 0.6 \times \dfrac{4.85}{2}$

$= 70.96$ kN

Centre of pressure from surface = $\dfrac{2}{3} \times 4.85$

$= 3.233$ m

Centre of pressure from top = $3.233 + 1.15$

$= 4.383$ m

Taking moments about the top

$R_B = 70.96 \times \dfrac{4.385}{6}$

$= 51.86$ kN

This is also the shear force at the bottom of the stiffener.

$$\text{Length of weld metal} = 255 + 255 + 100 + 100 - 12.5$$
$$= 697.5 \text{ mm}$$
$$\text{Area of weld metal} = 697.5 \times 5$$
$$= 3487.5 \text{ mm}^2$$
$$\text{Shear stress in weld} = \frac{51.86 \times 10^3}{3487.5 \times 10^{-6}}$$
$$= 14.87 \times 10^6 \text{ N/m}^2$$
$$= 14.87 \text{ MN/m}^2$$

52. Tank A, free surface effect $= \dfrac{\rho_1 i}{\rho \nabla}$

$$= \frac{i}{\nabla} \text{ since } \rho_1 = \rho_2$$
$$= \frac{12 \times 16^3 \times 1.025}{12 \times 5000}$$
$$= 0.840 \text{ m}$$

Tank B, free surface effect $= \dfrac{14 \times 15^3 \times 1.025}{12 \times 5000}$

$$= 0.807 \text{ m}$$

Tank C, free surface effect $= \dfrac{14 \times 16^3 \times 1.025}{12 \times 5000}$

$$= 0.980 \text{ m}$$

The tank with the lowest free surface effect is filled first and thus they should be filled in the order B, A, C.

Note: Since the difference in free surface effect depends upon the product ($l \times b^3$), this value could have been calculated for each tank instead of the complete free surface effect.

53.

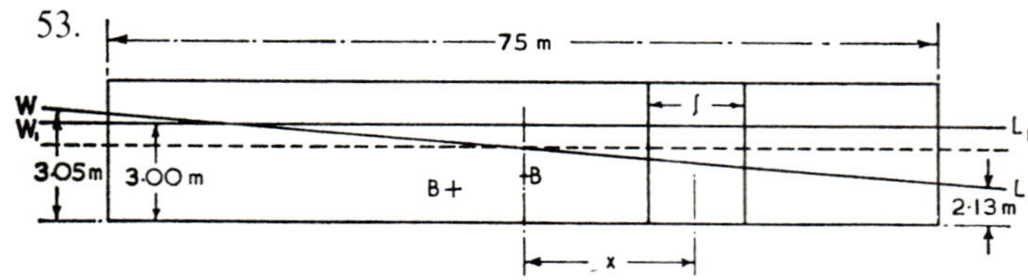

Fig. 136

$$\text{Original mean draught} = \frac{3.05 + 2.13}{2}$$
$$= 2.59 \text{ m}$$
$$\text{Volume of displacement} = 75 \times 8.5 \times 2.59$$
$$= 1651 \text{ m}^3$$

$$\text{LCB from midships} = \frac{37.5 \times \frac{0.46}{2} \times \frac{4}{3} \times 37.5}{75 \times 2.59}$$

$$= 2.22 \text{ m aft}$$

$$\text{Final volume of displacement} = 75 \times 8.5 \times 3.0$$
$$= 1912 \text{ m}^3$$

LCB at midships

Increase in volume of displacement

$$= 1912 - 1651$$
$$= 261 \text{ m}^3$$

The effect of this added volume is to bring the vessel to an even keel.

Let x = distance of Lcg of compartment forward of midships

Taking moments about midships:

$$1912 \times 0 = 1651 \times 2.22 - 261 \times x$$

$$x = \frac{1651 \times 2.22}{261}$$

$$= 14.04 \text{ m}$$

$$\text{Immersed cross-sectional area} = 8.5 \times 3$$
$$= 25.5 \text{ m}^2$$

$$\therefore \text{Length of compartment} = \frac{261}{25.5}$$

$$= 10.24 \text{ m}$$

54. $$\text{TPC} = A_w \times 0.01\,026$$

$$A_w = \frac{27.5}{0.01\,026} \text{ m}^2$$

$$\text{Change in draught} = \frac{\Delta \times 100}{A_w} \left(\frac{\rho_S - \rho_R}{\rho_R \times \rho_S} \right) \text{ cm}$$

$$25 = \frac{\Delta \times 100 \times 0.01\,026}{27.5} \left(\frac{1.026 - 1.000}{1.000 \times 1.026} \right)$$

$$= \frac{\Delta \times 0.026}{27.5}$$

$$\Delta = \frac{25 \times 27.5}{0.026}$$

$$= 26\,442 \text{ tonne}$$

If, with this displacement, the vessel moves into the river water, then:

$$\text{Change in draught} = \frac{26\,442 \times 1.026}{27.5}\left(\frac{1.026 - 1.012}{1.012 \times 1.026}\right)$$

$$= 13.30 \text{ cm}$$

Thus the maximum allowable draught in the river water

$$= 8.25 + 0.133$$
$$= 8.383 \text{ m}$$

$$\text{Actual draught} = 8.44 \text{ m}$$

$$\therefore \quad \text{Excess draught} = 0.057 \text{ m}$$

$$\text{TPC in river water} = 27.5 \times \frac{1.012}{1.026}$$

$$= 27.12$$

$$\therefore \quad \text{Excess mass} = 5.7 \times 27.12$$
$$= 154.6 \text{ tonne}$$

55.

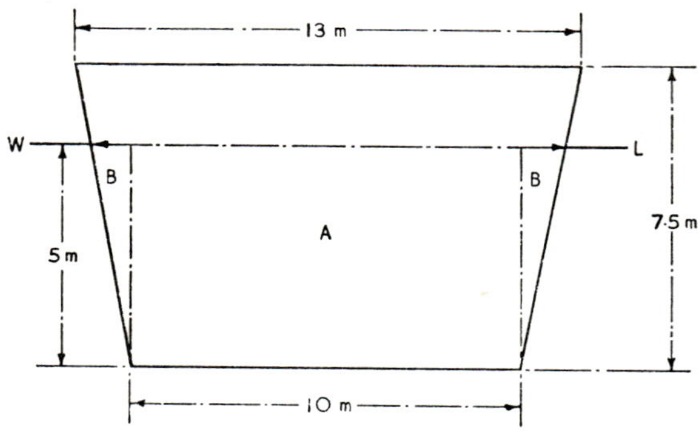

Fig. 137

$$\text{Breadth at water level} = 10 + \frac{3}{7.5} \times 5$$

$$= 12 \text{ m}$$

Divide into a rectangle A and two triangles B.

$$\text{Load on A} = \rho g A H$$
$$= 1.025 \times 9.81 \times 10 \times 5 \times 2.5$$
$$= 1256.91 \text{ kN}$$

$$\text{Centre of pressure from WL} = \tfrac{2}{3} \times 5$$
$$= 3.33 \text{ m}$$

$$\text{Load on B} = 1.025 \times 9.81 \times \tfrac{1}{2} \times 5 \times \frac{5}{3} \times 2$$
$$= 83.79 \text{ kN}$$

$$\text{Centre of pressure from WL} = \tfrac{1}{2} \times 5$$
$$= 2.5 \text{ m}$$
$$\text{Total load} = 1256.91 + 83.79$$
$$= 1340.7 \text{ kN}$$

Taking moments about the waterline:

$$\text{Centre of pressure from WL} = \frac{1256.91 \times 3.33 + 83.79 \times 2.5}{1256.91 + 83.79}$$
$$= \frac{4189.7 + 209.5}{1340.7}$$
$$= 3.281 \text{ m}$$

∴ Centre of pressure is 5.781 m from the top of the bulkhead.

56.

½ ord	½ ord³	SM	product
5.5	166.38	1	166.38
4.6	97.34	4	389.36
4.3	79.51	2	159.02
3.7	50.65	4	202.60
3.0	27.00	1	27.00
			944.36

$$h = \frac{23}{4}$$
$$= 5.75 \text{ m}$$

Second moment of area about centreline

$$= \tfrac{2}{9} \times 5.75 \times 944.36$$
$$= 1206.7 \text{ m}^4$$

$$\text{Free surface effect} = \frac{\rho_1 i}{\rho \nabla}$$

$$0.2 = \frac{\rho_1 \times 1206.7 \times 1.025}{1.025 \times 5350}$$

$$\text{Density of liquid } \rho_1 = \frac{0.2 \times 5350}{1206.7}$$

$$= 0.887 \text{ t/m}^3$$

57.

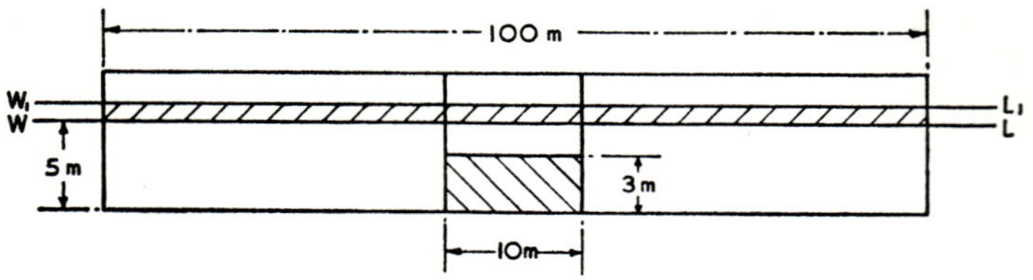

Fig. 138

Before bilging:

$$KB = \frac{5}{2}$$

$$= 2.5\,\text{m}$$

$$BM = \frac{B^2}{12 \times 5}$$

$$KG = 2.5 + \frac{B^2}{60} - 0.8$$

$$= 1.7 + \frac{B^2}{60}$$

After bilging:

$$\text{Increase in draught} = \frac{10 \times B \times 3}{100 \times B}$$

$$= 0.3\,\text{m}$$

$$\text{New draught} = 5.3\,\text{m}$$

$$KB_1 = \frac{100 \times 5.3 \times \frac{5.3}{2} - 10 \times 3 \times \frac{3}{2}}{100 \times 5.3 - 10 \times 3}$$

$$= \frac{1404.5 - 45}{100 \times 50}$$

$$= 2.719\,\text{m}$$

$$B_1M_1 = \frac{100 + B^3}{12 \times 100 \times B \times 5}$$

$$= \frac{B^2}{60}$$

$$\text{New metacentric height } GM_1 = KB_1 + B_1M_1 - KG$$

$$= 2.719 + \frac{B^2}{60} - \left(1.7 + \frac{B^2}{60}\right)$$

$$= 1.019\,\text{m}$$

58. (a) Show that $C = \dfrac{\Delta^{\frac{2}{3}} V^3}{sp}$

(b) Displacement $= 160 \times 22 \times 9.2 \times 0.765 \times 1.025$
$= 25\,393$ tonne

Theoretical speed $V_t = \dfrac{4.0 \times 96 \times 60}{1852}$

$= 12.44$ knots

Real slip $0.33 = \dfrac{12.44 - V_a}{12.44}$

$V_a = 12.44(1 - 0.33)$
$= 8.335$ knots

Wake fraction $0.335 = \dfrac{V - V_a}{V}$

$0.665V = V_a$

$V = \dfrac{8.335}{0.665}$

$= 12.53$ knots

Admiralty Coefficient $= \dfrac{25\,393^{\frac{2}{3}} \times 12.53^3}{2900}$

$= 585.9$

At 15 knots: shaft power $= 2900 \times \left(\dfrac{15}{12.53}\right)^3$

$= 4976$ kW

59. Wetted surface area of ship

$= 2.57\sqrt{24\,000 \times 180}$
$= 5341$ m^2

Wetted surface area of model

$= 5241 \times \left(\dfrac{6}{180}\right)^2$

$= 5.934$ m^2

Model:

$$R_t = 40 \text{ N in FW}$$

$$R_f = 0.492 \times 5.934 \times 3.6^{1.825}$$
$$= 30.24 \text{ N in FW}$$

$$R_r = 40 - 30.24$$
$$= 9.76 \text{ N in FW}$$
$$= 9.76 \times 1.025$$
$$= 10.004 \text{ N in SW}$$

Ship:

$$R_r \propto L^3$$

$$\therefore \quad R_r = 10.004 \times \left(\frac{180}{6}\right)^3$$

$$= 270\,110 \text{ N}$$

$$V \propto \sqrt{L}$$

$$\therefore \quad V = 3.6\sqrt{\frac{180}{6}}$$

$$= 19.72 \text{ knots}$$

$$R_f = 0.421 \times 5341 \times 19.72^{1.825}$$
$$= 518\,930 \text{ N}$$

$$R_t = 518\,930 + 270\,110$$
$$= 789\,040 \text{ N}$$

$$\text{ep}_n = 789\,040 \times 19.72 \times \frac{1852}{3600} \times 10^{-3}$$

$$= 8005 \text{ kW}$$

60. (a) Description

(b)

	mass	deflection	deviation
	3 tonne	64 mm S	64 mm
	3 tonne	116 mm S	52 mm
	6 tonne	3 mm S	113 mm
	3 tonne	54 mm P	57 mm
	3 tonne	113 mm P	59 mm
	6)18 tonne		6)345 mm
Mean	3 tonne		57.5 mm

$$\tan\theta = \frac{57.5}{5.5 \times 1000}$$

$$GM = \frac{m \times d}{\Delta \times \tan\theta}$$

$$= \frac{3 \times 12 \times 5500}{8000 \times 57.5}$$

$$= 0.430 \text{ m}$$

61. Change in mean draught

$$= \frac{\Delta \times 100}{A_w} \left(\frac{\rho_S - \rho_R}{\rho_R \times \rho_S}\right) \text{ cm}$$

$$= \frac{8000 \times 100 \times 1.024}{16 \times 100} \left(\frac{1.024 - 1.000}{1.000 \times 1.024}\right)$$

$$= 12 \text{ cm increase}$$

Shift in centre of buoyancy

$$= \frac{\rho_S - \rho_R}{\rho_S} \times FB$$

$$= \frac{1.024 - 1.000}{1.024} \times (3.0 - 0.6)$$

$$BB_1 = 0.05625 \text{ m aft}$$

$$\text{Change in trim} = \frac{8000 \times 0.05\,625}{65}$$

$$= 6.92 \text{ cm by the head}$$

$$\text{New trim} = 30 - 6.92$$
$$= 23.08 \text{ cm}$$

$$\text{Moment required} = 23.08 \times 65$$

$$\therefore \text{Distance moved by mass} = \frac{23.08 \times 65}{50}$$

$$= 30.00 \text{ m}$$

Total change in trim = 30 cm by the head

$$\text{Change forward} = +\frac{30}{110}\left(\frac{110}{2}+3\right)$$

$$= +15.8 \text{ cm}$$

Final level keel draught = 6.00 + 0.12 + 0.158
= 6.278 m

62.

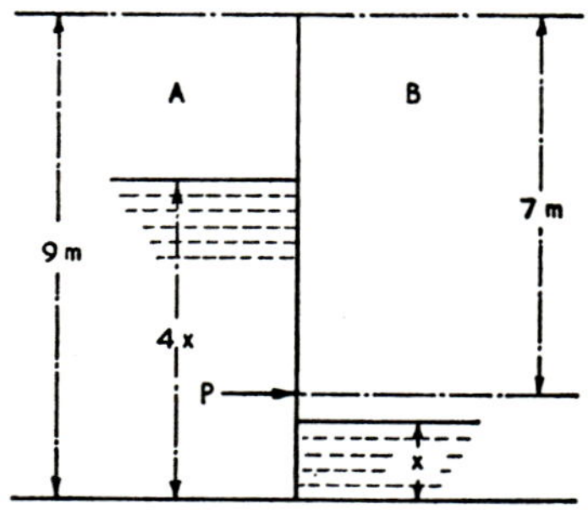

Fig. 139

Let x = height of water on side B

$4x$ = height of water on side A

Load on side A = $\rho g A H$
= $1.025 \times 9.81 \times 4x \times 14.5 \times 2x$
= $145.8 \times 8x^2$

Centre of pressure on side A

= $\frac{2}{3} \times 4x$ from surface
= $\frac{1}{3} \times 4x$ from bottom

Load on side B = $1.025 \times 9.81 \times x \times 14.5 \times 0.5x$
= $145.8 \times 0.5x^2$

Centre of pressure on side B

= $\frac{1}{3} \times x$ from bottom

Taking moments about the *bottom* of the bulkhead:
$2(145.8 \times 8x^2 - 145.8 \times 0.5x^2) =$

$$145.8 \times 8x^2 \times \frac{4x}{3} - 145.8 \times 0.5x^2 \times \frac{x}{3}$$

Dividing by $148.5x^2$:

$$2(8-0.5) = \frac{32x}{3} - \frac{0.5x}{3}$$

$$x = \frac{2 \times 7.5 \times 3}{31.5}$$

$$= 1.429 \text{ m}$$

$$\text{Height on side B} = 1.429 \text{ m}$$

$$\text{Height on side A} = 5.716 \text{ m}$$

$$\text{Resultant load} = 145.8x^2(8-0.5)$$
$$= 2233 \text{ kN}$$
$$= 2.233 \text{ MN}$$

63. There are several methods of approach with this question. Probably the most straightforward is to plot curves of ep (trial) and ep (service).

V	13.0	14.1	15.2	16.3
ep_n	1690	2060	2670	3400
ep_t	1909.7	2327.8	3017.1	3842
ep_s	2247.7	2739.8	3551.1	4522

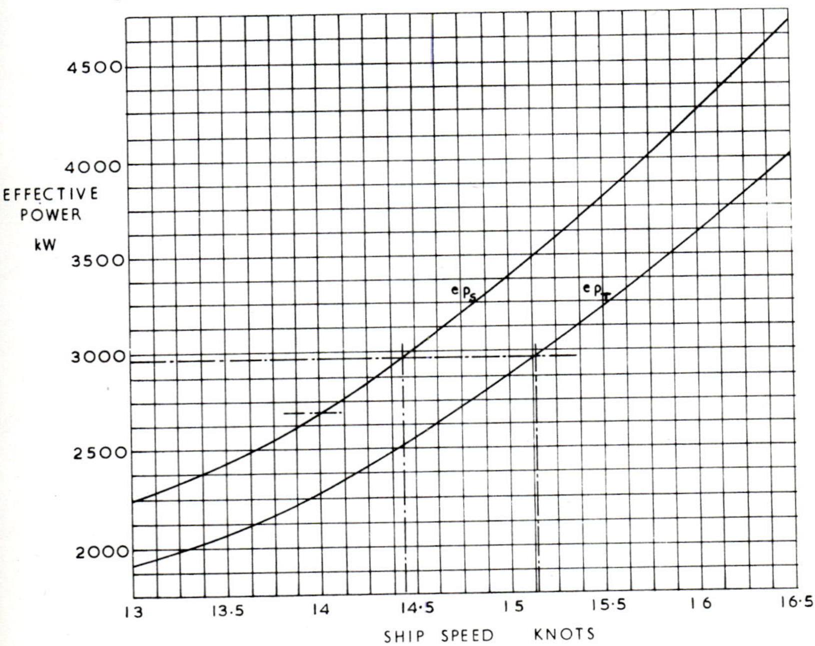

Fig. 140

At 14 knots, ep (service) = 2675 kW

$$\text{ep (max)} = \frac{2675}{0.9}$$

$$= 2970 \text{ kW}$$

(a) Required

$$\text{ip} = \frac{2970}{0.72 \times 0.965 \times 0.87}$$

$$= 4920 \text{ kW}$$

(b) At ep 2970 kW, from graph:

Service speed = 14.46 knot

Trial speed = 15.14 knot

64. $GG_1 = 0.20$ m

θ	$\sin\theta$	$GG_1 \sin\theta$	GZ	G_1Z	SM	product
0	0	0	0	0	1	0
15°	0.259	0.0518	0.38	0.43	4	1.72
30°	0.500	0.100	1.00	1.10	2	2.20
45°	0.707	0.1414	1.41	1.55	4	6.20
60°	0.866	0.1732	1.20	1.37	1	1.37
						11.49

(a)

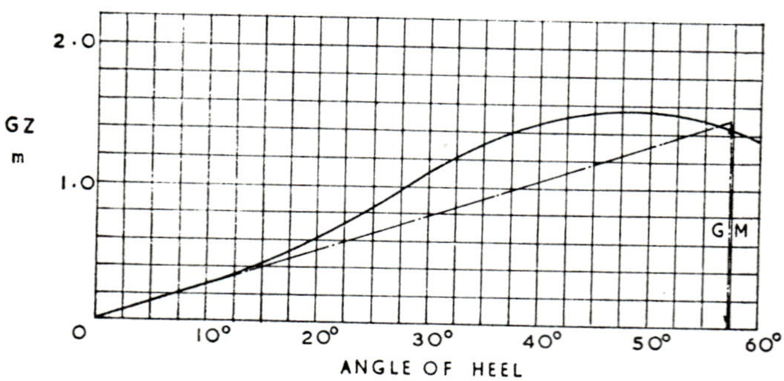

Fig. 141

(b) Dynamical stability $= \frac{1}{3} \times \frac{15}{57.3} \times 11.49 \times 9.81 \times 15\,000$

$$= 147\,500 \text{ kN m}$$

$$= 147.5 \text{ MJ}$$

65. Let

$$V = \text{original speed}$$
$$C = \text{original cons/day}$$
$$K = \text{original cons for the voyage}$$
$$\text{Cons/day} \propto \text{speed}^3$$
$$\therefore \quad \frac{C}{C+25} = \left(\frac{V}{V+1.5}\right)^3$$
$$\text{Voyage cons} \propto \text{speed}^2$$
$$\frac{K}{1.2K} = \left(\frac{V}{V+1.5}\right)^2$$
$$\sqrt{\frac{1}{1.2}} = \frac{V}{V+1.5}$$
$$V+1.5 = \sqrt{1.2}V$$
$$= 1.095\,V$$
$$V = \frac{1.5}{0.095}$$
$$\therefore \quad \text{Original speed} = 15.79 \text{ knots}$$
$$\frac{C}{C+25} = \left(\frac{15.79}{15.79+1.5}\right)^3$$
$$= 0.7617$$
$$C = 0.7617\,(C+25)$$
$$C(1-0.7617) = 0.7617 \times 25$$
$$C = \frac{0.7617 \times 25}{0.2383}$$
$$\therefore \quad \text{Original consumption} = 79.91 \text{ tonne/day}$$

66.

Width	SM	product	lever	product	lever	product
6.0	1	6.0	0	—	0	—
6.0	4	24.0	1	24.0	1	24.0
5.3	2	10.6	2	21.2	2	42.4
3.6	4	14.4	3	43.2	3	129.6
0.6	1	0.6	4	2.4	4	9.6
				90.8		205.6

$$\text{1st moment} = \frac{h^2}{3}\Sigma m$$
$$= \frac{3^2}{3} \times 90.8$$
$$= 272.4 \text{ m}^3$$

$$\begin{aligned} \text{Load on bulkhead} &= \rho g \times \text{1st moment} \\ &= 0.80 \times 9.81 \times 272.4 \\ &= 2138 \text{ kN} \\ &= 2.138 \text{ MN} \end{aligned}$$

$$\begin{aligned} \text{Centre of pressure} &= \frac{\text{2nd moment}}{\text{1st moment}} \\ &= \frac{h\Sigma_I}{\Sigma_m} \\ &= \frac{3 \times 205.6}{90.8} \\ &= 6.79 \text{ m from top of bulkhead} \end{aligned}$$

67.

V (knot)	12.50	13.25	14.00
ep (kW)	1440	1800	2230
QPC	0.705	0.713	0.708
dp (kW)	2042	2525	3150

$$\text{dp} = \frac{\text{ep}}{\text{QPC}}$$

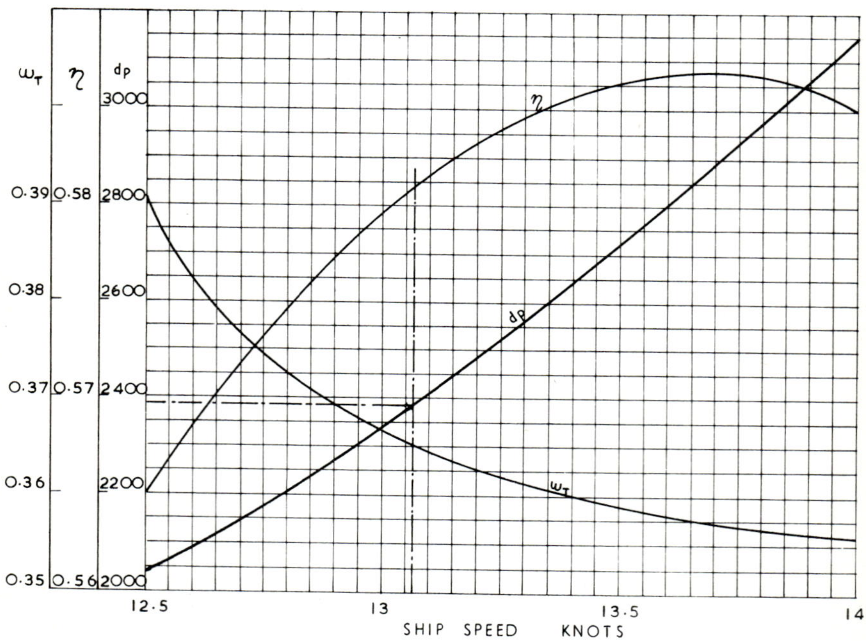

Fig. 142

From graph, when dp is 2385 kW,

$$\text{Ship speed} = 13.06 \text{ knot}$$

At this speed, $w_t = 0.365$ and propeller efficiency = 0.581

Thrust power $\quad \text{tp} = 2385 \times 0.581$

$$= 1385 \text{ kW}$$

$$\text{Ship speed} = 13.06 \times \frac{1852}{3600}$$

$$= 6.72 \text{ m/s}$$

$$0.365 = \frac{6.72 - v_a}{6.72}$$

$$v_a = 6.72(1 - 0.365)$$

$$= 4.266 \text{ m/s}$$

$$\text{tp} = T \times v_a$$

Propeller thrust $\quad T = \dfrac{1385}{4.266}$

$$= 325 \text{ kN}$$

68.

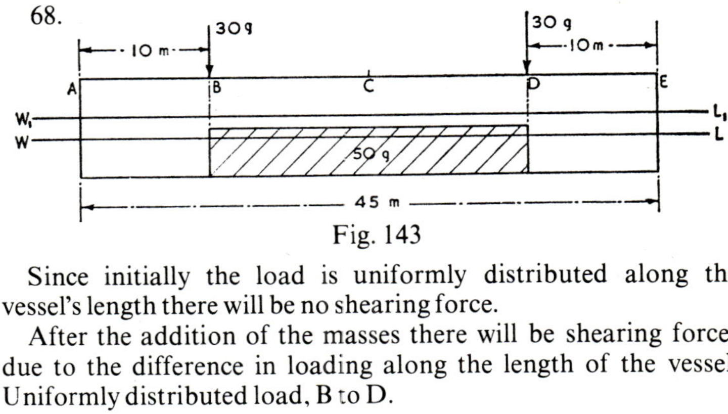

Fig. 143

Since initially the load is uniformly distributed along the vessel's length there will be no shearing force.

After the addition of the masses there will be shearing forces due to the difference in loading along the length of the vessel. Uniformly distributed load, B to D.

$$= \frac{50g}{25}$$

$$= 2g \text{ kN/m}$$

Additional buoyancy required

$$= \left(\frac{30 + 30 + 50}{45}\right) g$$

$$= 2.444g$$

Shearing force at A $\quad = 0$

Shearing force at left hand of B

$$= 2.444g \times 10$$

$$= 24.44g$$

Shearing force at right hand of B

$$= 24.44g - 30g$$
$$= -5.56g$$

Shearing force at C $= 2.444g \times 22.5 - 30g - 2.0g \times 12.5$

$$= 55g - 30g - 25g$$
$$= 0$$

Since the vessel is symmetrically loaded, these values will be repeated, but of opposite sign.

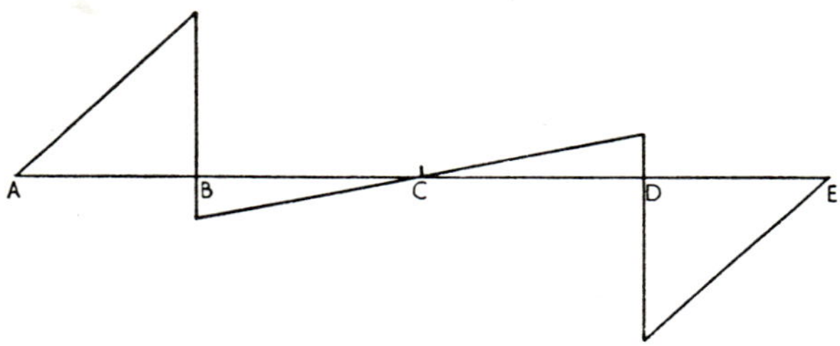

SHEARING FORCE DIAGRAM

Fig. 144

The maximum shearing force occurs at B and D.

Maximum shearing force $= 24.44g$

$$= 239.8 \text{ kN}$$

69. $R_t = T(1-t)$

$$= 378(1 - 0.24)$$
$$= 287.3 \text{ kN}$$

Ship speed $= 15.5 \times \dfrac{1852}{3600}$

$$= 7.974 \text{ m/s}$$

Effective power ep $= 287.3 \times 7.974$

$$= 2291 \text{ kW}$$

$$v_t = 4.87 \times 1.58$$
$$= 7.695 \text{ m/s}$$

$$0.30 = \frac{7.695 - v_a}{7.695}$$

$$v_a = 7.695(1 - 0.30)$$
$$= 5.386 \text{ m/s}$$

Taylor wake fraction $w_T = \dfrac{7.974 - 5.386}{7.974}$

$$= 0.324$$

Shaft power $\quad sp = \dfrac{3540}{0.97}$

$$= 3649.5 \text{ kW}$$

Propulsive coefficient $= \dfrac{ep}{sp}$

$$= \frac{2291}{3649.5}$$

$$= 0.628$$

$$ep = ep_n + \text{appendage allowance}$$

$$ep_n = \frac{2291}{1.15}$$

$$= 1992 \text{ kW}$$

Quasi-Propulsive Coefficient QPC $= \dfrac{ep_n}{dp}$

$$= \frac{1992}{3540}$$

$$= 0.563$$

70. $\quad$ MCT1 cm $= \dfrac{14\,000 \times 120}{100 \times 125}$

$$= 134.4 \text{ tonne m}$$

Let $\quad m$ = mass added

d = distance of mass from F

Change in trim required $= (8.5 - 7.9)\,100$

$$= 60 \text{ cm}$$

Since the after draught remains constant, the change in trim aft must be equal to the bodily sinkage.

Change in trim aft $= \dfrac{60}{125}\left(\dfrac{125}{2} + 3\right)$

$$= 31.44 \text{ cm}$$

Bodily sinkage $= \dfrac{m}{19}$

$$\therefore \quad \frac{m}{19} = 31.44$$

$$m = 31.44 \times 19$$

$$= 597.36 \text{ tonne}$$

But change in trim $$60 = \frac{m \times d}{\text{MCT1 cm}}$$

$$d = \frac{60 \times 134.4}{597.36}$$

$$= 13.50 \text{ m forward}$$

Thus 597.36 tonne must be added 16.50 m forward of midships.

71. When a mass is suspended from a derrick head, its centre of gravity may be taken at the derrick head.

(a) $$GG_1 = \frac{50 \times 18}{4000}$$

$$= 0.225 \text{ m up}$$

(b)The mass has been moved the same distance:

i.e. $$GG_1 = 0.225 \text{ m up}$$

(c) $$GG_1 = \frac{50 \times 12}{4000}$$

$$= 0.15 \text{ m up}$$

(d) $$GG_1 = \frac{50 \times 14}{4000}$$

$$= 0.175 \text{ m outboard}$$

72.

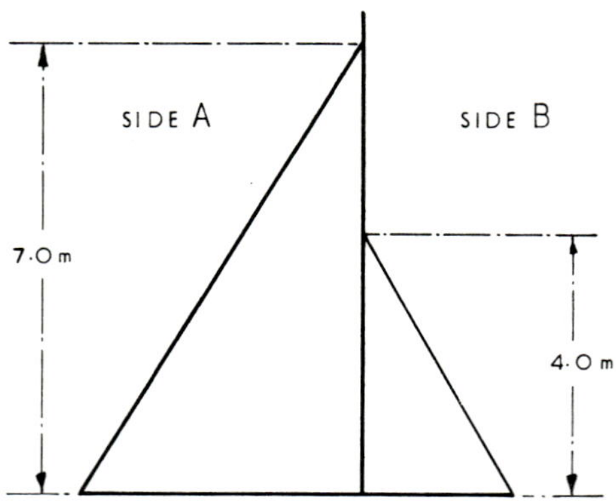

Fig. 145

Load on side A $= \rho g A H$

$$= 1.00 \times 9.81 \times 7 \times 8 \times \frac{7}{2}$$

$$= 1922.76 \text{ kN}$$

Centre of pressure from bottom $= \frac{7}{3}$ m

Load on side B $= 0.850 \times 9.81 \times 4 \times 8 \times \frac{4}{2}$

$= 533.66$ kN

Centre of pressure from bottom $= \frac{4}{3}$ m

(a) Resultant load $= 1922.76 - 533.66$

$= 1389.1$ kN

(b) Take moments about the bottom of the bulkhead:

$$\text{Resultant centre of pressure} = \frac{1922.76 \times \frac{7}{3} - 533.66 \times \frac{4}{3}}{1389.1}$$

$$= \frac{4486.44 - 711.54}{1389.1}$$

$= 2.718$ m from bottom

73.

Section	½ area	SM	Product for volume	lever	Product for moment
AP	0.4	1	0.4	+ 5	+ 2.0
1	7.6	4	30.4	+ 4	+ 121.6
2	21.4	2	42.8	+ 3	+ 128.4
3	33.5	4	134.0	+ 2	+ 268.0
4	40.8	2	81.6	+ 1	+ 81.6
5	45.5	4	182.0	0	+ 601.6
6	48.4	2	96.8	− 1	− 96.8
7	52.0	4	208.0	− 2	− 416.0
8	51.1	2	102.2	− 3	− 306.6
9	34.4	4	137.6	− 4	− 550.4
FP	0	1	—	− 5	—
			1015.8		− 1369.8

(a) New Displacement $= \frac{2}{3} \times 9.15 \times 1015.8 \times 1.025$

$= 6351.3$ tonne

Original Displacement $= 5750.0$ tonne

$\therefore$ Mass of water added $= 601.3$ tonne

(b) Moment of buoyancy about midships

$$= \frac{2}{3} \times 9.15^2 \times (1369.8 - 601.6) \times 1.025$$

$$= 43\,949 \text{ tonne m forward}$$

Original moment of buoyancy $= 5750.0 \times 4.6$

$$= 26\,450 \text{ tonne m forward}$$

$\therefore$ Moment of ballast about midships

$$= 43\,949 - 26\,450$$

$$= 17\,499 \text{ tonne m forward}$$

Centre of gravity of ballast from midships

$$= \frac{17\,499}{601.6}$$

$$= 29.09 \text{ m forward}$$

74. Mass of water pumped out $= 14 \times 12 \times 0.6 \times 1.025$
$= 103.3$ tonne

The centre of gravity of this water is 0.9 m above the keel. Taking moments about the keel:

$$\text{New } KG = \frac{7500 \times 6.7 - 103.3 \times 0.9}{7500 - 103.3}$$

$$= \frac{50\,250 - 93.0}{7396.7}$$

$$= 6.781 \text{ m}$$

$$\text{Free surface effect} = \frac{i}{\nabla}$$

$$= \frac{14 \times 12^3 \times 1.025}{12 \times 7396.7}$$

$$= 0.279 \text{ m}$$

Original $KM = 6.70 + 0.45$
$= 7.15$ m

New $GM = 7.15 - 6.781 - 0.279$
$= 0.090$ m

75. Change in mean draught

$$= \frac{\Delta \times 100}{A_w}\left(\frac{\rho_S - \rho_R}{\rho_R \times \rho_S}\right) \text{cm}$$

$$= \frac{9100 \times 100 \times 1.024}{16.8 \times 100}\left(\frac{1.024 - 1.000}{1.000 \times 1.024}\right)$$

$$= \frac{9100 \times 0.024}{16.8}$$

$$= 13.0 \text{ cm reduction}$$

Shift in centre of buoyancy

$$= \frac{\rho_S - \rho_R}{\rho_R} FB$$

$$= \frac{1.024 - 1.000}{1.000} \times (0.6 + 2.25)$$

$$= 0.0684 \text{ m}$$

$$\text{Change in trim} = \frac{9100 \times 0.0684}{122}$$

$$= 5.10 \text{ cm by the stern}$$

$$\text{Change forward} = -\frac{5.10}{120}\left(\frac{120}{2} + 0.6\right)$$

$$= -2.6 \text{ cm}$$

$$\text{Change aft} = +\frac{5.10}{120}\left(\frac{120}{2} - 0.6\right)$$

$$= +2.5 \text{ cm}$$

New draught forward = 6.70 – 0.13 – 0.026
= 6.544 m

New draught aft = 6.70 – 0.13 + 0.025
= 6.595 m

76.

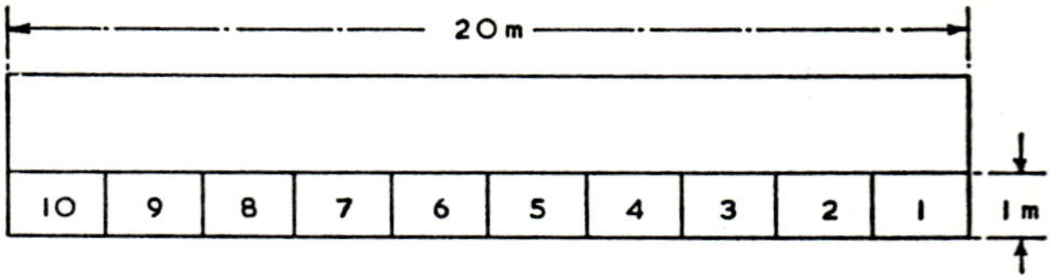

Fig. 146

$$\text{Mass added/tank} = 2 \times 10 \times 1 \times 1.025 = 20.50 \text{ tonne}$$

$$\text{Total mass added} = 41.0 \text{ tonne}$$

$$\text{mass/m} = \frac{20.5}{2} = 10.25 \text{ tonne}$$

$$\text{Weight/m} = 10.25g \quad \text{kN}$$

$$\text{Buoyancy required/m} = \frac{41.0g}{20} = 2.05g$$

Hence, in way of ballast,

$$\text{Excess load/m} = 10.25g - 2.05g = 8.20g \quad \text{kN}$$

(a) With No. 1 and No. 10 tanks filled:

$$\text{S.F. at aft end of vessel} = 0$$

$$\text{S.F. at fore end of No. 10} = -8.20g \times 2 = -16.40g \quad \text{kN}$$

$$\text{S.F. at midships} = -16.40g + 2.05g \times 8 = 0$$

$$\text{S.F. at aft end of No. 1} = +2.05g \times 8 = +16.40g$$

$$\text{S.F. at fore end of vessel} = +16.40g - 8.20g \times 2 = 0$$

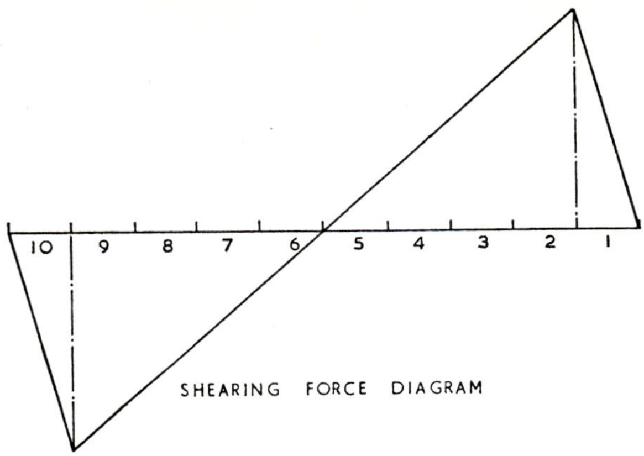

Fig. 147

(b) With No. 3 and No. 8 tanks filled:

$$\text{S.F. at aft end of vessel} = 0$$

$$\begin{aligned}\text{S.F. at aft end of No. 8} &= +2.05g \times 4\\ &= +8.20g\end{aligned}$$

$$\begin{aligned}\text{S.F. at fore end of No. 8} &= +8.20g - 8.20g \times 2\\ &= -8.20g\end{aligned}$$

$$\begin{aligned}\text{S.F. at midships} &= -8.20g + 2.05g \times 4\\ &= 0\end{aligned}$$

$$\begin{aligned}\text{S.F. at aft end of No. 3} &= +2.05g \times 4\\ &= +8.20g\end{aligned}$$

$$\begin{aligned}\text{S.F. at fore end of No. 3} &= +8.20g - 8.20g \times 2\\ &= -8.20g\end{aligned}$$

$$\text{S.F. at fore end of vessel} = 0$$

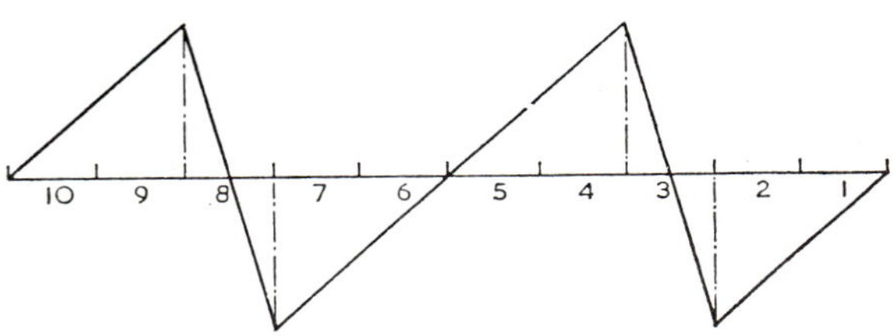

Fig. 148

(c) With No. 5 and No. 6 tanks filled:

$$\text{S.F. at aft end of vessel} = 0$$

$$\begin{aligned}\text{S.F. at aft end of No. 6} &= +2.05g \times 8\\ &= +16.40g\end{aligned}$$

$$\begin{aligned}\text{S.F. at fore end of No. 5} &= +16.40g - 8.20g \times 4\\ &= -16.40g\end{aligned}$$

$$\text{S.F. at fore end of vessel} = 0$$

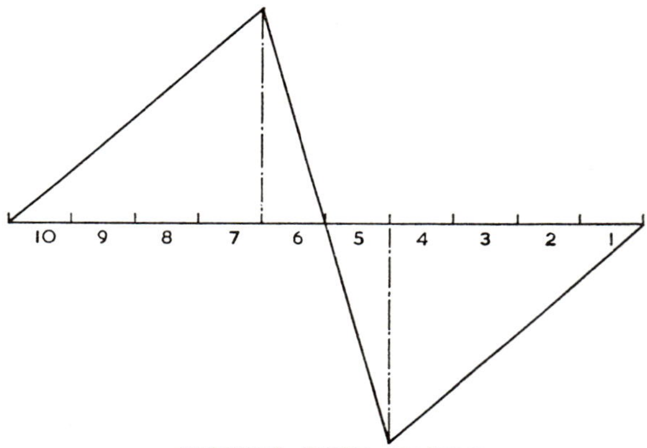

SHEARING FORCE DIAGRAM

Fig. 149

The maximum shearing force in case (b) is half of the maximum values in cases (a) and (c). Thus (b) is the best loaded condition.

77. $$\begin{aligned}\text{Final displacement} &= 216 + 81 + 54\\ &= 351 \text{ tonne}\\ &= L \times B \times d \times \rho\end{aligned}$$

$$\text{New draught} = \frac{351}{32 \times 5.5 \times 1.025}$$

$$= 1.946 \text{ m}$$

$$KB = \frac{1.946}{2}$$

$$= 0.973 \text{ m}$$

$$BM = \frac{32 \times 5.5^3}{12 \times 32 \times 5.5 \times 1.946}$$

$$= 1.295 \text{ m}$$

$$KM = 0.973 + 1.295$$
$$= 2.268 \text{ m}$$
$$\text{Final} \quad GM = 0.130 \text{ m}$$
$$\therefore \quad \text{Final} \quad KG = 2.138 \text{ m}$$

Let $x = Kg$ of machinery

$$351 \times 2.138 = 216 \times 1.8 + 81x + 54 \times 0.15$$
$$750.44 = 388.8 + 81x + 8.1$$
$$81x = 353.54$$

Kg of machinery $x = 4.365$ m

78. If a ship moves from sea water of 1.025 t/m³ into fresh water of 1.000 t/m³,

$$\text{change in mean draught} = \frac{\Delta}{40\,\text{TPC}} \text{ cm}$$
$$\therefore \quad 23 = \frac{\Delta}{40 \times 23}$$
$$\Delta = 23 \times 40 \times 23$$
$$21\,160 \text{ tonne}$$

The draught in the river water is the same as the allowable draught in the fresh water, but the displacement is 202 tonne greater, i.e. 21 362 tonne.

Let ρ_R = density of river water

Volume of displacement in fresh water

$$= \frac{21\,160}{1.000} \text{ m}^3$$

Volume of displacement in river water

$$= \frac{21\,362}{\rho_R} \text{ m}^3$$

But volume in fresh water = volume in river water

$$\frac{21\,160}{1.000} = \frac{21\,362}{\rho_R}$$
$$\rho_R = \frac{21\,362}{21\,160}$$

Density of the river water = 1.010 t/m³

79.

Speed of advance (m/s)	1.22	1.46	1.70	1.94
Thrust (N)	93.7	72.3	49.7	24.3
Thrust power (W)	114.3	105.6	84.5	47.1
Torque (Nm)	3.90	3.23	2.50	1.61
Delivered power (W)	196.1	162.4	125.7	80.9

$$\text{Thrust power} = \text{thrust} \times \text{speed of advance}$$
$$\text{Delivered power} = \text{torque} \times 2\pi \times \text{rev/s}$$

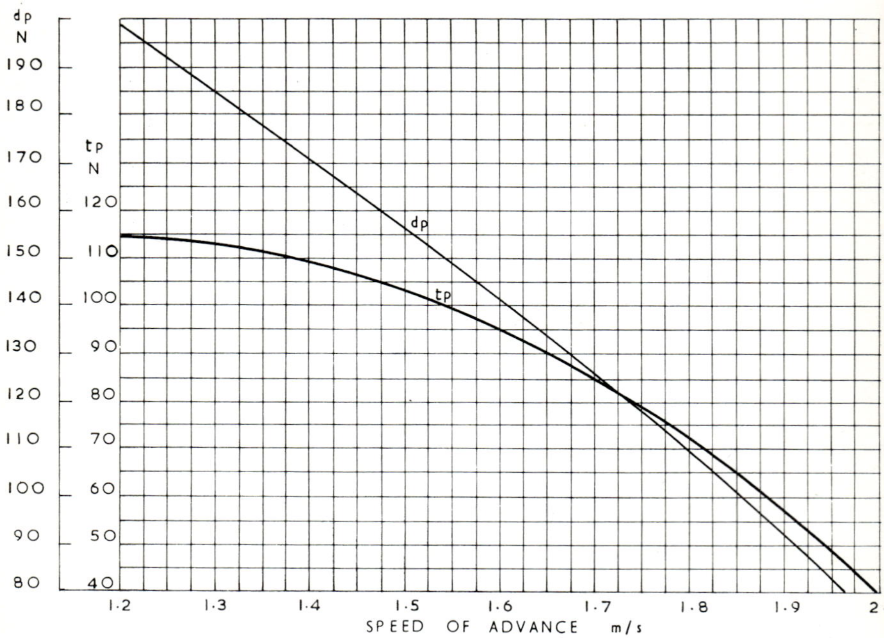

Fig. 150

For ship $\quad dp = 3000 \text{ kW in sea water}$

Equivalent for model $\quad dp = 3000 \times \left(\frac{0.3}{4.8}\right)^{3.5} \times \frac{1.00}{1.025}$

$$= 178.6 \text{ W}$$

From graph at this dp: $v_a = 1.348 \text{ m/s}$
and at this speed: $tp = 111.1 \text{ W}$

(a) For ship $\quad tp = 111.1 \times \left(\frac{4.8}{0.3}\right)^{3.5} \times \frac{1.025}{1.00}$

$$= 1866 \text{ kW}$$

(b) Propeller efficiency $= \frac{1866}{3000} \times 100$

$$= 62.2\%$$

80.

Draught	C_W	SM	Product for volume	(1) lever	Product for vert. moment	LCF	Product for (2) longl. moment
1.22	0.78	1	0.78	1	0.78	+ 1.30	+ 1.014
2.44	0.82	4	3.28	2	6.56	+ 1.21	+ 3.969
3.66	0.85	2	1.70	3	5.10	+ 0.93	+ 1.581
4.88	0.88	4	3.52	4	14.08	+ 0.50	+ 1.760
6.10	0.90	1	0.90	5	4.50	− 0.06	− 0.054
			10.18		31.02		8.270

(1) Levers taken from the keel

(2) Product of volume column and LCF

(a) Displacement of layer $= \frac{1.22}{3} \times 10.18 \times 128 \times 16.75 \times 1.025$

$= 9097.8$ tonne

(b) (i) Longitudinal moment $= \frac{1.22}{3} \times 8.27 \times 128 \times 16.75$

LCB of layer from midships $= \dfrac{\frac{1.22}{3} \times 8.27 \times 128 \times 16.75}{\frac{1.22}{3} \times 10.18 \times 128 \times 16.75}$

$= \dfrac{8.27}{10.18}$

$= 0.812$ m forward

(ii) Vertical moment $= \frac{1.22^2}{3} \times 31.02 \times 128 \times 16.75$

VCB of layer from keel $= \dfrac{\frac{1.22^2}{3} \times 31.02 \times 128 \times 16.75}{\frac{1.22}{3} \times 10.18 \times 128 \times 16.75}$

$= \dfrac{1.22 \times 31.02}{10.18}$

$= 3.717$ m

INDEX